Advanced Studies in Nearly Zero-Energy Buildings and Optimal Design

Advanced Studies in Nearly Zero-Energy Buildings and Optimal Design

Guest Editors

Cristina Carpino
Miguel Chen Austin
Dafni Mora
Natale Arcuri

MDPI

Basel • Beijing • Wuhan • Barcelona • Belgrade • Novi Sad • Cluj • Manchester

Guest Editors
Cristina Carpino
Department of Mechanical,
Energy and Management
Engineering
University of Calabria
Cosenza
Italy

Miguel Chen Austin
Faculty of Mechanical
Engineering
Technological University
of Panama
Panama City
Panama

Dafni Mora
Faculty of Mechanical
Engineering
Technological University
of Panama
Panama City
Panama

Natale Arcuri
Department of Mechanical,
Energy and Management
Engineering
Università della Calabria
Cosenza
Italy

Editorial Office
MDPI AG
Grosspeteranlage 5
4052 Basel, Switzerland

This is a reprint of the Special Issue, published open access by the journal *Buildings* (ISSN 2075-5309), freely accessible at: https://www.mdpi.com/journal/buildings/special_issues/Energy_Design.

For citation purposes, cite each article independently as indicated on the article page online and as indicated below:

Lastname, A.A.; Lastname, B.B. Article Title. *Journal Name* **Year**, *Volume Number*, Page Range.

ISBN 978-3-7258-8089-8 (Hbk)
ISBN 978-3-7258-8090-4 (PDF)
https://doi.org/10.3390/books978-3-7258-8090-4

Cover image courtesy of Miguel Alejandro Chen Austin

Contents

Preface . . . . **vii**

Cristina Carpino, Evangelia Loukou, Miguel Chen Austin, Birgitte Andersen, Dafni Mora and Natale Arcuri
Risk of Fungal Growth in Nearly Zero-Energy Buildings (nZEB)
Reprinted from: *Buildings* **2023**, *13*, 1600, https://doi.org/10.3390/buildings13071600 . . . . . . **1**

Katherine Chung-Camargo, Jinela González, Miguel Chen Austin, Cristina Carpino, Dafni Mora and Natale Arcuri
Advances in Retrofitting Strategies for Energy Efficiency in Tropical Climates: A Systematic Review and Analysis
Reprinted from: *Buildings* **2024**, *14*, 1633, https://doi.org/10.3390/buildings14061633 . . . . . . **21**

Jaime Resende and Helena Corvacho
Optimisation of Nearly Zero Energy Building Envelope for Passive Thermal Comfort in Southern Europe
Reprinted from: *Buildings* **2024**, *14*, 2757, https://doi.org/10.3390/buildings14092757 . . . . . . **49**

Cristina Carpino, Miguel Chen Austin, Dafni Mora and Natale Arcuri
Retrofit Measures for Achieving NZE Single-Family Houses in a Tropical Climate via Multi-Objective Optimization
Reprinted from: *Buildings* **2024**, *14*, 566, https://doi.org/10.3390/buildings14030566 . . . . **75**

Lissette De León, Dafni Mora, Cristina Carpino, Natale Arcuri and Miguel Chen Austin
A Reference Framework for Zero Energy Districts in Panama Based on Energy Performance Simulations and Bioclimatic Design Methodology
Reprinted from: *Buildings* **2023**, *13*, 315, https://doi.org/10.3390/buildings13020315 . . . . **97**

Ángel Benigno González-Avilés, Carlos Pérez-Carramiñana, Antonio Galiano-Garrigós and María Isabel Pérez-Millán
Energy Benefits of Tourist Accommodation Using Geodesic Domes
Reprinted from: *Buildings* **2024**, *14*, 505, https://doi.org/10.3390/buildings14020505 . . . . **125**

Xiao Teng, Zhenjiang Shen and Dara Citra Saraswati Tutuko
Evaluating the Impact of Insulation Materials on Energy Efficiency Using BIM-Based Simulation for Existing Building Retrofits: Case Study of an Apartment Building in Kanazawa, Japan
Reprinted from: *Buildings* **2025**, *15*, 570, https://doi.org/10.3390/buildings15040570 . . . . **150**

Paolo Zangheri, Delia D'Agostino, Roberto Armani and Paolo Bertoldi
Review of the Cost-Optimal Methodology Implementation in Member States in Compliance with the Energy Performance of Buildings Directive
Reprinted from: *Buildings* **2022**, *12*, 1482, https://doi.org/10.3390/buildings12091482 . . . . . . **170**

Byung Chang Kwag, Gil Tae Kim and In Tae Hwang
Integration of Photovoltaic Systems for Energy Self-Sufficient Low-Rise Multi-Family Residential Buildings in Republic of Korea
Reprinted from: *Buildings* **2024**, *14*, 2522, https://doi.org/10.3390/buildings14082522 . . . . . . **189**

Marwa Alsuwaidi, Noura Almahri, Latifa Alazeezi, Taif Alkhzaimi, Lindita Bande, Anwar Ahmad, et al.
The Design of a Wellbeing Center with an Advanced Parametric Roof in Jabel Hafeet Mountain Located in Al Ain City, United Arab Emirates
Reprinted from: *Buildings* **2024**, *14*, 974, https://doi.org/10.3390/buildings14040974 . . . . **211**

Keon Ho Lee and Young-Hak Song
Analysis of Energy Reduction and Energy Self-Sufficiency Improvement Effects by Applying a Bidirectional Reflectance PV Array with Integrated External Shading at a School Building
Reprinted from: *Buildings* **2023**, *13*, 2915, https://doi.org/10.3390/buildings13122915 . . . . . . **231**

Stefania Perrella, Fabio Bisegna, Piero Bevilacqua, Daniela Cirone and Roberto Bruno
Solar-Assisted Heat Pump with Electric and Thermal Storage: The Role of Appropriate Control Strategies for the Exploitation of the Solar Source
Reprinted from: *Buildings* **2024**, *14*, 296, https://doi.org/10.3390/buildings14010296 . . . . **245**

Preface

The urgency of addressing climate change and achieving global sustainability goals places the building sector under intense scrutiny. Buildings are central to energy consumption and carbon emissions, making the transition to Nearly Zero-Energy Buildings (nZEBs) and Net-Zero-Energy Buildings (NZEBs) not just an environmental aspiration but a critical regulatory and technological imperative. It is with this motivation that we introduce this Special Issue, "Advanced Studies in Nearly Zero-Energy Buildings and Optimal Design."

This Special Issue compiles impactful contributions received following our initial call for papers, offering a comprehensive review of the current state of the art in high-performance building design and operation. Our primary aim was to create a forum for sharing innovative, evidence-based research that moves beyond theoretical concepts to explore practical, implementable solutions. The scope of the collected works is broad yet focused, encompassing advanced optimization techniques, the integration of renewable energy systems, innovative materials, and crucial performance assessments across diverse geographical and climatic conditions.

This collection is addressed to a wide audience: dedicated researchers and academics seeking foundational knowledge, industry engineers and architects requiring practical design data, and policymakers looking for robust insights to inform future legislative action. The success of this Special Issue is a direct result of the high quality and dedication of the contributing authors, who represent international leaders in sustainable built environments.

We extend our sincere gratitude to them, as well as to the expert reviewers whose meticulous and timely feedback ensured that rigorous scientific standard was maintained throughout this publication. Their collective assistance and commitment were instrumental in shaping this valuable resource.

Cristina Carpino, Miguel Chen Austin, Dafni Mora, and Natale Arcuri
Guest Editors

buildings
MDPI

Perspective

Risk of Fungal Growth in Nearly Zero-Energy Buildings (nZEB)

Cristina Carpino [1,*], Evangelia Loukou [2], Miguel Chen Austin [3], Birgitte Andersen [2], Dafni Mora [3] and Natale Arcuri [1]

1 Department of Mechanical, Energy and Management Engineering, University of Calabria, Via Pietro Bucci 46C, 87036 Cosenza, Italy; natale.arcuri@unical.it
2 Department of the Built Environment, Aalborg University, A.C. Meyers Vænge 15, A, 1-348, 2450 Copenhagen, Denmark; elou@build.aau.dk (E.L.); bian@build.aau.dk (B.A.)
3 Faculty of Mechanical Engineering, Universidad Tecnológica de Panamá, Panama City 0819-07289, Panama; miguel.chen@utp.ac.pa (M.C.A.); dafni.mora@utp.ac.pa (D.M.)
* Correspondence: cristina.carpino@unical.it

Abstract: Research on nearly zero-energy buildings has addressed mainly the aspects of energy saving or technical and economic optimization, while some studies have been conducted on comfort and indoor air quality. However, the potential problems that may arise in low-energy buildings during the operational phase, and especially the risk of fungal growth, which can deteriorate the indoor environment and pose a health risk to the occupants, are yet to be extensively investigated. The present work intends to analyze previous research on microbial contamination in zero-energy buildings in order to identify the possible risks that may lead to fungal formation and the possible strategies to prevent the proliferation of molds. The methodology is based on a systematic literature review and subsequent critical analysis to outline perspectives on this topic. The main results indicate that high envelope insulation and inadequate ventilation are the leading causes of fungal growth in energy-efficient buildings. The need for more detailed regulation in this area is also highlighted. The study's outcomes underline the need for more attention to be paid to the design and management of zero-energy buildings, aiming to achieve the reduction in energy demands while ensuring the occupants' well-being.

Keywords: nearly zero-energy buildings (nZEBs); fungal growth; mold risk; indoor air quality; microbial contamination; mechanical ventilation systems

1. Introduction

The green transition puts pressure on increasing energy savings through better building insulation and decreased ventilation losses [1]. Different sustainability goals can often contradict each other if not applied consciously, for example, when the health and well-being of occupants are overlooked or compromised by sustainable development practices [2].

Research has shown that indoor fungal growth can cause Building-Related Symptoms (BRS), especially in the case of damp, poorly maintained, or water-damaged buildings [3,4]. Ebbehøj et al. [3] examined 25 employees working in severely moisture-damaged buildings before and after renovation. The reported symptoms included eye, nose, and throat irritations, headache, fatigue, dizziness, and concentration difficulties. Based on a national survey conducted on multi-family buildings in Sweden, building dampness, mold, and insufficient ventilation are among the most critical risk factors associated with headaches and ocular symptoms [5]. Wang et al. [6] investigated the correlation between asthma, lung function, and specific volatile organic compounds (VOCs) found in homes affected by dampness and microbial growth. The survey involved 159 adults from three northern European cities (Reykjavik, Uppsala, and Tartu). The results showed an increased concentration of specific VOCs in the presence of certain mold species or indoor dampness.

Therefore, the association between VOCs and lung disease may be stronger in homes with mold and dampness problems [6]. Furthermore, laboratory experiments conducted by Maundrill et al. [7] demonstrated that long-term fungal contamination represents the most dangerous degradation mechanism for building materials. Mechanical tests were performed to measure the decay of 290 fibrous plaster samples subjected to moisture and fungal treatments over a two-year period.

As Ginestet et al. [8] explained, the leading causes of indoor fungal development are increased air moisture and water on building elements. These conditions can be triggered due to insufficient or lack of ventilation, faulty or ineffective heating, sensitive constructions and building elements, as well as structural problems, and accidental incidents leading to water damage. In this regard, several studies have investigated the relationship between fungal growth risk and building construction. Silveira et al. [9] and Xue et al. [10] analyzed the influence of thermal insulation and the solar orientation of external walls in different climates on the surface condensation risk and, thus, fungal growth. The conclusions indicate that the presence of thermal insulation decreases the risk of surface condensation. In contrast, insufficient ventilation and low surface temperatures are critical in creating ideal conditions for fungal growth [9]. Xue et al. [10] concluded that the wall orientation and insulation positioning are closely connected to the indoor moisture content, condensation, and mold risk. High heat flux at the exterior surface due to solar radiation can reduce the moisture content of the building envelope. On the other hand, exterior insulation can increase the condensation risk [10,11]. Morelli et al. [12] investigated the wind–vapor barrier ratio to avoid mold growth, in correlation with the internal humidity class and exterior climate, through 1-D hygrothermal simulations.

The risk of mold growth varies according to climatic conditions. The influence of the external climate is due to the association of fungi with the plants present outdoors. Furthermore, the humidity level is also influenced by the external climate and the occupants' behavior. For example, in hot climates, there is a known habit of opening windows more frequently and for extended periods, implementing high air changes. In contrast, in cold climates, there is a tendency to keep the windows closed, causing an increase in internal humidity, which increases the risk of condensation on inner surfaces when considering the lower temperatures.

Additionally, there is an increasing demand for the use of less energy-intensive and more sustainable building materials. Wood or other new, biogenic materials are, therefore, substituting traditional elements like brick and concrete in many cases [12–14]. However, organic materials are more sensitive to moisture, which makes them more susceptible to fungal growth and deterioration [2,12,13].

Moreover, based on the analysis by Lu et al. [15] and simulations on mass timber walls by Defo and Lacasse [16], the mold growth risk may rise in the future due to increased rain loads under the effect of climate change. The results depend on the specific climatic characteristics of each region evaluated in near- and far-future scenarios. The impact of climate change on the actual performance of energy-efficient houses has also been highlighted by Ozarisoy [17] and D'Agostino et al. [18].

The need for energy savings has changed the construction style to airtight and highly insulated buildings. The obligations of nearly Zero-Energy Buildings (nZEBs) are towards energy efficiency (reduction in thermal transmittance and infiltration) and moisture control (preventing forced air exfiltration). A reduction in heat losses is achieved through the airtightness of the building envelope, which minimizes uncontrolled air movement through infiltration.

The generally imposed requirements for obtaining zero-energy buildings are the heating and cooling demand limitations and the integration of energy from renewable sources. However, the concept has evolved over time, resulting in several generations of nZEBs [19], and it continues evolving towards even more ambitious goals. In fact, the "zero" target has been surpassed by "positive"-energy buildings [20], which aim to obtain a positive energy balance, i.e., the energy produced is higher than what is consumed.

A great opportunity is offered by the conversion of existing buildings into nZEBs. Considerable interest has been developed in this area in recent years, as improving the performance of the building stock has become a priority to achieve the zero-carbon objectives. However, as demonstrated by the comprehensive review conducted by Tetteh et al. [21], in the current framework for renovation, "energy efficiency" retrofitting has become the highest research priority, while "occupant behavior" and "indoor environmental quality" (comfort, satisfaction, air quality, etc.) are diminishing in prominence.

As the characteristics of buildings vary depending on context and specific use, different optimization strategies have been created to coordinate multiple needs. The optimization includes both the building envelope [22] and the systems, in particular mechanical ventilation and heat recovery [23], as well as high-performance HVAC technologies like geothermal heat pumps [24].

Furthermore, questions about the actual usability and levels of well-being experienced within these high-efficiency buildings are rising [25,26]. After examining the air quality of 25 nZEB cases, Wu et al. [27] warned that overheating, contaminant accumulation, thermal discomfort, and Building Sickness Syndrome are frequently reported problems and that there are several indoor environmental quality (IEQ) risk factors, as well as other complex issues that need to be addressed in nearly zero-energy buildings as they can seriously affect the occupants' health and well-being. Steinemann et al. [28] examined the issue of how green buildings, although they may improve energy efficiency and sustainability, may not necessarily enhance occupants' health and well-being by improving indoor air quality. Furthermore, certain green practices and green products may actually degrade indoor air quality. A negative impact on health has also been reported due to excessive overheating in zero-energy buildings in a Mediterranean climate [29]. Therefore, the "green" label does not always imply good indoor air quality.

The occupants' behavior significantly influences achieving the right balance between energy consumption and thermo-hygrometric comfort [30]. The maintenance and operation of the systems are also relevant. For example, when there is mechanical ventilation, the periodic replacement of the filters is essential to ensure adequate indoor air quality [31]. Studies based on monitoring campaigns and post-occupancy surveys have tried to evaluate the performance of zero-energy buildings during the operation phase and the occupants' satisfaction level. However, the fungal growth risk in nZEBs has not been tackled yet. In this regard, a gap in the literature has been identified.

1.1. Motivation and Objective

Based on the examined literature, it emerges that the fungal formation risk in zero-energy buildings and high-performance buildings has not been clearly identified. The present research intends to draw attention to the problem of mold formation and air quality in low-consumption buildings. Indoor microbial contamination threatens human health, while its detection poses a challenge, especially when the affected area is hidden. The perspective provided by this work helps to improve the understanding of this phenomenon and outlines possible intervention strategies to prevent and assess the adverse effects associated with fungal growth in nZEBs.

The present work intends to address the following research questions:

- Is there a relevant body of literature dealing with the risk of fungal formation in energy-efficient buildings, and particularly in nearly zero-energy buildings?
- What are the possible causes of mold growth in nZEBs?
- What solutions and intervention strategies can be adopted to prevent the formation of fungi and ensure a trade-off between energy efficiency and air quality in nZEBs?

The originality of the proposed work can be expressed in three points. First, it draws attention to the emerging challenges of nearly zero-energy buildings and directs future studies toward approaches that combine energy-saving needs with the occupants' well-being. Secondly, the risk of hidden mold contamination in nZEBs is investigated, with particular focus on the operation of mechanical ventilation. Finally, the need for integrated design and interdisciplinary collaboration (e.g., designers, health experts, building automation and control systems specialists, and legislators) is highlighted to define measures aiming to obtain healthy nZEBs.

1.2. Structure Description

The article is organized into four main sections. In the first phase of the study (Section 2), an extensive survey of the existing literature was conducted using the VOSviewer software to identify literature trends during the last decade. Subsequently, the literature search used specific keywords to focus on works produced in the last three years.

In the second step, a thorough literature analysis was conducted, focusing, on the one hand, on the possible causes of mold formation in nearly zero-energy buildings and generally in energy-efficient buildings (Section 3). On the other hand, possible suggestions to avoid the problem of mold growth in nZEBs are provided (Section 4). The latter is accompanied by the description of the available tools for the detection and identification of the problem, if present, and for the prediction of mold formation to adopt suitable solutions already at the design stage.

Finally, the main findings are discussed, and possible perspectives are outlined (Section 5). Figure 1 illustrates the adopted methodology.

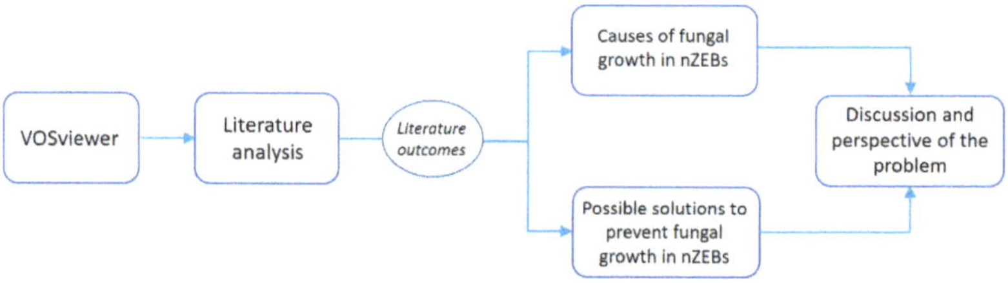

Figure 1. Methodology outline.

2. Methodology

2.1. Research Trend over the Past Years

To analyze the direction and focus researchers have been moving towards, overlay visualizations were developed using the VOSviewer software v1.6.17. The overlay map (Figure 2) based on keywords shows a concentrated occurrence of the keyword "indoor air quality" (with the larger size bubble). The keywords "fungi", "mold", "ventilation", and "thermal comfort" appear with the same occurrence (or same bubble size), which could be interpreted as having the same importance among the documents found. However, these keywords appeared to be used around 2014 (purple). After that, the keywords "indoor air quality", "ventilation", and "indoor environment" were employed around 2017, but "thermal comfort", "hygrothermal performance", and "mould growth" appeared towards 2018–2019 (light green). Lately, between 2020 and 2023 (yellow), the keywords "COVID-19", "condensation", "energy performance", and "air pollutants" have appeared.

This tendency shows that the research direction is moving towards assessing the energy and hygrothermal performance based on condensation and mold growth, which is the specific subject of this perspective paper.

The overlay map based on terms in the title and abstract (Figure 3) shows two distinctive clusters around "concentration" (on the left-hand side) and "performance" (on the right-hand side). The color contrast between these two clusters revealed a focus shift from fungal concentration aspects and health effects (having a peak period around 2014) towards performance assessment in buildings (around 2017) and solutions via humidity control, materials, and other technologies, such as air purifiers (beyond 2020).

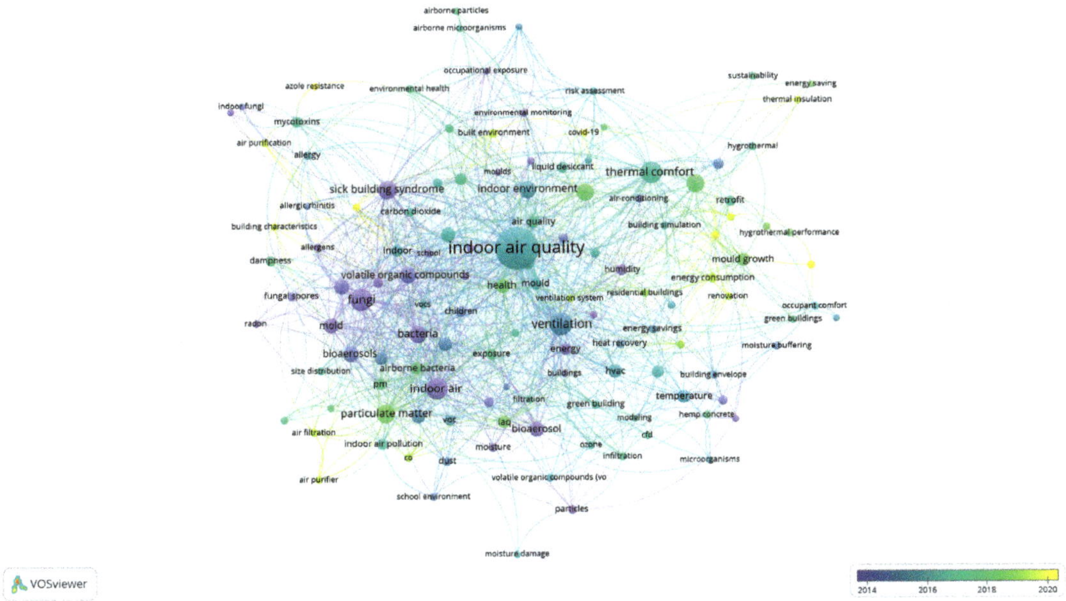

Figure 2. Overlay map based on keywords occurrence. The keyword occurrence threshold is five, where only 130 keywords met such threshold.

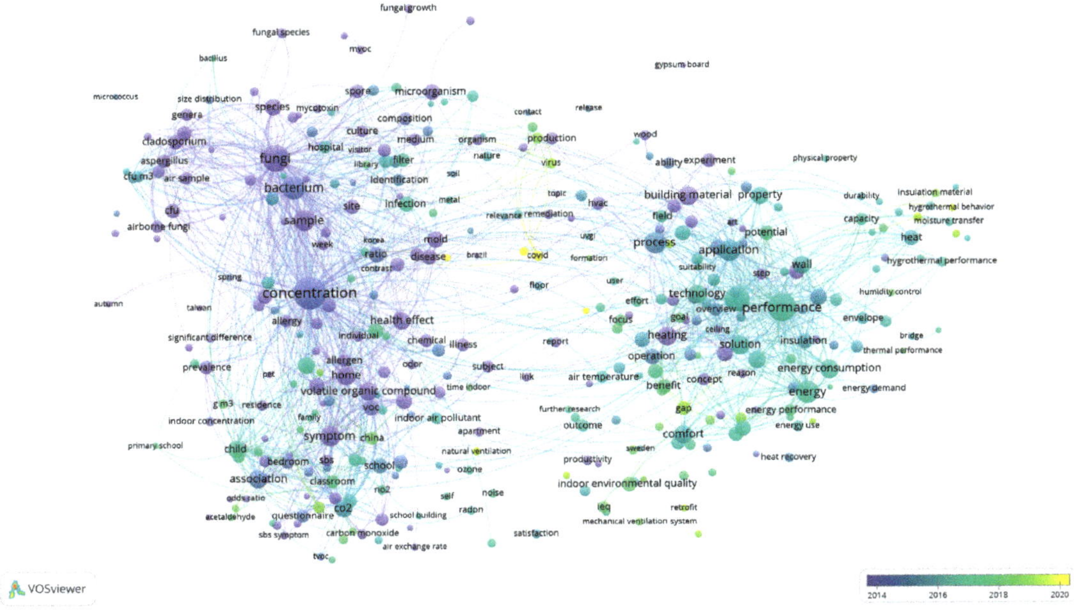

Figure 3. Overlay map based on the terms in the title and abstract. The terms occurrence threshold is ten, where only 622 terms met such threshold.

However, no mention of "nZEB" and "NZEB" was found in the 1128 documents employed in the mapping, which means these terms or keywords did not appear more than the recommended thresholds (five times for the keywords and ten times for terms in title and abstract). The keyword "nzeb" appears with a threshold of two. Only when a threshold of one is selected do the keywords "zero carbon-ready building", "zero energy building", and "zero-energy building" appear. This leaves a gap in the literature on how mold growth control affects the achievement of zero-energy buildings.

The literature search for the development of the overlay maps was limited to the years from 2010 to 2023 within the SCOPUS database to retrieve information from the literature concerning indoor air quality and mold growth in buildings. About 1000 references were encountered; only 113 were strictly related to this review's main subject. The first selection criterion corresponds to selecting only the documents where indoor air quality and mold growth are directly linked to aspects within zero-energy buildings (ZEBs). Due to a low number of studies, a unique selection of documents was performed for those directly related to each aspect of ZEBs. In these, mold growth can threaten indoor air quality and energy consumption, i.e., the design, construction, systems and efficiency measures, operation and maintenance, and regulations. The final list of documents chosen to be analyzed agrees with the observations in Figures 2 and 3.

2.2. Literature Search

After exploring the research trend through the VOSviewer visual maps, a systematic literature review was conducted on the specific topic in the context of the last three years. The investigation was carried out by searching the Scopus database. In the first phase, the association of words "nZEB" and "fungi", or "nZEB" and "mould", was used, but very few results were obtained. Subsequently, the acronym "nZEB" was replaced with the full name "nearly Zero Energy Buildings" and the more general one "Zero Energy Buildings" to broaden the scope of research. However, the results remained limited. Consequently, to extrapolate a greater number of articles, the search was extended using the more general combination "Zero Energy Buildings" and "contaminants", and "Zero Energy Buildings" and "Indoor Air Quality." However, by limiting the field to the category of zero-energy buildings, the findings from the literature were reduced. Finally, the search was further enlarged by inserting the keyword "buildings" combined with the words "fungi", "mold", "dampness", and "moisture." Furthermore, a check was performed by removing the filter for the time period. However, only a few works completed prior to the period of the last three years were deemed particularly interesting to include in the expanded, final literature list. The literature was subsequently analyzed in detail, and the results are reported below.

3. Expected Problems of Energy-Efficient Buildings

The authors deduced the risk areas of energy-efficient buildings through the literature analysis. The domains that require attention are related to the buildings' construction details and practices, mechanical ventilation and its operation, regulations about the prevention of fungal growth, and the occupants' behavior and operation practices. For example, according to research conducted by Kang and Nagano [32], there is evidence to support the presence of organic contaminants inside nearly zero-energy residential buildings in Japan. Kang and Nagano [32] investigated a building where several energy efficiency measures were implemented in the design, including sealing of the envelope, the installation of a highly efficient heating system with heat pumps, and the integration of solar collectors. A ventilation system was set up, using an earth tube and a sensible heat recovery ventilator, supplying fresh air. The results showed that the concentrations of fungi and bacteria vary according to the season and the occupants' lifestyle, while higher concentrations are found in summer. Furthermore, it was found that the environment inside the earth tube favors the growth of fungi.

3.1. Buildings' Construction Details and Techniques

Several studies have pointed out that highly insulated and airtight buildings are prone to increased moisture levels indoors, which can lead to an accumulation of indoor air pollutants if sufficient natural or mechanical ventilation is not ensured [33,34]. Specific parameters can increase the risk of indoor dampness and lead to the right conditions for fungal growth already before the completion of the construction phase [8]. New buildings may face problems of built-in moisture in the envelope due to construction techniques that use materials with high moisture content, water damage incidents, or poorly executed construction details, combined with the lack of regulations and control [8,34]. Ingebretsen et al. [35] analyzed the microclimatic conditions in air cavities and roofs by collecting wood moisture, air temperature, and humidity data for three case buildings in Norway. Even though significant differences are registered between case buildings and the position of the sensors, there was a risk for mold formation in all analyzed buildings. In particular, the authors [35] highlight the critical issues related to ventilated cavities. Insufficient cavity ventilation can prevent structures from drying out, increasing humidity levels and leading to mold growth. Excessive ventilation, on the other hand, can cool the inside layer of the cladding, causing condensation and, thus, mold formation. The air cavities showed higher annual temperature amplitudes for long periods than the external air, which occurs to a greater extent for the ZEB Laboratory [35]. Furthermore, poorly designed or faulty installed wind and air barriers in external walls can lead to water condensation in the construction [34]. Finally, the wide use of economic interior finish solutions with poor hygroscopic properties is worsening the situation [34].

When it comes to interventions in existing buildings, it may be challenging to upgrade them to reach the energy performance of new buildings (nZEBs) [36,37]. Window replacement or the application of internal insulation can lead to indoor dampness if the solutions are not well thought out and designed with focus on the moisture performance of the building [34]. The results of a study by Morelli and Møller [38] show that in some cases, the relative humidity behind the internal insulation would reach 95% during the heating season. The same problems have been documented for interventions in historic buildings that require internal thermal insulation [39]. As Lu et al. [15] highlighted, these actions expose the structures to an increased risk of humidity accumulation due to condensation and the impossibility of drying, causing mold growth and material decay.

3.2. *Mechanical Ventilation Risks of Poor Maintenance or Faulty Operation*

High-performance buildings are less susceptible to high humidity compared to traditional buildings [40]. It is believed that the implementation of mechanical ventilation prevents mold development indoors [8]. At the same time, in heavily polluted areas, mechanical ventilation has substituted natural ventilation to reduce the entrance of external pollutants [2]. Therefore, in new buildings, the installation of mechanical ventilation is often mandatory [8]. However, not many studies have verified the effectiveness of such solutions in energy-efficient buildings [33]. Additionally, the ventilation system may often malfunction, or its performance can be tampered with by the occupants' behavior or intervention, e.g., reduction in ventilation to save energy for heating.

As outdoor air is the primary source of fungal spores indoors, mechanical ventilation is more effective than natural ventilation in creating a better indoor environment while decreasing the relative humidity levels that significantly affect fungal growth. Yet, that is only if the filters are changed regularly (once or twice per year), as fungi can grow in the filters and create an indoor source of spores. Niculita-Hirzel et al. [33] hypothesized that mechanical ventilation is not more effective in reducing indoor fungal growth. Lin et al. [41] provided observed data on the diffusion and retention of biological contaminants by HVAC systems, indicating that they can operate efficiently in removing dispersed fungal spores in the absence of internal or external sources. On the other hand, in the presence of mold, the diffusion of spores within the room is very fast. Additionally, integrating a mechanical ventilation system during the design stage has proven more efficient than a system added during renovation.

In air-handling units (AHUs), fungal growth has often been isolated from the blower wheel fan blades, ductwork, cooling coil fins, and insulation [42]. More specifically, the blower wheel fan blades are the most contaminated locations. On a positive note, they are easily accessible for sampling. Extensive sampling from AHUs by Wilson et al. [43] revealed *Cladosporium* spp. (*Cladosporium cladosporioides* or *Cladosporium sphaerospermum*) to be the most commonly isolated from blower wheel fan blades, ducts, and cooling coil fins. At the same time, species from *Penicillium chrysogenum*, *Aspergillus* spp., and *Paecilomyces* spp. were found at the cooling coil fins and insulation. The article has shown that mold growth is very common in AHUs, as the right conditions of water activity and substrate can be provided. If the filters are improperly maintained or not changed regularly, they can serve as a microbiological niche. The spores can enter AHUs, either from outdoors or indoors, through the return air system. No clear correlations were drawn between the risk of contamination and the operating conditions of the units.

Poor or a lack of maintenance increases the risk of mold growth inside duct surfaces in HVAC systems. That is because duct cleaning often follows a visual inspection only based on the dust quantity rather than on mold content [44], and the spore aerosolization rate increases with the colony age [45]. Few studies have focused on fungal aerosolization and resuspension inside ventilation ducts. Li [46] studied this phenomenon with *Aspergillus niger* over the elbow and reducer duct with circular and rectangular sections, while Liu et al. [47] studied hydrophobic particles in a straight-duct uniform section.

Aerosolization increases when the spores grow at the center of rectangular ducts rather than at the corners. Similarly, it increases with the reducer's constriction angle and elbow angle [46]. Ventilation and the regular use of air conditioning are critical factors in potentially reduced aerosolization [47]. The resuspension rate increases with particle size [47] and air-flow velocity [47,48], while a slight roughness in ducts' walls decreases this rate. However, humidity and temperature variations do not impact the resuspension rate significantly [47]. Higher mold levels of floor dust were found in rooms with mechanical ventilation systems than in rooms with extractor fans only [49]. However, another large study showed insufficient effectiveness in only using exhaust fans to remedy mold problems [50]. The filtration efficiency of an air-handling unit installed in a low-energy office building was analyzed by Pavard et al. [42], showing that the microbial concentration of fresh air entering the AHU was higher than that in the air extracted from the office. Bakker et al. [51] demonstrated that the low efficiency of the upstream filters of an air conditioning plant determines an increase in bacteria and fungi introduced through aerosolization into the environment.

Previous studies [52] have reported the relationship between indoor humidity levels and ventilation rates, with different associations. Indeed, some indicate that indoor humidity decreases as ventilation rates increase, while others point out the opposite effect. A low air exchange rate could increase the risk of health symptoms [49] with higher prevalence when using mechanical rather than natural ventilation [53]. Furthermore, the actual ventilation rates do not always comply with the minimum demands set by the standards.

Even though mechanical ventilation is supposed to prevent the increase in indoor dampness, incorrect design and sizing can compromise the effectiveness of such systems [34]. Alaidroos and Mosly [54] tackled the issue of maintaining adequate indoor air quality levels in buildings with high mechanical ventilation rates in hot and humid climates. The results demonstrated that increasing the ventilation rate causes an increased risk of mold growth due to the high outdoor humidity levels of the specific climate in which the building is located. Moreover, the ventilation rate is more significant for the mold index than the cooling setpoint temperature. On the other hand, Almatawah et al. [55] evaluated the concentration of airborne microbes (bacteria and fungi) in the external and internal environment for a case study in Kuwait's hot and arid desert climate. The concentration of fungi was very high in summer and winter, demonstrating a very high level of contamination.

In many European countries, the air-tightness standards can be pretty strict [34]. When sufficient ventilation is not ensured due to inadequate air change rates, such dwellings may have increased concentrations of fungal particles and indoor pollutants, decreased indoor air quality, high indoor dampness, condensation on buildings' surfaces, and fungal growth [2,13,34]. Relying solely on mechanical ventilation can also create problems during the construction phase, when the ventilation is not yet operational, or in case of an electric power cut. Finally, the systems often rely on the occupants' behavior to operate as designed, while most systems do not control humidity levels [2].

3.3. Building Regulations

Even though there is definitive proof of the adverse health effects of fungal growth in buildings, indoor mold standards are inconsistent and lack clear and restrictive guidelines [2]. For example, ventilation requirements are dictated by the energy savings and occupants' comfort levels. However, in many cases, these requirements might be contradictory or not weigh the same. Airtightness has increased in high-performance buildings; nevertheless, the standards regarding ventilation rates have not changed compared to traditional buildings [52]. An approach toward humidity control and fungal growth risk is missing [2,52]. Coulburn and Miller [56] conducted an integrative review to investigate the prevalence, risk factors, and impact associated with mold in Australian housing. The authors draw attention to the extent and prevalence of diagnosed mold in buildings constructed following the regulations. The limited focus of regulations and standards related to indoor microbial pollutants like bacteria, viruses, and fungi is probably due to the difficult assessment process for these pollutants [57,58].

Some studies have focused on analyzing the influence of temperature and relative humidity on the concentration levels of pollutants indoors. De Jonge and Laverge [59] performed dynamic modeling of volatile organic compounds (VOCs) using the zero-energy house "E-cube", which is characterized by high air tightness and is equipped with demand-control ventilation. The results highlight the significant influence of temperature and humidity on the internal concentration levels of VOCs, which exceed the permitted levels set by the guidelines. This underlines the importance of adequately designing the mechanical ventilation systems and related control strategies, considering the dynamic behavior of VOCs. Finally, sometimes the standards and regulations are not oriented to avoid the risk of increased humidity, for example, in the case of small- and medium-sized residential high-performance buildings in China [52].

3.4. Energy Savings in Heating and Fuel Poverty

Fuel poverty has a socio-economic aspect related to inadequate heating, the conditions and maintenance level of buildings, as well as financial limitations that may, for example, be connected to the overcrowding of houses. When buildings are over-occupied, indoor relative humidity increases, while the design ventilation rates may become insufficient [8]. On the other hand, inadequate heating can lead to cold surfaces and, consequently, to water condensation and mold development. In many countries, the recommended indoor air temperature has decreased by 1 or 2 degrees due to the energy crisis that started in 2022. Even though this decrease can lead to significant savings in energy consumption, there is a significant risk of mold formation on cold surfaces due to moisture condensation [34].

In the comprehensive, holistic review carried out by Coulburn and Miller [56] for mold-affected housing in Australia, risk factors linked to poor housing conditions, poor-quality rental accommodation, and socioeconomic aspects have been reported by most processed studies. The works reviewed show the association between conditions of poverty and the reporting of indoor mold problems. Households highlight the difficulty of keeping the home warm or cool due to the poor quality of the construction and the need to save on bills. The authors [56] also pointed out the high prevalence of poor-quality housing and indoor mold in socioeconomically disadvantaged groups.

The investigation carried out by Sharpe et al. [60] in the UK found energy poverty to be affecting a third of the survey participants (671 households). Energy-poor behaviors associated with insufficient or no heating due to the high costs increase the risk of mold contamination. Furthermore, the mediation analysis conducted on the collected data showed that fuel-poor occupants might not benefit from energy efficiency measures on buildings due to the residents' ineffective heating and ventilation practices.

4. Suggestions

4.1. Necessary Steps to Avoid Fungal Growth Problems in nZEBs

There is a need for regulations that include assessment procedures (commissioning) [8,61], prevention/mitigation strategies that are planned already at the design process and implemented during the construction phase [2,8,34], and an operation period [2,8]. It is also essential to raise awareness concerning better practices for occupants [2,34,61]. Such regulations must be localized to match the specific context of the building location, climate, construction type, energy usage, and fungal species, among other factors [62].

On the one hand, the literature has shown that the risk for mold growth, visible or not, increases in environments with relative humidity (RH) above 80% [63]. Conversely, significantly lower risk has been found in environments with RH lower than 40% (for example, when working with *C. cladosporioides*) [64], specifically in materials such as hempcrete and straw [63,65,66]. Thus, correct material selection during the construction phase should reduce the risk. Furthermore, this could be achieved with the use of clay or lime plaster by pre-treating the materials with anti-fungal substances [63] or by using non-porous [47] or water- and moisture-protected materials [63,67]. Other materials, such as glass wool, cement blocks, and compressed earth brick, retain less moisture than hemp-based biomaterials [65]. Finally, adding insulation layers could also help reduce the mold index to near zero [68], whereas an insulation thickness of at least 40 mm appeared to be enough to reduce the condensation risk, regardless of the presence of thermal bridges [69].

On the other hand, for indoor environments with high relative humidity (RH), during the operation phase, the literature recommends mitigation measures such as control of the relative humidity fluctuations [70] rather than controlling indoor air temperature fluctuations [64]. The same applies to highly insulated, airtight buildings with reduced air infiltration and hygrothermal buffering, where it becomes necessary to control the fluctuation in indoor RH. That cannot be achieved through the supply of outdoor air by a mechanical ventilation system, and therefore there is need for other humidity-buffering technologies [70], like air dehumidification devices [34]. However, in case of microbial contamination in air-handling units (AHUs) or other areas of the dwelling, it is necessary to clean the affected areas using anti-microbial agents and following the standards and industry recommendations, as well as using high-efficiency particulate air (HEPA) vacuum cleaning, unless the contamination is so extensive that the replacement of the components or materials might be a better practice [43].

Quality assurance and moisture safety commissioning processes are necessary for nZEB renovations, consisting of fixed regulations and inspections of all implemented solutions before, during, and after the completion of construction. During construction, it is crucial to ensure that all materials, structures, and products are protected against moisture and weather conditions. At the same time, all surfaces need to be dry and water-free before being covered [61].

When accounting for all the mitigation measures, an increase in the overall energy consumption of the building could occur, thus increasing the risk of non-compliance with the energy efficiency regulation requirements for nZEBs or Passive Haus. However, supposing that the recommendations to avoid the mold growth risk during the construction phase are followed, the energy efficiency requirements for nZEBs may only be affected by the initial investment costs and the periodical commissioning for indoor environmental quality assurance. The commissioning procedures or assessment methods for mold detection during the building's operation phase are presented hereafter.

4.2. Detection of Mold- and Moisture-Related Problems through Questionnaires

Questionnaires are among the most widespread tools for collecting data on the characteristics of buildings and how the interior spaces are experienced by the occupants, as well as their perception of air quality, comfort levels, and critical conditions [71]. Several authors have experimented with the use of questionnaires to investigate problems related to humidity and mold in homes, highlighting the difficulty of reporting the presence of fungal formations and, consequently, diagnosing the causes. This problem is reported, for example, by Marasinghe et al. [72]. The authors conducted a survey to test the validity of the data reported through self-administered questionnaires regarding the characteristics of the buildings and the presence of dampness. Inspections and measurements in the houses were conducted to verify the correspondence between the reported data and the actual situation. The questionnaires proved to be a useful data collection tool concerning the construction characteristics of the buildings. However, regarding the dampness indicators, less agreement was observed between the information reported by the occupants and that detected by the inspectors. The findings of this research thus highlight the need to use integrative tools, such as digital photos, moisture-sensitive devices, or more detailed information, to increase the validity of the questionnaires used to investigate mold and humidity problems in dwellings.

Wang et al. [73] conducted an extensive questionnaire-based survey campaign (40,279 questionnaires) in eight Chinese cities to investigate the perceived indoor air quality and home environment. The results disclosed a relationship between mold and odors and climatic conditions. In particular, visible mold or damp stains and mold odor are more common in cities of south China, characterized by higher mean ambient temperature. Also, these phenomena are more frequent in old buildings. Moreover, the study highlighted other associated factors for mold and dampness in the home, including pets and pests (rats/mice, cockroaches), low ventilation levels, and inadequate cleaning. A correlation was also found between dampness/mold, allergic rhinitis or asthma in adults, and the perception of odors. An observational assessment tool was developed by Park and Cox-Ganser [74] for evaluating indoor dampness and mold. The tool is based on visual and olfactory inspections and uses a standardized evaluation form supplemented by a questionnaire about the participants' respiratory diseases. Water stains, visible mold, mold odor, and moisture are detected, and their magnitude is scored based on intensity or size. All building components are included in the assessment (walls, ceiling, floor, windows, furniture, ventilation systems, pipes, and materials). The results showed significant associations between the individual exposure index and building-related respiratory symptoms. The proposed dampness and mold assessment sheet can be used for an initial assessment of building moisture and mold presence or following a water damage incident. Nevertheless, indoor microbial assessment cannot rely entirely on questionnaires, as occupants are often unaware of the mold-related problems in their homes, especially when it comes to hidden fungal growth.

Figure 4 shows the questions generally contained in questionnaires aimed at identifying the presence of fungi in indoor environments.

4.3. Detection Methods for Visible and Hidden Fungal Growth

Several detection methods for the assessment of indoor microbial contamination have been studied over the years and are commonly used by inspection companies. These methods can be used during the construction or operation phase to characterize dampness and mold-associated growth in buildings. Such investigations should be held when there are signs or suspicions of microbiological growth or after water damage incidents (e.g., pipe leaks, flooding, water penetration). It is essential to raise awareness and educate occupants, owners, and operational staff about fungal growth risks so that they are qualified to prevent, assess the signs, and remedy indoor fungal growth. These methods could be classified as direct and indirect:

(a) Standardized Direct Methods:

Direct microbial assessment methods comprise various sampling protocols [43], depending on the purpose and source of the sample, combined with a variety of quantitative or qualitative analysis techniques: e.g., culture [44], direct microscopic spore count

(DMSC) [44], β-N-acetyhexosaminidase (NAHA) dosing [44], and qPCR [44]. Even though there are no standardized protocols or generally accepted guidelines to be followed, the research community has been actively working toward this direction.

(b) Indirect Methods:

There are several driving forces to moisture transport, such as capillary pressure, moisture content, vapor density gradient, partial water vapor pressure, relative humidity, and chemical composition, which can be used for assessing hygrothermal dynamics. Selecting the driving forces to find accurate hygrothermal results is a challenging task. However, the pressure inside buildings is assumed to be constant, so the literature suggests that using the vapor pressure difference as a potential guide for isobaric cases can give relatively better results [67].

Among the indirect detection methods are modeling approaches that allow a relationship to be established between relative humidity and mold growth, e.g., the Sedlbauer model [75], the Sautour relationship (considered an improvement to the former) [75], and the VTT model [76]. Other approaches are implementing the temperature factor as an indicator of condensation risk (or mold growth), i.e., ISO 13788 [77], ASHRAE 160-2009 [78]; comparing only the relative humidity of indoor air [79]; comparing barrier temperatures [79], i.e., BS5250; and comparing the air vapor pressures [79]. However, standardized indicators have proven to not consistently provide reliable results in predicting the risk of mold growth [50].

(c) Other Proposed Methods:

Some innovative methods and techniques are under investigation and are intended to attain more consistent and reproducible results. For example, the proposed fungal spore source strength tester (FSSST) can assess the aerosolization potential of fungal spores from contaminated surfaces. It can be effectively used in the field to measure the highest possible level of spore concentration released from a mold source in an indoor environment [80]. Another creative method is the electronic nose (e-nose), developed by Suchorab et al. [81], to detect fungal-contaminated materials and surfaces at an early stage by using MOS-type sensors.

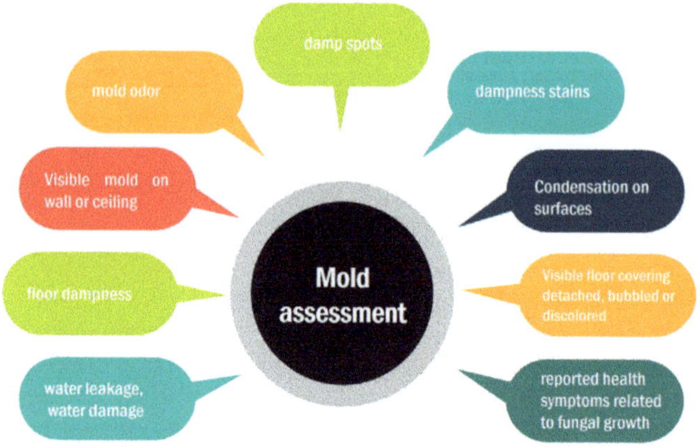

Figure 4. Questions to assess fungal formations in buildings.

4.4. Prediction Models

In evaluating the mold growth risk, the accuracy of the prediction models plays an important role. These models have evolved, improving their reliability and adopting numerous parameters that can contribute to the development of fungi in indoor environments.

A new dynamic mold growth model was developed by Boardman et al. [82], namely the Dose–Response Simple Isopleth for Mold, which can predict the moisture risk in wooden buildings. A mold germination isopleth is calculated based on the critical relative humidity, which depends on the surface temperature and relative humidity conditions. The model's predictions have been compared with field studies, proving its reliability. The model can be effectively used to evaluate the risk of mold starting from temperature and relative humidity data measured in the field or provided by hygrothermal simulations.

Moon and Augenbroe [83] have highlighted the need to construct a practical indicator for predicting mold growth. The authors draw attention to the disparity between the theoretical analysis of mold risk, based on simulations, and the risk in actual conditions, comprising uncontrolled and unconsidered factors intervening with mold formation. Four categories are identified as the causal factors of unpredicted mold growth: spore source, substrate conditions, HVAC maintenance and operation, and local building details. The research proposes a method based on a mold germination graph, considering additional building parameters in idealized situations. These are then used to define a practical performance indicator by introducing the additional causal factors evaluated for each case.

Using reliable and consistent data regarding the hygrometric properties of materials in modeling and simulating vapor transfer through structures makes predictions more robust. The required data can only be obtained through laboratory experiments. With this aim, Olaoye et al. [84] demonstrated how a low-cost hygrothermally controlled test room could be used instead of the more expensive climatic chamber to successfully conduct tests on the hygrometric properties of materials. In particular, a flexible membrane was tested, for which the water vapor resistivity and diffusion properties were determined.

4.5. Smart Technologies to Prevent Fungal Formation

Alternatively, low-consumption, humidity-buffering technologies with sensors have been shown to prevent fungal formation. Regarding ventilation strategies, a ventilation solution that uses an artificial thermal bridge with a dewpoint sensor that is designed to activate the ventilation system when the conditions become critical in risk areas of the inner envelope helps avoid constant ventilation and high energy consumption, along with the risk of mold development [1]. Another ventilation technique based on indoor CO_2 levels combined with a hygrothermal sensor is analyzed in [85], aiming to maintain good environmental conditions with low energy consumption while mitigating the condensation risk [85]. However, the authors in [79] suggest a deflection of wind-driven rain as the most effective strategy for decreasing the risk of mold growth. Other solutions could be the improvement of the water tightness of the envelope and cladding (red matt clay brick cladding walls) ventilation. Advances in sensor technology and smart control systems can help in monitoring indoor air quality and ensuring healthy and comfortable conditions for occupants [86].

5. Discussion

Nearly zero-energy buildings are an effective solution for reducing greenhouse gas emissions and decarbonizing the built environment. Indeed, the advancement in technologies has made it possible to achieve very-high-performance buildings with minimal energy requirements. Energy-efficient buildings are highly insulated and airtight to achieve low energy consumption, which can lead to an augmented moisture level if sufficient ventilation is not ensured. Indoor dampness is the main trigger for the formation of microbial growth. Therefore, indoor spaces can become fertile grounds for the germination of spores, the growth of molds, and the accumulation of pollutants indoors, often without the occupants being aware of the problem.

Most nearly zero-energy buildings are equipped with mechanical ventilation, which can create good indoor environmental conditions by removing humidity and indoor pollutants and preventing outdoor contaminants from entering the building. However, the situation can transform if the mechanical ventilation is not designed or dimensioned correctly, if it malfunctions, is not operated properly, or its designed operation is tempered by the occupants' accidental or intentional interference. Furthermore, if the mechanical ventilation system is not maintained correctly and checked regularly, there is a high risk that fungal growth can initiate inside the ducts and other components of the system.

Another area that requires special attention is the construction techniques and the sensitive building elements. This does not apply only to energy-efficient buildings, but airtight dwellings are more sensitive to trapped moisture than traditional buildings with high infiltration and exfiltration rates. Increased moisture and dampness can originate from errors or faulty installation, materials or other elements with high moisture content, water damage incidents, etc. Special attention must be given to isolated interventions that might not be well designed or thought through, i.e., adding insulation, implementing low-quality interior finishes, replacing windows, etc.

Awareness of the problem is required on all fronts, from designers and construction teams to the building owners, occupants, as well as operating and maintenance staff. Designers are called to perform accurate assessments in the planning phase to devise the most appropriate solutions to ensure healthy internal conditions. Construction workers must have expertise in the field, especially in the installation of systems, which play a crucial role in nZEBs. The occupants must be suitably educated and informed to be able to correctly use zero-energy buildings, without compromising their efficiency and without, on the other hand, generating unhealthy internal conditions. Finally, the operating and maintenance staff must follow a well-structured maintenance plan to detect signs of dampness and fungal growth and know how to assess or remediate fungal contamination incidents, if necessary.

Altogether, the literature underlines the importance of the proper design of mechanical ventilation systems and related control strategies [59]; detailed condensation risk analyses during early design stages [13]; climate-specific design, especially for new construction solutions [13,87]; and the monitoring of hygrothermal conditions in the building envelope for low-energy dwellings to verify whether the moisture safety requirements are met and to predict moisture-related risks during the building's service life [88]. Additionally, planning and periodically reviewing HVAC systems' maintenance and cleaning activities, including all system components, e.g., ducts, filters, blower wheel fan blades, cooling coil fins, air-handling unit water tanks, etc. [2,55]. Finally, system monitoring may be a way to ensure its correct operation.

Greater attention to this problem is also required from legislators and those responsible for drafting the technical standards. There is, in fact, a lack of regulations and guidelines in this area. The current regulations envisage only "indirect" types of control for the formation of fungi in buildings through control parameters such as temperature and relative humidity. However, precise limits of these parameters have yet to be established, with a particular focus on zero-energy buildings. Some regulations provide reference values without, however, specifying when, how, and how often to verify these values, which can constantly vary during the operation phase of the building. Therefore, implementing standards and regulations for the operation phase, focusing on indoor microbial pollutants, is crucial. Adopting a "precautionary approach" is also suggested before future measures on energy efficiency requirements are introduced, as they can increase the risk of surface and interstitial condensation with the consequent formation of mold [56].

Automation and control systems could provide a possible aid for the prevention of mold growth. These, if properly integrated into the building, should be able to decouple the management of the house from occupants' habits. Through an apparatus of sensors and actuators, automation systems or the most recent evolution of intelligent and cognitive systems can act to maintain adequate levels of temperature and internal humidity, as well as detecting the presence of contaminants to intervene promptly.

6. Conclusions

This study aims to raise awareness about the risk of indoor fungal contamination in nearly zero-energy buildings. Indisputably, indoor environmental quality cannot be compromised to decrease the energy consumption, tampering with the occupants' health and well-being. It is the authors' opinion that indoor mold formation and the construction practices for energy-efficient buildings have not been clearly associated, creating a pressing need to address the problem and define prevention strategies. From a theoretical perspective, this analysis presents an overview of the state of the art on the issue and identifies the most recurring risk areas that require focus and further research. The adopted measures must be considered and planned throughout all the stages of the building's lifetime, starting from the design to the construction phase and the operation period, which requires a multidisciplinary and integrated approach. From the practical perspective, this work suggests solutions and measures to help prevent dampness and mold growth incidents, as well as diagnostic techniques and detection methods in case of suspicions or signs of fungal growth.

Based on the results of the study, the following answers can be provided to the research questions:

- A gap in the literature on fungal development in zero-energy buildings has been identified. Even though a few studies have directly addressed the topic, the problem of indoor fungal contamination has not yet received enough focus to spark a generalized reaction from legislative institutes or motivate the industry to take action. There is enough proof that indoor fungal growth can cause or exacerbate adverse health problems for exposed occupants. Still, no organized and widespread efforts to tackle the problem have taken effect.
- The number of specific studies on mold growth in zero-energy buildings is small, so the complete picture of the causes is still unclear. Nevertheless, the literature analysis has revealed some clear risk areas and trigger factors for indoor mold contamination. The conclusions can be summarized into four areas: the building construction and sensitive areas that can result from errors or poorly designed solutions; insufficient or inadequate ventilation due to problematic operation or lack of proper/regular maintenance; lack of regulations and standards to address the problem and impose preventive measures; and occupants' behavior and daily habits, very often due to a lack of knowledge or awareness of correct practices.
- Consequently, to prevent or mitigate the risk of fungal formation in nZEBs it is necessary to adopt a suitable, multidisciplinary approach for the design and construction of energy-efficient buildings, with direct focus on this topic. That would entail a moisture control design and quality assurance practices during the construction phase. During the operation phase, it is essential to ensure that all systems function properly and receive regular inspections and maintenance. Of course, updating or implementing regulations to address the problem, together with informing and educating both professionals and occupants, are necessary steps toward these goals. Finally, using smart, automated control technologies can ensure healthy indoor conditions.

The following lines of future research have been identified:

- Expand the available data on the topic to assess the extent of the problem and provide further proof. This requires performing numerical and simulation analyses to better understand the potential risks and underlying conditions leading to mold formation. Then, a large-scale field study is needed to collect a large amount of data from energy-efficient buildings and evaluate the actual risk and causes.
- Define the necessary steps and activities for training, knowledge, and information dissemination for professionals, operating staff, owners, and occupants. Some critical aspects for preventing dampness- and mold-related problems are proper building and system operation, regular inspections and maintenance, correct occupational practices, etc. In this regard, commissioning and retrofitting processes need to include the aspects of moisture safety. Fungal detection methods need to be standardized, and intervention mechanisms and protocols must be appointed if suspicions of microbial contamination arise.

- Adjust the regulations to include a direct consideration and assessment of fungal growth risks in buildings. Clear guidelines and prevention measures are necessary, adjusted to the special conditions of different climates.
- The design and development of innovative technologies and automated control systems to measure and regulate indoor environmental parameters, including microbial products.
- Further investigation regarding material properties and especially new, biogenic materials that have not been researched enough yet.

This study does have some limitations. Even though the interaction is evident, the occupants' role and behavior need to be more extensively covered, as they play a critical role in mold-related problems in buildings. Therefore, further investigation is required in future studies. Furthermore, the range of detection methods and predictive models presented in the article is not exhaustive, since the purpose of the study is to define the leading causes of fungal growth in nZEBs and suggest possible solutions. For more in-depth knowledge of these tools, please refer to the suggested literature. Additionally, remediation strategies for fungal removal have not been addressed. Such techniques are essential to restore safe and healthy indoor conditions, ensure the conservation of building materials and elements, and interrupt the degradation process after the detection of fungal growth. Some damages could be irreversible if not addressed (for example, stains and detachments of surface finishes or the encrustation and corrosion of the system's pipes). This subject requires specific attention and extensive research to uncover the necessary actions to be implemented.

Author Contributions: Conceptualization, C.C. and E.L.; methodology C.C., E.L. and M.C.A.; software, M.C.A.; investigation, E.L.; resources, N.A.; writing—original draft preparation, C.C., E.L. and M.C.A.; writing—review and editing, C.C. and E.L.; visualization, B.A. and D.M.; supervision, B.A., D.M. and N.A.; project administration, M.C.A., D.M. and N.A.; funding acquisition, M.C.A., D.M. and N.A. All authors have read and agreed to the published version of the manuscript.

Funding: This research was funded by the Panamanian Institution Secretaría Nacional de Ciencia, Tecnología e Innovación SENACYT, under the project code IDDS22-30, as well as supported by the Sistema Nacional de Investigación (SNI). The authors would also like to thank Landsbyggefonden and Grundejernes Investeringsfond for their financial support.

Data Availability Statement: Not applicable.

Conflicts of Interest: The authors declare no conflict of interest. The funders had no role in the study's design; in the collection, analyses, or interpretation of data; in the writing of the manuscript; or in the decision to publish the results.

Abbreviations

nZEBs	nearly Zero-Energy Buildings
VOCs	Volatile Organic Compounds
HVAC	Heating, Ventilation, and Air Conditioning
IEQ	Indoor Environmental Quality
AHU	Air Handling Unit
RH	Relative Humidity
HEPA	High-Efficiency Particulate Air filter

References

1. Krus, M.; Sedlbauer, K.; Sinnesbichler, H. *Sensor-Free Control of the Ventilation for the Prevention of Mould Growth by an Artificial Thermal Bridge with a Dew Point Switch*; Fraunhofer-Institute for Building Physics: Munich, Germany, 2005.
2. Brambilla, A.; Sangiorgio, A. Mould growth in energy efficient buildings: Causes, health implications and strategies to mitigate the risk. *Renew. Sustain. Energy Rev.* **2020**, *132*, 110093. [CrossRef]
3. Ebbehøj, N.E.; Hansen, M.Ø.; Sigsgaard, T.; Larsen, L. Building-related symptoms and molds: A two-step intervention study. *Indoor Air* **2002**, *12*, 273–277. [CrossRef] [PubMed]

4. Yang, S.; Pernot, J.G.; Jörin, C.H.; Niculita-Hirzel, H.; Perret, V.; Licina, D. Energy, indoor air quality, occupant behavior, self-reported symptoms and satisfaction in energy-efficient dwellings in Switzerland. *Build. Environ.* **2020**, *171*, 106618. [CrossRef]
5. Yang, Q.; Wang, J.; Norbäck, D. The home environment in a nationwide sample of multi-family buildings in Sweden: Associations with ocular, nasal, throat and dermal symptoms, headache, and fatigue among adults. *Indoor Air* **2021**, *31*, 1402–1416. [CrossRef]
6. Wang, J.; Janson, C.; Gislason, T.; Gunnbjörnsdottir, M.; Jogi, R.; Orru, H.; Norbäck, D. Volatile organic compounds (VOC) in homes associated with asthma and lung function among adults in Northern Europe. *Environ. Pollut.* **2023**, *321*, 121103. [CrossRef] [PubMed]
7. Maundrill, Z.C.; Dams, B.; Ansell, M.; Henk, D.; Ezugwu, E.K.; Harney, M.; Stewart, J.; Ball, R.J. Moisture and fungal degradation in fibrous plaster. *Constr. Build. Mater.* **2023**, *369*, 130604. [CrossRef]
8. Ginestet, S.; Aschan-Leygonie, C.; Bayeux, T.; Keirsbulck, M. Mould in indoor environments: The role of heating, ventilation and fuel poverty. *A French perspective. Build. Environ.* **2020**, *169*, 106577. [CrossRef]
9. Silveira, V.D.C.; Pinto, M.M.; Westphal, F.S. Influence of environmental factors favorable to the development and proliferation of mold in residential buildings in tropical climates. *Build. Environ.* **2019**, *166*, 106421. [CrossRef]
10. Xue, Y.; Fan, Y.; Wang, Z.; Gao, W.; Sun, Z.; Ge, J. Facilitator of moisture accumulation in building envelopes and its influences on condensation and mould growth. *Energy Build.* **2022**, *277*, 112528. [CrossRef]
11. Kon, O.; Caner, İ. The Effect of External Wall Insulation on Mold and Moisture on the Buildings. *Buildings* **2022**, *12*, 521. [CrossRef]
12. Morelli, M.; Rasmussen, T.V.; Therkelsen, M. Exterior wood-frame walls—Wind–vapour barrier ratio in denmark. *Buildings* **2021**, *11*, 428. [CrossRef]
13. Torres-Rivas, A.; Palumbo, M.; Haddad, A.; Cabeza, L.F.; Jiménez, L.; Boer, D. Multi-objective optimisation of bio-based thermal insulation materials in building envelopes considering condensation risk. *Appl. Energy* **2018**, *224*, 602–614. [CrossRef]
14. Bruno, R.; Arcuri, N.; Carpino, C. Study of innovative solutions of the building envelope for passive houses in Mediterranean areas. *Energy Procedia* **2017**, *140*, 80–92. [CrossRef]
15. Lu, J.; Marincioni, V.; Orr, S.A.; Altamirano-Medina, H. Climate resilience of internally-insulated historic masonry assemblies: Comparison of moisture risk under current and future climate scenarios. *Minerals* **2021**, *11*, 271. [CrossRef]
16. Defo, M.; Lacasse, M.A. Effects of climate change on the moisture performance of tallwood building envelope. *Buildings* **2021**, *11*, 35. [CrossRef]
17. Ozarisoy, B. Energy effectiveness of passive cooling design strategies to reduce the impact of long-term heatwaves on occupants' thermal comfort in Europe: Climate change and mitigation. *J. Clean. Prod.* **2022**, *330*, 129675. [CrossRef]
18. D'Agostino, D.; Parker, D.; Epifani, I.; Crawley, D.; Lawrie, L. How will future climate impact the design and performance of nearly zero energy buildings (NZEBs)? *Energy* **2022**, *240*, 122479. [CrossRef]
19. López-Ochoa, L.M.; Las-Heras-Casas, J.; Olasolo-Alonso, P.; López-González, L.M. Towards nearly zero-energy buildings in Mediterranean countries: Fifteen years of implementing the Energy Performance of Buildings Directive in Spain (2006–2020). *J. Build. Eng.* **2021**, *44*, 102962. [CrossRef]
20. Barrutieta, X.; Gainza, J.; Irulegi, O.; Hernández, R. The zero building: An exemplary nearly zero energy office building (NZEB) and its potential to become a positive energy building (PEB). *Arch. Sci. Rev.* **2021**, *66*, 214–225. [CrossRef]
21. Tetteh, M.O.; Darko, A.; Chan, A.P.C.; Jafari, A.; Brilakis, I.; Chen, W.; Nani, G.; Yevu, S.K. Scientometric mapping of global research on green retrofitting of existing buildings (GREB): Pathway towards a holistic GREB framework. *Energy Build.* **2022**, *277*, 112532. [CrossRef]
22. Chegari, B.; Tabaa, M.; Simeu, E.; Moutaouakkil, F.; Medromi, H. An optimal surrogate-model-based approach to support comfortable and nearly zero energy buildings design. *Energy* **2022**, *248*, 123584. [CrossRef]
23. Liu, P.; Alonso, M.J.; Mathisen, H.M. Global sensitivity analysis and optimal design of heat recovery ventilation for zero emission buildings. *Appl. Energy* **2023**, *329*, 120237. [CrossRef]
24. D'Agostino, D.; Minelli, F.; Minichiello, F. New genetic algorithm-based workflow for multi-objective optimization of Net Zero Energy Buildings integrating robustness assessment. *Energy Build.* **2023**, *284*, 112841. [CrossRef]
25. Ozarisoy, B.; Altan, H. Limitations: Developing an Evidence-Based Energy Policy Framework to Asset Robust Energy Performance Evaluation and Certification Schemes. In *Handbook of Retrofitting High Density Residential Buildings: Policy Design and Implications on Domestic Energy Use in the Eastern Mediterranean Climate of Cyprus;* Springer International Publishing: Cham, Switzerland, 2022; pp. 617–640. [CrossRef]
26. Lim, A.Y.; Yoon, M.; Kim, E.H.; Kim, H.A.; Lee, M.J.; Cheong, H.K. Effects of mechanical ventilation on indoor air quality and occupant health status in energy-efficient homes: A longitudinal field study. *Sci. Total Environ.* **2021**, *785*, 147324. [CrossRef]
27. Wu, Z.; Jin, Y.; Huang, L.; Zhang, N.; Luo, X.; Wang, X. Appraisal of indoor environment quality (IEQ) in nearly Zero energy Building (nZEB): A literature review. *Sci. Technol. Built Environ.* **2023**, *29*, 251–267. [CrossRef]
28. Steinemann, A.; Wargocki, P.; Rismanchi, B. Ten questions concerning green buildings and indoor air quality. *Build. Environ.* **2017**, *112*, 351–358. [CrossRef]
29. Vidal, I.R.; Otaegi, J.; Oregi, X. Thermal comfort in nzeb collective housing in Northern Spain. *Sustainability* **2020**, *12*, 9630. [CrossRef]
30. De Masi, R.F.; Gigante, A.; Festa, V.; Ruggiero, S.; Vanoli, G.P. Effect of hvac's management on indoor thermo-hygrometric comfort and energy balance: In situ assessments on a real nzeb. *Energies* **2021**, *14*, 7187. [CrossRef]

31. Liu, Z.; Zhou, Q.; Yin, H.; Xu, W.; Yang, X.; Gao, J. Indoor environmental quality and energy consumption real-time assessment: A field measurement of a nearly zero-energy building in cold region of China. *Energy Build.* **2021**, *246*, 111093. [CrossRef]
32. Kang, Y.; Nagano, K. Field measurement of indoor air quality and airborne microbes in a near-zero energy house with an earth tube in the cold region of Japan. *Sci. Technol. Built Environ.* **2016**, *22*, 1010–1023. [CrossRef]
33. Niculita-Hirzel, H.; Yang, S.; Jörin, C.H.; Perret, V.; Licina, D.; Pernot, J.G. Fungal contaminants in energy efficient dwellings: Impact of ventilation type and level of urbanization. *Int. J. Environ. Res. Public Health* **2020**, *17*, 4936. [CrossRef]
34. Di Giuseppe, E. *Nearly Zero Energy Buildings and Proliferation of Microorganisms. A Current Issue for Highly Insulated and Airtight Building Envelopes*; Springer: Berlin/Heidelberg, Germany, 2013.
35. Ingebretsen, S.B.; Andenæs, E.; Gullbrekken, L.; Kvande, T. Microclimate and Mould Growth Potential of Air Cavities in Ventilated Wooden Façade and Roof Systems—Case Studies from Norway. *Buildings* **2022**, *12*, 1739. [CrossRef]
36. Romero, M.J.; Aguilar, F.; Vicente, P.G. Analysis of design improvements for thermal bridges formed by double-brick façades and intermediate slabs for nZEB residential buildings in Spain. *J. Build. Eng.* **2021**, *44*, 103270. [CrossRef]
37. Recart, C.; Dossick, C.S. Hygrothermal behavior of post-retrofit housing: A review of the impacts of the energy efficiency upgrade strategies. *Energy Build.* **2022**, *262*, 112001. [CrossRef]
38. Morelli, M.; Møller, E.B. Energy savings and risk of mold growth in apartments renovated with internal insulation. *Sci. Technol. Built Environ.* **2019**, *25*, 1199–1211. [CrossRef]
39. Carpino, C.; Loukou, E.; Andersen, B.; Settino, J.; Arcuri, N. Biodeterioration in historic buildings. Indoor environmental conditions and risk of fungal growth. In Proceedings of the MEKO TC4 International Conference on Metrology for Archaeology and Cultural Heritage, Consenca, Italy, 19–21 September 2022; pp. 62–67.
40. Wallner, P.; Munoz, U.; Tappler, P.; Wanka, A.; Kundi, M.; Shelton, J.F.; Hutter, H.-P. Indoor environmental quality in mechanically ventilated, energy-efficient buildings vs. *Conventional buildings. Int. J. Environ. Res. Public Health* **2015**, *12*, 14132–14147. [CrossRef] [PubMed]
41. Lin, W.R.; Ho, Y.H.; Lee, W.K.; Cheng, H.M.; Wang, P.H. Spatiotemporal distribution and the passive dispersal of fungal spores through HVAC systems. *Aerobiologia* **2022**, *38*, 13–21. [CrossRef]
42. Pavard, G.; Joubert, A.; Andrès, Y.; Le Cann, P. Analysis of Particulate and Microbiological Filtration Performance of Air Handling Unit Filters in a Low-Energy Office Building over 12 Months. *Buildings* **2022**, *12*, 1475. [CrossRef]
43. Wilson, S.C.; Palmatier, R.N.; Andriychuk, L.A.; Martin, J.M.; Jumper, C.A.; Holder, H.W.; Straus, D.C. Mold contamination and air handling units. *J. Occup. Environ. Hyg.* **2007**, *4*, 483–491. [CrossRef]
44. Bi Mve, M.J.B.; Cloutier, Y.; Lacombe, N.; Lavoie, J.; Debia, M.; Marchand, G. Comparison of methods to evaluate the fungal biomass in heating, ventilation, and air-conditioning (HVAC) dust. *Environ. Monit. Assess.* **2016**, *189*, 8. [CrossRef]
45. Teertstra, W.R.; Tegelaar, M.; Dijksterhuis, J.; Golovina, E.A.; Ohm, R.A.; Wösten, H.A.B. Maturation of conidia on conidiophores of Aspergillus niger. *Fungal Genet. Biol.* **2017**, *98*, 61–70. [CrossRef]
46. Li, X. Aerosolization Potential of Fungal Spores from the Colony Growing on HVAC Ducts. *Polish J. Environ. Stud.* **2022**, *31*, 4177–4187. [CrossRef]
47. Liu, Z.; Niu, H.; Rong, R.; Cao, G.; He, B.J.; Deng, Q. An experiment and numerical study of resuspension of fungal spore particles from HVAC ducts. *Sci. Total Environ.* **2020**, *708*, 134742. [CrossRef] [PubMed]
48. Gopalakrishnan, S.; Arigela, R.; Gupta, S.K.; Raghunathan, R. Dynamic response of passive release of fungal spores from exposure to air. *J. Aerosol Sci.* **2019**, *133*, 37–48. [CrossRef]
49. Meyer, H.W.; Suadicani, P.; Nielsen, P.A.; Sigsgaard, T.; Gyntelberg, F. Moulds in floor dust—A particular problem in mechanically ventilated rooms? *A study of adolescent schoolboys under the Danish moulds in buildings program. Scand. J. Work Environ. Health* **2011**, *37*, 332–340. [CrossRef] [PubMed]
50. Kempton, L.; Kokogiannakis, G.; Cooper, P. Mould risk evaluations in residential buildings via site audits and longitudinal monitoring. *Build. Environ.* **2021**, *191*, 107584. [CrossRef]
51. Bakker, A.; Siegel, J.A.; Mendell, M.J.; Prussin, A.J.; Marr, L.C.; Peccia, J. Bacterial and fungal ecology on air conditioning cooling coils is influenced by climate and building factors. *Indoor Air* **2020**, *30*, 326–334. [CrossRef] [PubMed]
52. Tang, S.; Zhi, C.; Fan, Y.; Ye, W.; Su, X.; Zhang, X. Unhealthy indoor humidity levels associated with ventilation rate regulations for high-performance buildings in China. *Build. Environ.* **2020**, *177*, 106839. [CrossRef]
53. Seppänen, O.A.; Fisk, W.J. Summary of Human Responses to Ventilation. *Indoor Air* **2004**, *14*, 102–218. Available online: https://pubmed.ncbi.nlm.nih.gov/15330778/ (accessed on 1 May 2023). [CrossRef]
54. Alaidroos, A.; Mosly, I. Preventing mold growth and maintaining acceptable indoor air quality for educational buildings operating with high mechanical ventilation rates in hot and humid climates. *Air Qual. Atmos. Health* **2022**, *16*, 341–361. [CrossRef]
55. Almatawah, Q.A.; Al-Rashidi, M.S.; Yassin, M.F.; Varghese, J.S. Microbiological contamination of indoor and outdoor environments in a desert climate. *Environ. Monit. Assess.* **2022**, *194*, 355. [CrossRef]
56. Coulburn, L.; Miller, W. Prevalence, Risk Factors and Impacts Related to Mould-Affected Housing: An Australian Integrative Review. *Int. J. Environ. Res. Public Health* **2022**, *19*, 1854. [CrossRef]
57. Wei, G.; Yu, X.; Fang, L.; Wang, Q.; Tanaka, T.; Amano, K.; Yang, X. A review and comparison of the indoor air quality requirements in selected building standards and certifications. *Build. Environ.* **2022**, *226*, 109709. [CrossRef]
58. Sousa, J. Association between home characteristics and occupant's behaviours and concentrations of bacteria, fungi and endotoxins. *J. Build. Eng.* **2022**, *45*, 103409. [CrossRef]

9. De Jonge, K.; Laverge, J. Modeling dynamic behavior of volatile organic compounds in a zero energy building. *Int. J. Vent.* **2021**, *20*, 193–203. [CrossRef]
0. Sharpe, R.A.; Thornton, C.R.; Nikolaou, V.; Osborne, N.J. Fuel poverty increases risk of mould contamination, regardless of adult risk perception & ventilation in social housing properties. *Environ. Int.* **2015**, *79*, 115–129. [CrossRef]
1. Pihelo, P.; Kalamees, T. Commissioning of moisture safety of nZEB renovation with prefabricated timber frame insulation wall elements. *Wood Mater. Sci. Eng.* **2021**, *16*, 110–117. [CrossRef]
2. Nath, S.; Dewsbury, M.; Künzel, H.; Watson, P. Mould Growth Risks for a Clay Masonry Veneer External Wall System in a Temperate Climate. *Atmosphere* **2022**, *13*, 1755. [CrossRef]
3. Raamets, J.; Lokko, L.; Ruus, A.; Kalamees, T.; Muoni, K. Assessment of moisture and mould of hempcrete and straw panels. *J. Phys. Conf. Ser.* **2021**, *2069*, 012194. [CrossRef]
4. Wu, H.; Wong, J.W.C. Temperature versus Relative Humidity: Which Is More Important for Indoor Mold Prevention? *J. Fungi* **2022**, *8*, 696. [CrossRef] [PubMed]
5. Kabore, A.; Maref, W.; Plamondon, C.O. Hygrothermal performance of the building envelope with low environmental impact: Case of a hemp concrete envelope. *J. Phys. Conf. Ser.* **2021**, *2069*, 012006. [CrossRef]
6. Tlaiji, G.; Pennec, F.; Ouldboukhitine, S.; Ibrahim, M.; Biwole, P. Hygrothermal performance of multilayer straw walls in different climates. *Constr. Build. Mater.* **2022**, *326*, 126873. [CrossRef]
7. Verma, S.K.; Anand, Y.; Gupta, N.; Jindal, B.B.; Tyagi, V.V.; Anand, S. Hygrothermal dynamics for developing energy-efficient buildings: Building materials and ventilation system considerations. *Energy Build.* **2022**, *260*, 111932. [CrossRef]
8. Cho, H.M.; Yang, S.; Wi, S.; Chang, S.J.; Kim, S. Hygrothermal and energy retrofit planning of masonry façade historic building used as museum and office: A cultural properties case study. *Energy* **2020**, *201*, 117607. [CrossRef]
9. Arregi, B.; Garay, R.; Garrido-Marijuan, A. Assessment of thermal performance and surface moisture risk for a rear-ventilated cladding system for façade renovation. *IOP Conf. Ser. Earth Environ. Sci.* **2020**, *410*, 012102. [CrossRef]
0. Hall, M.R.; Casey, S.P.; Loveday, D.L.; Gillott, M. Analysis of UK domestic building retrofit scenarios based on the E.ON Retrofit Research House using energetic hygrothermics simulation—Energy efficiency, indoor air quality, occupant comfort, and mould growth potential. *Build. Environ.* **2013**, *70*, 48–59. [CrossRef]
1. Colclough, S.; Hegarty, R.O.; Murray, M.; Lennon, D.; Rieux, E.; Colclough, M.; Kinnane, O. Post occupancy evaluation of 12 retrofit nZEB dwellings: The impact of occupants and high in-use interior temperatures on the predictive accuracy of the nZEB energy standard. *Energy Build.* **2022**, *254*, 111563. [CrossRef]
2. Marasinghe, S.A.; Sun, Y.; Hou, J.; Wang, P.; Zhang, Q.; Norbäck, D. Validity of questionnaire survey in evaluating residence's characteristics and dampness in contemporary human dwellings in Tianjin, China. *Sci. Technol. Built Environ.* **2022**, *29*, 241–249. [CrossRef]
3. Wang, J.; Zhang, Y.; Li, B.; Zhao, Z.; Huang, C.; Zhang, X.; Deng, Q.; Lu, C.; Qian, H.; Yang, X.; et al. Effects of mold, water damage and window pane condensation on adult rhinitis and asthma partly mediated by different odors. *Build. Environ.* **2023**, *227*, 109814. [CrossRef]
4. Park, J.H.; Cox-Ganser, J.M. NIOSH Dampness and Mold Assessment Tool (DMAT): Documentation and Data Analysis of Dampness and Mold-Related Damage in Buildings and Its Application. *Buildings* **2022**, *12*, 1075. [CrossRef]
5. Zeng, L.; Chen, Y.; Ma, M.; Du, B.; Gao, J.; Cao, G.; Li, J. Prediction of mould growth rate within building envelopes: Development and validation of an improved model. *Build. Serv. Eng. Res. Technol.* **2023**, *44*, 63–79. [CrossRef]
6. Menneer, T.; Mueller, M.; Sharpe, R.A.; Townley, S. Modelling mould growth in domestic environments using relative humidity and temperature. *Build. Environ.* **2022**, *208*, 108583. [CrossRef]
7. *ISO 13788:2012*; Hygrothermal Performance of Building Components and Building Elements—Internal Surface Temperature to Avoid Critical Surface Humidity and Interstitial Condensation—Calculation Methods. ISO: Geneva, Switzerland, 2012. Available online: https://www.iso.org/standard/51615.html (accessed on 29 April 2023).
8. *ASHRAE 160-2009*; Criteria for Moisture-Control Design Analysis in Buildings. ASHRAE: Peachtree Corners, GA, USA, 2016. Available online: www.ashrae.org (accessed on 29 April 2023).
9. Starakiewicz, A.; Miasik, P.; Krasoń, J.; Lichołai, L. Methods for determining mold development and condensation on the surface of building barriers. *Buildings* **2020**, *10*, 4. [CrossRef]
30. Sivasubramani, S.K.; Niemeier, R.T.; Reponen, T.; Grinshpun, S.A. Fungal spore source strength tester: Laboratory evaluation of a new concept. *Sci. Total Environ.* **2004**, *329*, 75–86. [CrossRef]
31. Suchorab, Z.; Frąc, M.; Guz, Ł.; Oszust, K.; Łagód, G.; Gryta, A.; Bilińska-Wielgus, N.; Czerwiński, J. A method for early detection and identification of fungal contamination of building materials using e-nose. *PLoS ONE* **2019**, *14*, e0215179. [CrossRef]
32. Boardman, C.R.; Glass, S.V.; Lepage, R. Dose-response simple isopleth for mold (DR SIM): A dynamic mold growth model for moisture risk assessment. *J. Build. Eng.* **2023**, *68*, 106092. [CrossRef]
33. Moon, H.J.; Augenbroe, G.L.M. Towards a practical mould growth risk indicator. *Build. Serv. Eng. Res. Technol.* **2004**, *25*, 317–326. [CrossRef]
34. Olaoye, T.S.; Dewsbury, M.; Kunzel, H. A method for establishing a hygrothermally controlled test room for measuring the water vapor resistivity characteristics of construction materials. *Energies* **2021**, *14*, 4. [CrossRef]
35. Shin, M.-S.; Rhee, K.-N.; Lee, E.-T.; Jung, G.-J. Performance evaluation of CO_2-based ventilation control to reduce CO_2 concentration and condensation risk in residential buildings. *Build. Environ.* **2018**, *142*, 451–463. [CrossRef]

86. Schieweck, A.; Uhde, E.; Salthammer, T.; Salthammer, L.C.; Morawska, L.; Mazaheri, M.; Kumar, P. Smart homes and the control of indoor air quality. *Renew. Sustain. Energy Rev.* **2018**, *94*, 705–718. [CrossRef]
87. Strang, M.; Leardini, P.; Brambilla, A.; Gasparri, E. Mass timber envelopes in passivhaus buildings: Designing for moisture safety in hot and humid australian climates. *Buildings* **2021**, *11*, 478. [CrossRef]
88. Fedorik, F.; Alitalo, S.; Savolainen, J.P.; Räinä, I.; Illikainen, K. Analysis of hygrothermal performance of low-energy house in Nordic climate. *J. Build. Phys.* **2021**, *45*, 344–367. [CrossRef]

buildings

MDPI

Systematic Review

Advances in Retrofitting Strategies for Energy Efficiency in Tropical Climates: A Systematic Review and Analysis

Katherine Chung-Camargo [1], Jinela González [1], Miguel Chen Austin [1,2,3,*], Cristina Carpino [1,4], Dafni Mora [1,2,3] and Natale Arcuri [4]

1 Research Group Energy and Comfort in Bioclimatic Buildings (ECEB), Faculty of Mechanical Engineering, Universidad Tecnológica de Panamá, Panama City 0819-07289, Panama; katherine.chung@utp.ac.pa (K.C.-C.); jinela.gonzalez@utp.ac.pa (J.G.); cristina.carpino@unical.it (C.C.); dafni.mora@utp.ac.pa (D.M.)

2 Centro de Estudios Multidisciplinarios en Ciencias, Ingeniería y Tecnología (CEMCIT-AIP), Panama City 0819-07289, Panama

3 Sistema Nacional de Investigación (SNI), Clayton 0816-02852, Panama

4 Department of Mechanical, Energy and Management Engineering, University of Calabria, V. P. Bucci 46/C, 97036 Arcavacata di Rende, CS, Italy; natale.arcuri@unical.it

* Correspondence: miguel.chen@utp.ac.pa

Abstract: The global construction industry significantly contributes to energy consumption and greenhouse gas emissions, necessitating immediate action for sustainable development. Recognizing the impact of buildings on emissions, the United Nations has set ambitious energy-related goals for 2030. Retrofitting buildings emerges as a strategic method for reducing energy consumption, offering lower environmental impact and life cycle costs. However, retrofitting is a complex process influenced by diverse factors such as policies, available resources, techniques, building-specific data, and uncertainties. Thus, this paper reviews the existing literature on retrofitting strategies for tropical and humid climates to identify effective approaches for enhancing energy efficiency, thermal comfort, and overall building performance in these regions. Through comprehensive analyses, including bibliometric analysis using VOSviewer version 1.6.18 and systematic assessments, this study investigates various retrofitting strategies. This study categorizes tropical climates into Af (Tropical Rainforest Climate) and Aw (Tropical Savanna Climate) based on the Köppen climate classification. It reveals distinct emphases, with Af climates concentrating on office buildings and Aw climates prioritizing residential structures. Passive strategies were predominantly favored in office buildings, with glazing being the most commonly implemented approach. Residential structures, on the other hand, adopted a combination of passive strategies such as phase change materials along with active methods like appliance replacement. Educational buildings tended to rely on passive strategies, including roof covers, shading, and glazing. The absence of specific cost values underscores the importance of establishing baseline metrics, revealing significant challenges in retrofit techniques. This study further highlights an opportunity to explore passive methods in educational buildings, stressing the need for comprehensive guidelines, especially in institutional settings. Moreover, it emphasizes the urgency for ambitious regulations to address carbon emissions and optimize energy efficiency in tropical climates.

Keywords: building; energy efficiency; humid; NZEB; retrofit; tropical climate; tropics

1. Introduction

The building sector holds a significant position in energy consumption and greenhouse gas (GHG) emissions. Presently, it constitutes 35% of the global energy usage and is responsible for 38% of the energy-related CO_2 emissions [1,2]. Building operations contribute 28% of the annual CO_2 emissions, while building materials and construction add another 11% annually [3]. Emissions from the construction industry are of particular concern, highlighting the imperative to transition towards green buildings and low-carbon construction materials to mitigate GHG emissions. Endeavors aimed at reducing emissions and curbing energy consumption within the building sector are essential for attaining global sustainability objectives. The building sector's impact on GHG emissions significantly contributes to climate change [2,4,5]. Buildings play a crucial role in the United Nations' Sustainable Development Goals (SDGs) report [6] by contributing directly and indirectly to various SDG targets [7,8]. Implementing sustainable building practices is vital for realizing the SDGs concerning health, sustainable consumption, sustainable cities, and other related objectives, requiring further action to achieve energy-related goals by 2030, as outlined in the United Nations' Sustainable Development Goals report [6].

Recognizing the significance of enhancing energy efficiency, international regulatory bodies, including the Energy Performance of Buildings Directive (EPBD), mandated nearly zero energy consumption for buildings after 2020. The revised directive will help achieve the goal of reducing emissions by at least 60% in the building sector by 2030 compared with 2015 and attaining climate neutrality by 2050 [9].

Building retrofitting, an effective strategy to lower energy consumption and identify energy-saving opportunities based on building conditions, types, and functions, boasts relatively lower environmental impact and life cycle costs than redevelopment [10,11]. However, retrofitting is a multifaceted process influenced by factors like policies and regulations [12], available resources, preferred techniques [13], building-specific data, human elements, and uncertainties [14]. This comprehensive procedure involves energy audits, performance evaluations, identification of energy conservation benefits, economic analyses, risk assessments, and measurement and verification of energy savings [15].

Given the substantial upfront investment required to retrofit a building against annual energy savings, building owners often hesitate because of uncertainties about the investment's value [10]. Hence, conducting a Cost–Benefit Analysis (CBA) becomes crucial to assess the economic and financial implications across various retrofitting levels, enabling informed and effective decision-making [16].

Building retrofitting has a different focus depending on the weather and country. Some focus on measures established for windows (United States [17]), HVAC systems (United States [17], Thailand [18], Vietnam [19]), financial incentives (European Union [20], India [21], Japan [22], Canada [23]), subsidies (European Union [24], India [21], Japan [22]), targets for green building construction (China [25]), improving cooling systems (Singapore [26], Indonesia [27], Philippines [28], Cambodia [29], Bangladesh [30], Sri Lanka [31]), energy performance (Singapore [26], Cambodia [29], Bangladesh [30]), thermal comfort (Malaysia [32], Sri Lanka [31]), improving insulation (Thailand [18], Philippines [28]), and building envelopes (Indonesia [27]).

Also, there exists substantial research focusing on various facets of building energy management, including model calibration [33,34], simulation [35], retrofit solution selection [36], life cycle cost (LCC) computation [37], and establishing optimal decision-making models [38,39]. However, more studies need to systematically study the different strategies for retrofitting in hot and humid weather. Such a study could serve as a comprehensive solution for building owners to assess the feasibility of energy retrofit projects, particularly in tropical regions where climate significantly influences energy consumption patterns and retrofitting strategies [40].

For instance, because of varying energy consumption patterns between different types of buildings, especially institutions, and their relatively constrained financial resources [41], it becomes imperative to investigate the cost–benefit viability of energy retrofitting projects specifically tailored for institutional buildings in tropical climates.

This research stands out for its meticulous analysis and synthesis of the available literature concerning retrofitting strategies focused on buildings in tropical and humid climates. By concentrating on enhancing energy efficiency, thermal comfort, and overall building performance within these challenging environmental conditions, this study aims to uncover the most effective methods, metrics, approaches, and techniques. Its originality lies in its focused exploration of retrofitting strategies uniquely suited to tropical and humid climates, offering a specialized perspective on addressing the sustainability challenges of these regions. Through its comprehensive review, this research aims to provide valuable insights and recommendations for professionals and research, contributing to the advancement of sustainable building practices in tropical and humid climates.

The present study expands upon retrofit research conducted in tropical climates. Section 2 delves into a comprehensive literature examination, incorporating bibliometric and scientometric analyses. Section 3 presents the outcomes obtained by implementing the previously outlined methodology. Lastly, Section 4 delineates the most pertinent retrofitting strategies suitable for tropical climates and accentuates, compares, and analyzes various strategies while discussing future research directions from this study.

2. Materials and Methods

The methodology implemented is divided into two parts as follows: bibliometric and systematic analysis. The next section explains each part in detail. In addition, the Preferred Reporting Items for Systematic Reviews and Meta-Analyses (PRISMA) guide is attached as Supplementary Materials for this study.

2.1. Literature Search Strategy

Regarding data collection, to identify the different applications based on retrofit in buildings, an examination of the literature was carried out using the search strategy presented in Figure 1. The selection of the final papers on this research included four steps as follows:

a. To gather as many studies as possible, the use of scientific databases such as Google Scholar, Science Direct, and Springer were selected.
b. The search was performed in each of the scientific databases. The main co-word combinations and Boolean operators were retrofit AND building AND "energy efficiency"; retrofit AND building AND "energy efficiency" AND (NZEB OR nZEB); retrofit AND building; retrofit AND (NZEB OR nZEB), which returned a total of 34,553 documents.
c. An exclusion criterion was applied, limiting the research to English and a period of the last five years (2018–2024) and using the Boolean operators, which returned a total of 16,649 documents.
d. The papers were analyzed using a combination of additional co-word combinations and Boolean operators to include tropic/tropical climate/humid; the combinations were the following: retrofit AND building AND "energy efficiency" AND (tropics OR "tropical climate" OR humid); retrofit AND building AND "energy efficiency" AND (NZEB OR nZEB) AND (tropics OR "tropical climate" OR humid); retrofit AND building AND (tropics OR "tropical climate" OR humid); retrofit AND (NZEB OR nZEB) AND (tropics OR "tropical climate" OR humid), which returned a total of 2211 documents.

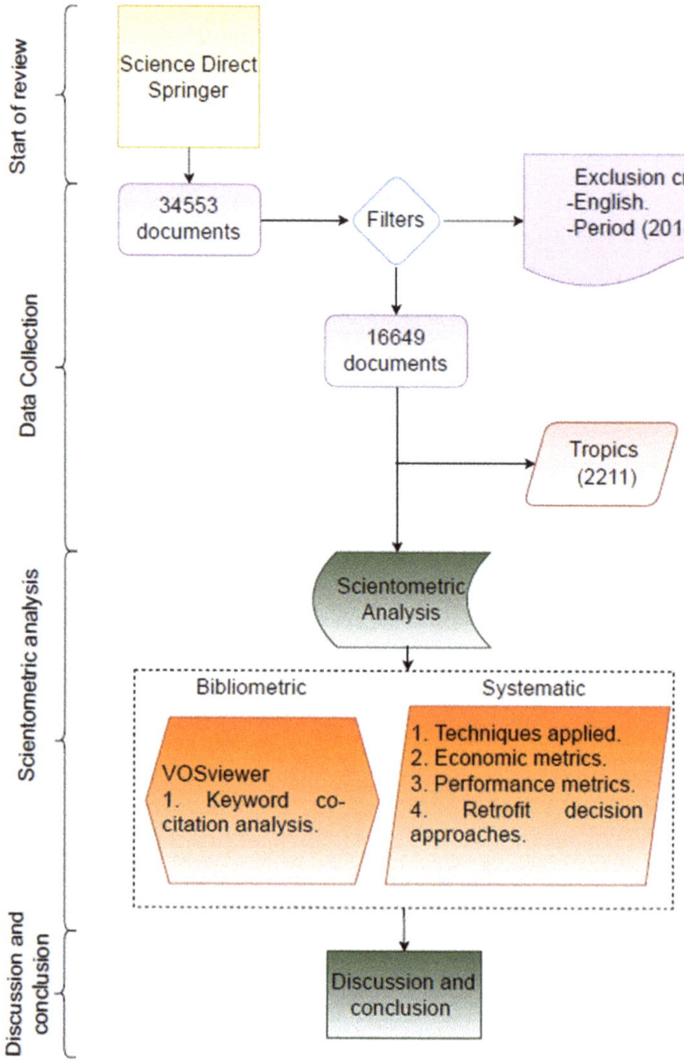

Figure 1. Research methodological process.

2.2. *Scientometric Analysis*

The final documents were analyzed using bibliometric and systematic analyses, both of which are explained in detail hereafter.

2.2.1. Bibliometric Analysis

A large number of documents from Section 1 in the data collection are presented. A bibliometric mapping tool named "VOSviewer" was used to analyze the information. VOSviewer has the advantage of presenting an informative visualization. VOSviewer is a free computer program that is especially useful for displaying large bibliometric maps in a way that is easy to interpret [42,43].

The bibliometric analysis was conducted, the map was created based on bibliographic data, the RIS file was uploaded, and the keyword co-occurrence analysis was selected.

Tables 1 and 2 list the query used for searching and the quantity of documents. Table 3 shows the most common keywords in the retrofit analysis, where some words were similar and thus grouped under similar keywords, as listed in Table 4. Once the keywords were grouped, a thesaurus file was used to merge the keywords.

Table 1. Query used for searching and the quantity of documents.

Keywords	Number of Documents	Period
retrofit AND building AND "energy efficiency"	10,299	2000–2024
retrofit AND building AND "energy efficiency" AND (NZEB OR nZEB)	753	2010–2024
retrofit AND building	22,672	2000–2024
retrofit AND (NZEB OR nZEB)	829	2010–2024
retrofit AND building AND "energy efficiency" AND (tropics OR "tropical climate" OR humid)	1444	2000–2024
retrofit AND building AND "energy efficiency" AND (NZEB OR nZEB) AND (tropics OR "tropical climate" OR humid)	166	2012–2024
retrofit AND building AND (tropics OR "tropical climate" OR humid)	2117	2000–2024
retrofit AND (NZEB OR nZEB) AND (tropics OR "tropical climate" OR humid)	178	2012–2023

Table 2. The query used to search for several documents in the last five years.

Keywords	Number of Documents	Period
retrofit AND building AND "energy efficiency"	5107	2018–2024
retrofit AND building AND "energy efficiency" AND (NZEB OR nZEB)	504	2018–2023
retrofit AND building	10488	2018–2024
retrofit AND (NZEB OR nZEB)	550	2018–2024
retrofit AND building AND "energy efficiency" AND (tropics OR "tropical climate" OR humid)	812	2018–2024
retrofit AND building AND "energy efficiency" AND (NZEB OR nZEB) AND (tropics OR "tropical climate" OR humid)	130	2018–2023
retrofit AND building AND (tropics OR "tropical climate" OR humid)	1131	2018–2024
retrofit AND (NZEB OR nZEB) AND (tropics OR "tropical climate" OR humid)	138	2018–2023

Table 3. List of the most keywords that occurred in the retrofit search.

Keywords	Occurrence
energy efficiency	762
thermal comfort	298
retrofit	282
energy retrofit	183
building retrofit	179
climate change	160
energy consumption	152
sustainability	129
nzeb	126
buildings	123
residential buildings	121
renewable energy	116
energy saving	111
optimization	108

Table 4. Related terms are grouped under the stated keyword.

Keyword	Terms Grouped with the Keyword
buildings	building; building energy efficiency; building stock
energy efficiency	energy saving; energy performance; energy savings; building energy performance
nzeb	zero energy building; net-zero energy building; nearly zero energy building; net zero energy building; net-zero; nearly zero energy buildings; net zero energy buildings; net-zero energy buildings; zero energy buildings; nearly zero energy; nzebs; net zero energy; zeb; nearly zero energy building (nzeb); nearly zero-energy building (nzeb); nearly-zero energy buildings; net zero energy building (nzeb); net-zero buildings; near zero energy buildings; nearly zero energy buildings (nzebs); nearly zero energy buildings (nzebs); net-zero emissions buildings; net-zero energy rural house; nzeb target; near zero energy building; nearly-zero and positive energy paradigms; net zero energy buildings (nzebs); net-zero, ghg emissions; zero energy buildings (zebs); zero energy solar household; zero-carbon building; zero-energy targets; net-zero energy
PEB	positive energy buildings (pebs)
residential buildings	residential building; residential
retrofit	energy retrofit; building retrofit; seismic retrofit; retrofitting; building renovation; building energy retrofit; building retrofitting; existing buildings; existing building; renovation energy retrofits; refurbishment; green retrofit; building retrofits; energy-efficient retrofit; energy retrofitting; deep energy retrofit; green retrofitting; retrofits; building refurbishment; energy renovation; retrofit delivery; retrofit measures; retrofit scenarios; deep retrofit; retrofitting strategies; deep renovation; energy efficiency retrofit; home energy retrofit; rehabilitation; domestic retrofit; energetic retrofit; existing residentials; housing retrofit; passive energy-saving retrofit; retrofit double glazing; retrofit guidelines; retrofit market; sustainable building renovation; sustainable building upgrade; zero energy building renovation; building energy renovation; building energy retrofits; building envelope retrofits; cost-optimal retrofit; façade retrofit; retrofitting measures; social housing retrofit; sustainable retrofit; thermal retrofit; deep retrofitting; renovation strategies; residential building retrofit; retrofit interventions
thermal comfort	adaptive thermal comfort; indoor thermal comfort; outdoor thermal comfort
tropical climate	tropics; hot and humid climate; hot and humid climates; hot-humid climates; hot/warm and humid climates

2.2.2. Systematic Analysis

A complete review of the preselected articles was needed to select sources that provided information related to retrofitted buildings and techniques applied. The final papers were selected by reviewing the title, abstract, and keywords.

The criteria employed in the literature selection included the following. (i) Retrofit in buildings in the last five years (2018–2023). (ii) Retrofit in buildings of any type. (iii) Retrofit in humid or tropical climates.

Finally, information on different retrofitting strategies in buildings and technical strategies applied in retrofitting were classified.

3. Results

The results of this study are presented in two parts. Section 2.2.1 explains the bibliometric analysis, and Section 2.2.2 explains the systematic analysis.

3.1. Bibliometric Analysis

Keyword Co-Citation Analysis

The number of documents listed in Table 1 is based on the primary co-word combinations and Boolean operators mentioned in the reviewed articles, excluding the specified exclusion criteria. The majority of these documents are from 2000 onwards. On the contrary, Table 2 includes exclusion criteria related to a recent five-year period.

Table 3 displays the most frequently appearing keywords along with their respective co-occurrence counts based on the completion of data collection and document exportation using Vosviewer. "Energy efficiency" emerges as the most prevalent keyword, exhibiting the highest frequency among all keywords and demonstrating connections to all other clusters. Subsequently, "thermal comfort" ranks as the second most frequent term, followed by "retrofit" in third place.

In the visual representation, 7318 keywords were considered, of which 1115 met the specified threshold criteria. The visualization in Figures 2 and 3 involves 1000 links and showcases 995 items. Figure 2 illustrates the interconnectedness of keywords and their links, with different colors denoting distinct clusters within the network. Meanwhile, Figure 3 depicts the network's links and co-occurrences of keywords across a timescale, highlighting recent years in yellow.

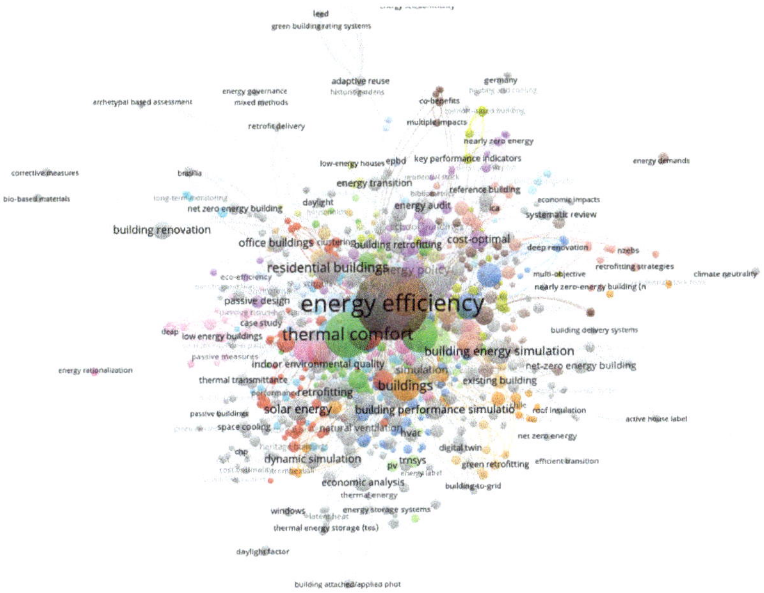

Figure 2. Network illustration for keywords used in the reviewed research.

In Figure 4, the keywords "tropical climate" and "tropics" exhibit the least occurrence, linked to various distinct clusters. Moreover, the timescale indicates that these keywords are more recent, suggesting a tendency towards novelty or recency.

As observed in the preliminary findings, various keywords derived from the highly co-occurring pairs in Table 3 were categorized under specific terms, as showcased in Table 4. This process is termed a thesaurus in Vosviewer and aims to standardize the data and enhance its practicality and realism for analysis. This approach yielded a new set of 6325 keywords, with only 123 meeting the defined threshold, thus generating 101 links.

Upon implementation, the resultant figures, Figures 5 and 6, depict the final representation, with "retrofit" emerging as the most prominent term. Additionally, Figure 7 groups keywords under "tropical climate." In Figure 7a, there is a prevailing trend indicating recent studies that are emphasizing tropical climates. Figure 7b illustrates the relationships between various domains such as retrofit, thermal comfort, nzeb, and energy efficiency.

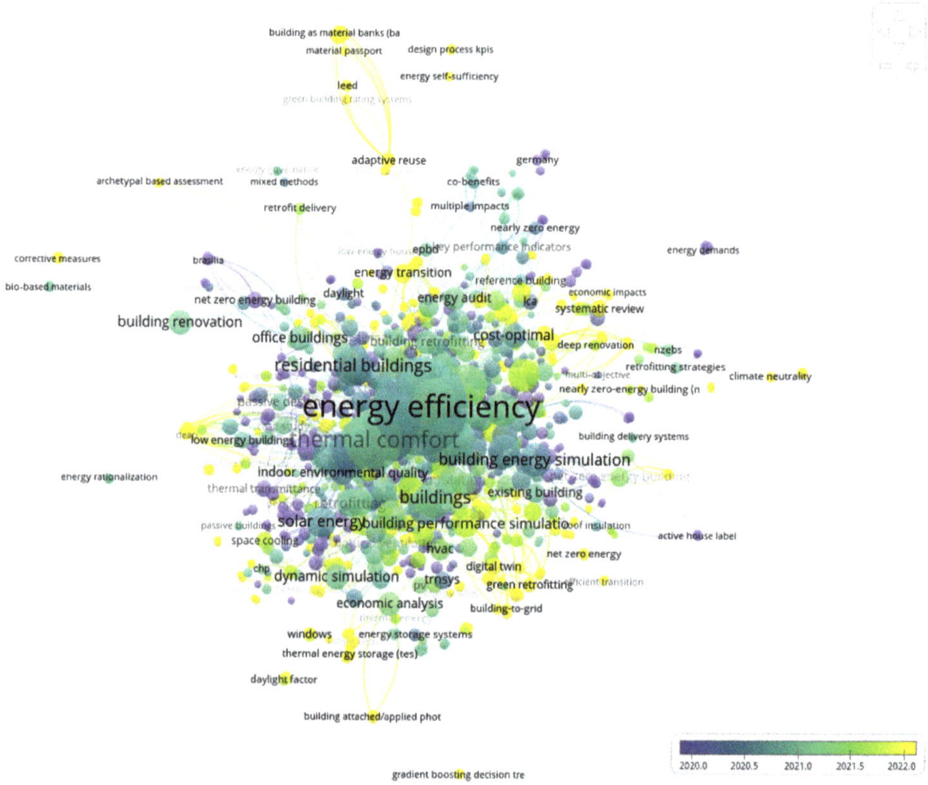

Figure 3. Network links and co-occurrences of the keywords on a timescale.

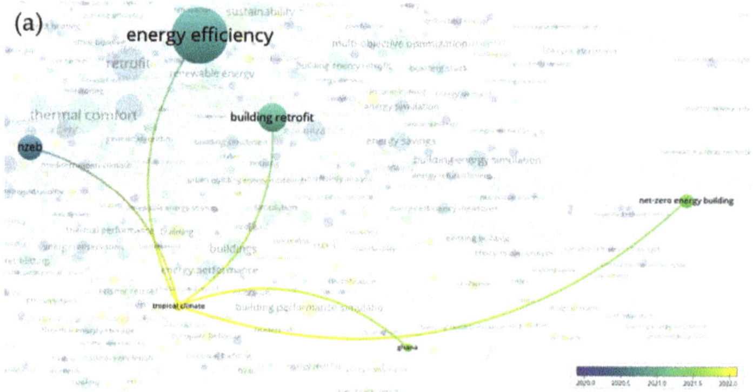

Figure 4. *Cont.*

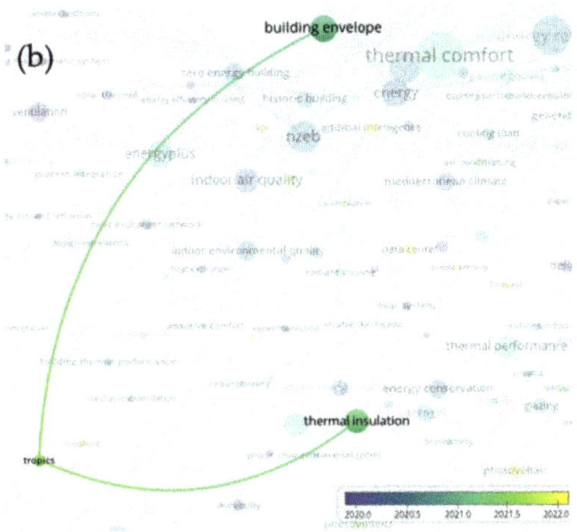

Figure 4. Co-occurrence of the links for the keywords (**a**) "tropical climate" and (**b**) "tropics".

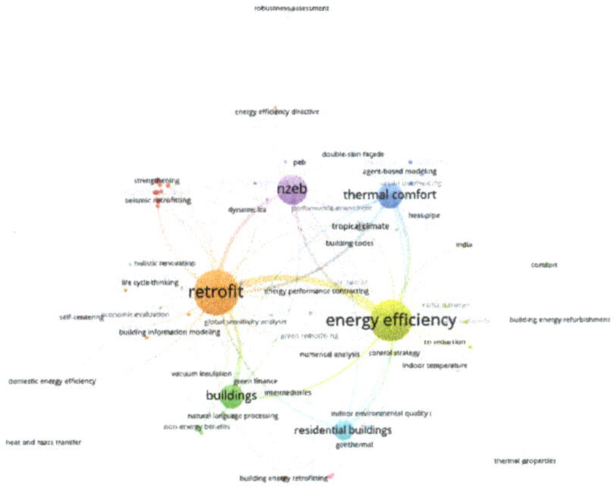

Figure 5. Network illustration for final keywords.

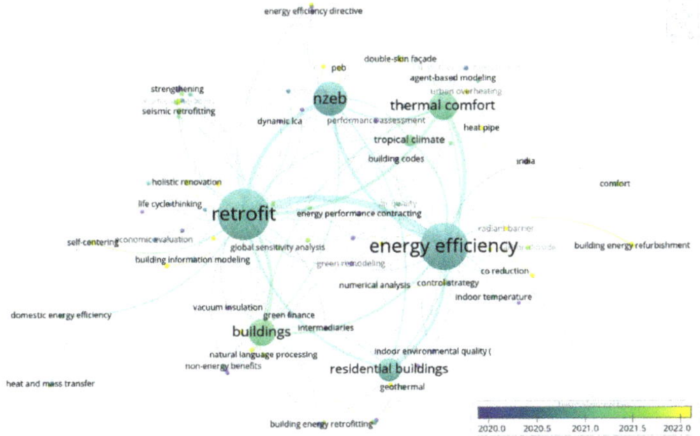

Figure 6. Networks links and co-occurrences for final keywords on timescale.

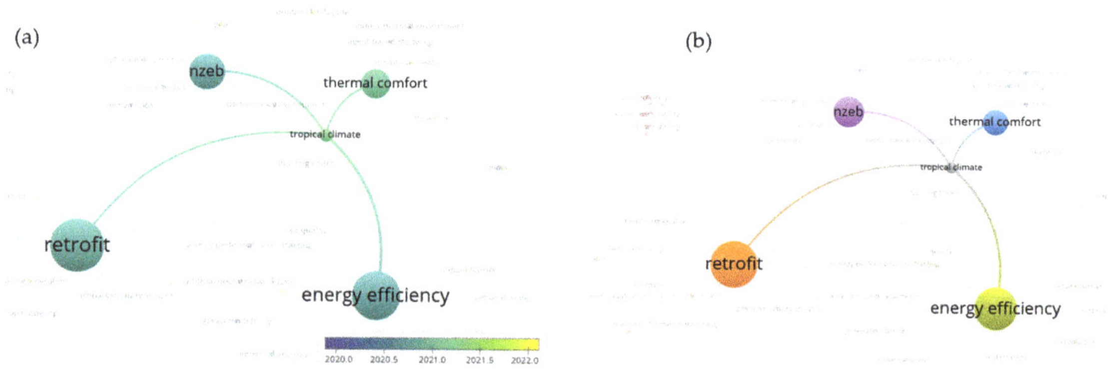

Figure 7. Keyword "tropical climate" (**a**) co-occurrence on a timescale and (**b**) co-occurrence of the link.

3.2. Systematic Analysis

Building energy efficiency retrofit entails optimizing technical, environmental, or economic parameters to identify the most suitable saving strategy. Although the analyzed methods may have different approaches, they must focus on meeting the established objective. An analysis of the literature revealed that enhanced comprehension can be categorized into techniques applied in retrofitting, studies involving economic metrics, performance metrics, and retrofit decision approaches.

Three subsections were developed regarding techniques applied in retrofitting. The first section focuses on building components (envelope). The second delves into utilizing systems or requirements for greater efficiency, while the third concentrates on renewable energy systems. The features of all the studies reviewed are summarized in Tables 5 and A1 (a follow-up of Table 5).

Economic metrics receive extensive evaluation because of their ability to determine the profitability of alternative approaches. Performance metrics are a crucial variable in energy-efficient assessments. Retrofit decision approaches reveal methods for identifying correct energy-saving approaches and play an essential role as tools to measure, monitor, and analyze the behavior of implemented strategies or technologies. The decision on which of these approaches to use depends on the specific objectives of the study.

3.2.1. Techniques Applied in Retrofitting

Building Component (Envelope)

A parametric analysis was carried out by Vidhya et al. [44], who investigated different saving strategies in their study focused on various methods to enhance thermal comfort in a school situated in Chennai, India. They concluded that natural ventilation, reducing infiltrations, and adding shading on windows can decrease the temperature by 3.2 °C in summer and 3.4 °C in winter. On the other hand, in the study developed by [45], natural ventilation was also implemented as an energy-saving strategy by changing the upper windows from fixed to pivot. They discovered that natural ventilation does not provide the best thermal comfort. For this reason, a cooling system was therefore incorporated. Then, the comfort hours were increased to 35%. Implementing a dynamic external shading device was also studied in [46], where a 24% cooling energy saving was achieved.

In the study presented by A. Shandilya [47], after comparing different saving strategies in a single-family house located in India, the results demonstrated that adding shading control and applying insulation to the walls and roof reduced the cooling energy demand by 70%, even when keeping single-glazed windows. The floor insulation was not included, as it decreased the comfort hours during the summer. Thus, it is important to consider the most possible parameters to determine the best strategies.

Replacing windows was an energy-saving strategy adopted in [48], where they evaluated different options to find the right type of window (louvered windows, plenum windows, or double-glazed and triple-glazed windows). According to the results, after evaluating life cycle cost, embodied energy, global warming potential (GWP), and energy use, the option with the best energy performance was triple-glazed windows. Nevertheless, considering the other indicators, the optimal choice was double-glazing windows because of their GWP and embodied energy values, 532 MJ kg/m^2 embodied energy and 101 kg/m^2 CO_2 embodied carbon, respectively. Equally, double-glazed windows reduced cooling energy consumption by 27%. In Sao Paulo, Brazil [49], researchers studied an office building and performed a sensitivity analysis to select the proper value for each of the following variables: the window opening effective area (WOEA), external shading angle, glazing thermal properties (U-value and heat gain), and wall absorptance solar radiation. They concluded that a double-glazed window with a U-value of 2.8 W/m^2 was preferred, employing a light color for the external walls with an absorptance (α) of 0.2, 93% WOEA, 45° for the external shading device vertical shadow angle (VSA), and 37% glazing solar heat gain. These improvements resulted in an energy saving from 8.7 kWh/m^2 per year to 21.4 kWh/m^2, depending on window orientation.

Window enhancement is important in achieving an efficient building. The authors of [50] demonstrated that an existing window could join a new one, becoming a double-glazed window. They developed a study in Singapore, where 8mm of sun energy grey glazing type and solar control film were adapted to 10 mm of clear glazing (the previous one installed). They were separated by 17 mm of air. This refurbishment reduced the U-value from 5.649 to 1.998 W m^{-2} K^{-1}, representing a 41.4% cooling energy saving. A study developed by S. Sebayang [51] revealed a window improvement in an educational building in Singapore. They implemented a combination of new glazing properties and vertical fins; the first reduced the overall thermal transfer value (OTTV) of the wall from 32.27 W/m^2 to 29 W/m^2, and the last produced 28.02 W/m^2, where the OTTV was 26.23 W/m^2. The results showed that the joint implementation of both strategies decreased the indoor air temperature by 0.835 °C.

Another investigation developed a similar study. Somasundaram et al. [52] adapted a second glazing to an existing glazing. Nevertheless, this system did not work the same way as the original double-glazed window; the authors suggested that this could be due to the Window-to-Wall Ratio (WWR), which was only 8%. Despite this, the cooling energy consumption was reduced by 5.9%. In [53], another study was carried out in which the WWR was established at 20%, where 3% and 7.5% were achieved as annual energy consumption savings of air conditioning and total energy saving, respectively. Based on these outcomes, a better result was achieved by implementing a greater WWR. In contrast, Chandrasekaran et al. [46] demonstrated that for larger percentages such as 40%, 50%, and 60%, energy savings are not positively correlated with the percentages. Their results showed that for 40%, more energy is saved than a WWR of 50% and 60%, which was because of a higher heat input. For this reason, it was recommended that the WWR be maintained between 20% and 40%. Alwi et al. [54] determined that a 30% WWR exhibited better performance, reducing solar gains by 6%. In [55], a parametric analysis found that having 32% of the façade as an opened surface was the optimal choice.

The window size or state (adapted to a shading system, opened, or closed) impacts natural ventilation and heat input and influences daylighting entry. Consequently, there must be a balance between all those variables to maintain an optimal comfort zone and low energy demands. One solution that was presented is daylighting control, which established a daylighting control to keep the illuminance level at 500 lux. This solution reduced artificial lighting use and limited heat gains, which positively influenced energy cooling demand and total energy consumption. Using these saving strategies, the first one resulted in a reduction of 20%, the second one in a decrease of 14.3%, and the WWRs remained in the initial percentages (from 23% to 38%). These outcomes were better than those provided by the other saving strategies studied, such as WWR in 20% (for all external windows), which only resulted in 5.6% reductions [56].

In another study, controlled systems, such as catching data from sunlight, were used to help achieve energy efficiency through dynamic façades, specifically, kinetic shading devices (louvers), to avoid incremental solar heat gain. A single-glazed blue-tinted device was used and was found to be the best case. The annual sunlight exposure decreased by 65% [57].

In the study by Gupta et al. [58], they added a green roof over an original roof, reflecting coating and bamboo-based shading systems. A parametric analysis was performed to find the best materials. The results indicated that by applying those technologies, cooling energy savings were achieved by 18.5% to 23%. In addition, bamboo-based shading devices helped to reduce greenhouse gas emissions because of their natural properties. It was demonstrated that these shading systems performed better than horizontal window shading in tropical climates.

Phase change material (PCM) technology is an alternative to building retrofitting. Based on [59], which analyzed different countries with tropical climates, PCMs are feasible for this climate type, given their influence on reducing energy consumption. The outcomes showed an energy savings from 16.58% to 68.63%. Also, the authors emphasized PCM layer thickness, which was positively correlated with energy saving. The investigation conducted in [60] proved this correlation through an analysis developed in a residential building located in Malaysia, in which they implemented Infinite R™ as the PCM material. Three layers of thicknesses, 6 mm, 12 mm, and 18 mm, were tested and located in the inner part of the external and internal walls; each one was evaluated with different melting temperatures, from 27 °C to 30 °C, and solidification temperatures, in the range of 26 °C to 29 °C. After analyzing each possible solution, the outcomes illustrated that the best PCM combination corresponded to 27 °C for melting and 26 °C for the solidification process. Implementing 18mm as the thickness of the layer, given the enhancement in the thermal comfort, increased the thermal comfort time (TCT) to 78%. Furthermore, natural ventilation improved performance because of the capacity to remove heat inside the room on hot nights.

On the other hand, a study developed by M.J. Abden et al. [61] in Darwin, Australia, showed slightly different results from the investigation reported in [60]. First, the PCM was composed of form-stable PCM (FSPCM), implementing methyl stearate and diatomite. The thickness of FSPCM was 25 mm, and the melting temperature was 27 °C. The FSPCM was joined with thermal insulation called expanded polystyrene (EPS) with a thickness of 60 mm. These materials were incorporated below the wall and ceiling. According to the outcomes, this combination represented a major decrease in total energy consumption by up to 10.3%, and the intensity of thermal discomfort was reduced by 22.1%. In this case, natural ventilation was ineffective because of tropical environmental conditions. The investigation developed by Kameni et al. [62] in an office building in Madagascar identified that adapting insulation below a wall improved thermal comfort; it also resulted in around 12% and up to 10% in CO_2 savings.

PCM was also studied in a residential building located in Darwin, Australia [63]. In this case, n-octadecane combined with gypsum was adopted as the PCM with a 2 cm thickness. The result showed that the optimal melting temperature was 24 °C, achieving 7.6% cooling load reduction and 4.76% total energy consumption. These outcomes differ slightly from those of the previous study in Darwin, which may be due to the melting temperature and thickness selected, as both were inferior to the values in the other article. The authors also mentioned that an insulation material could help even more.

Most retrofit studies are developed through energy simulation software, and experimental studies are less common. However, some of them were found in the literature review. According to [64], a way to improve thermal comfort in tropical climates is employing roof covers with high-density polyethylene. Their results presented a reduction of 70–88% in the convective heat flux, but they also mentioned that the roof covers did not perform as well as a mechanical ventilation method. Different roof envelope materials, such as covers or insulators, can be used.

Adapting advanced technology to an existing installation to achieve energy efficiency was used in the study by [65], in which the indoor air quality was improved through a hybrid air treatment cooling system (HATCS). They developed an oxygen generation process implementing water-splitting methods using solar energy. Consequently, ozone-based treatment was produced to eliminate bacteria and viruses. Then, an air scrubbing mechanism was applied to the HVAC system. This air treatment helped to avoid outdoor air intake, thus reducing energy consumption by cooling load by 25%, which saved energy. Also, this strategy enhanced indoor air quality by up to 19%.

Use of Systems or Equipment with Greater Efficiency

According to Litardo et al. [56], considerable energy savings can be achieved by replacing obsolete equipment with more efficient equipment. In their investigation, it was concluded that using the HVAC system proposed by the Energy Star program, which meets the greatest efficiency standards, can provide 77.07 kWh/m^{2}y.

In the study performed in [66], appliance replacement, such as lighting, air conditioners, refrigerators, water heaters, clothes dryers, and washers, represented the most feasible saving strategy, considering the LCC. This could be due to the continual usage of these appliances, particularly the lighting and air conditioner, which achieved 83,640 kWh and 498,486 kWh in energy reductions during its lifetime.

Renewable Energy System

E. Ohene et al. [67] conducted a study through parametric analysis employing simulation software. Their study implemented retrofit strategies in a residential building located in Ghana to achieve a net-zero energy building (NZEB). The researchers found that passive strategies like taking advantage of daylight to reduce interior light, improving natural ventilation, applying window overhangs, and minimizing infiltration reduced heat gains effectively, and 48–58% energy savings were achieved. Those strategies helped to reduce energy consumption, which initially was 137 kWh/m^{2}y, whereas, after refurbishing, it was 68 kWh/m^{2}y, which turned the building into an NZEB. Daylight is often implemented as an energy-saving method; however, it is important to consider heat inputs. The study by Z. Amin et al. [60] achieved an NZEB after a building's refurbishment by implementing daylighting controls in an educational building in Ecuador. The photovoltaic (PV) system provided 66,590 kWh/y, and the energy consumption was only 48,498 kWh. The energy demand before refurbishment was 97,958 kWh/y. Thanks to the energy-saving strategies, an optimized PV system was installed. A similar method was also developed in a case study in Panama [68]. Nevertheless, the retrofit was performed after the photovoltaic system installation because of the necessity of achieving an NZEB.

In [69], an NZEB was achieved after establishing saving strategies to decrease energy consumption in an office building in India. Those strategies replaced the single-glazed type with the double-glazed type, and insulation was applied to the walls and roof. Thus, the U-value became 0.46. This photovoltaic system was installed on the rooftop and supplied the whole building's energy consumption.

Another application of PV systems is in replacing wall or roof envelopes; these technologies are called building-integrated photovoltaic (BIPV) curtain walls and BIPV membrane roofs. Jhumka et al. [70] studied an office building retrofit in Mauritius, where BIPV replaced the original roof and south façade. However, the study evaluated the heat transfer through the new envelopes and showed that both systems, including a curtain wall and membrane envelope, increased the heat transfer reverberating in the indoor air temperature. For that reason, the BIPV membrane roof was insulated; this system reduced the cooling load by 15% and represented 172% in energy savings, in contrast to the BIPV curtain wall, which did not generate a great decrease in energy consumption.

A photovoltaic system can be installed considering some adjustments, such as implementing plants beneath the system. This has been carried out to avoid rising temperatures, thus increasing efficiency. C. Kaewpraek et al. [71] performed this application in a residential building in Thailand. According to the results of their investigation, a green rooftop combined with a PV system improved the building's energy performance and helped reduce CO_2 emissions. According to the study developed by [72], combining photovoltaic systems with green roofs allowed a temperature reduction of the module, and this represented a module efficiency improvement of 3 to 11%.

Table 5. Retrofit strategies in tropical climates.

City, Country	Climate Type	Type of Strategy	Building Typology	Retrofit Purpose	Retrofit Strategies Studied	Retrofit Strategies with the Best Performance	Year	Ref.
Bangalore, India; Kolkata, India; Tanzania	Aw	Passive	Residential	Reduce energy consumption	Phase change material—thickness of 5 mm to 40 mm with an interval of 5 by 5 and melting temperature from 21 °C to 31 °C	Phase change material—thickness of 5 cm to 40 cm with an interval of 5 by 5 and melting temperature from 21 °C to 31 °C	2019	[59]
Bangalore, India	Aw	Passive	Office	NZEB	Wall insulation and photovoltaic system	Wall insulation and photovoltaic system	2023	[69]
Brasilia, Brazil	Aw	Passive	Office	Reduce energy consumption	Appliance improvement, natural ventilation, thermal mass	Appliance improvement, natural ventilation, thermal mass	2020	[73]
Chennai, India	Aw	Passive	Educational	Reduce indoor air temperature and energy consumption	Roof cover, light color painting, shading with trees, WWR increase, ceiling fan, reflective glass, double glazing	Roof cover, light color painting, WWR increase, ceiling fan, reflective glass, double glazing	2023	[44]
Chennai, India	Aw	Passive	Office	Reduce energy consumption	Fixed and dynamic external shading device	Dynamic external shading device	2022	[46]
Darwin, Australia	Aw	Passive	Residential	Reduce energy consumption	Phase change materials (form-stable PCM) and expanded polystyrene, separated and combined	FSPCM and EPS combined	2022	[61]
Darwin, Australia	Aw	Passive	Residential	Low cooling demands and improved thermal comfort	Phase change material (n-octadecane microencapsulated) 1cm and 2cm thickness	Phase change material (n-octadecane) 2 cm thickness	2022	[63]
Ecuador	Aw	Passive	Educational	Reduce energy consumption	Daylighting control, WWR reduction, solar shading, triple glazing, HVAC system replacement	Daylighting control	2021	[56]
Ghana	Aw	Passive	Residential	NZEB	Daylighting, envelope airtightness, sun shading, natural ventilation, building-applied photovoltaics	Photovoltaic system	2022	[67]
India	Aw	Passive	Residential	Reduce energy consumption and improve thermal comfort	Triple glazing, thermal envelope insulation (roof and walls)	Triple glazing, thermal envelope insulation (roof and walls)	2020	[47]
Indonesia	Af	Passive	Educational	Improve thermal comfort, reduce the indoor air temperature	New glazing properties and vertical fins	New glazing properties and vertical fins	2023	[51]

Table 5. *Cont.*

City, Country	Climate Type	Type of Strategy	Building Typology	Retrofit Purpose	Retrofit Strategies Studied	Retrofit Strategies with the Best Performance	Year	Ref.
Indonesia	Af	Passive	Office	Low cooling demand	Double-glazed windows, WWR, temperature set point	Double-glazed windows, WWR, temperature set point	2023	[74]
Madagascar	Af	Passive	Office	Improve thermal comfort	Phase change material, adding insulation to walls and roof, solar protection, external shading	Insulation and external shading	2020	[62]
Malaysia	Af	Passive	Residential	Provide thermal comfort	Roof cover	Roof Cover	2020	[64]
Malaysia	Af	Passive	Residential	Provide thermal comfort	Phase change materials	Phase change material	2021	[60]
Malaysia	Af	Passive	Office	Apply multi-criteria decision-making to select a type of window	Louvered, plenum, double-glazed, and triple-glazed window	Double-glazed window	2022	[48]
Mauritius	Aw	Passive	Office	Low cooling demands	BIPV curtain wall and BIPV membrane roof	BIPV membrane roof	2023	[70]
Mumbai, India	Aw	Passive	Educational	Reduce energy consumption	Reflecting coating for external walls and green roof	Reflecting coating for external walls and green roof	2022	[67]
Mumbai, India	Aw	Passive	Educational	Low cooling demands	A green roof, reflecting coating, bamboo-based shading device	Green roof, reflecting coating, bamboo-based shading device	2022	[58]
Panama	Aw	Passive	Residential	NZEB	Temperature setpoint, cooling operation, occupancy, wall insulation	Temperature setpoint, cooling operation, occupancy, wall insulation	2022	[75]
Rio de Janeiro, Brazil	Aw	Passive	Office	Improve thermal comfort	Pivot windows	Pivot windows	2021	[45]
Rio de Janeiro, Brazil	Aw	Active	Residential	NZEB	Thermal energy storage for chiller and demand limiting	Thermal energy storage for chiller and demand limiting	2021	[76]
Sao Paulo, Brazil	Af	Passive	Office	Low cooling demands	Window opening area, shading, glazing properties (solar heat gain, U-value), wall absorptance	Overall, 93% of the opening area, 45° of external shading angle, light color for wall, 2.8 U-value	2020	[49]
Singapore	Af	Passive	Office	Low cooling demands and reduced energy consumption	Doble glazing	Double-glazed window	2019	[53]
Singapore	Af	Passive	Office	Low cooling demands	Doble glazing	Double-glazed window	2020	[52]
Singapore	Af	Passive	Office	Reduce energy consumption	Adding a secondary glazing (lowE, Sunenergy grey)	Adding a secondary glazing (lowE, Sunenergy grey)	2020	[50]
Singapore	Af	Active	Residential	Reduce energy consumption	Appliance replacement, window replacement, window and wall retrofit	Appliance replacement	2019	[66]
Singapore	Af	Active	Office	Reduce energy consumption	Hybrid air treatment cooling system	Hybrid air treatment cooling system	2019	[65]

3.2.2. Economic Metrics

Cost indicators are highly important in decision-making when evaluated based on existing weights at a specific decision-making moment. The most prevalent is the life cycle cost, defined in Section 3.2.4. On the other hand, based on the energy approach, an alternative method is to express the relationship between energy and cost. According to [66], the Dynamic Generation Cost (DGC) is adapted to study the energy saving generated by a specific strategy and consider its costs.

The DGC is calculated as shown in Equation (1):

$$DGC = \frac{\sum_{k=0}^{N} \frac{I_k + C_k}{(1+d)^k}}{\sum_{k=0}^{N} \frac{E_k}{(1+d)^k}} \quad (1)$$

where k represents the year evaluated, I is the investment cost, C is the lifetime cost, E corresponds to the energy saving, and d is the depreciation.

Equation (2) should be considered when evaluating the relationship between the life cycle cost and energy generation. The Levelized Cost of Energy considers *CAPEX*, which is the capital expenditure; *OM*, which is the annual operating maintenance cost; *FC*, which is the fuel cost; *TC*, which is the tax cost; and *EG*, which is the energy generation per year. This indicator is employed when a generation system, such as a photovoltaic system, is installed.

$$LCOE = \frac{CAPEX + OM + FC + TC}{EG} \quad (2)$$

3.2.3. Performance Metrics

Occupant behavior is an important variable to study in an energy-efficient assessment because it can interfere with a building's energy performance. Some occupants' actions are often related to the HVAC system since they can manipulate the set point temperature. In [56], the air conditioning temperature was fixed at 24 °C, after being 21–22 °C; this change helped to improve energy savings.

When thermal comfort impacts occupant behavior, this indicator should be assessed, similar to the study by Kameni et al. [62]. They found that integrating saving strategies could help to reduce thermal discomfort. According to their results, after employing the PCM strategy, the predicted mean vote (PMV) value was established from −0.21 to 1.08. In the study developed by [47], this indicator helped to choose the energy-saving strategies because thermal comfort was evaluated, and it was noticed that one strategy, which recorded a great energy performance, negatively impacted thermal comfort during the summer season. N. Ardiani et al. [74] showed that achieving low energy demand and ideal thermal comfort may require several different energy-saving techniques; for this reason, a thorough investigation is necessary.

After this literature review was developed, assessing the indicators employed by the researchers was appropriate. These indicators are an important tool for measuring, monitoring, and analyzing the behavior of the strategies or technologies applied. These depend on the objective of the study. As shown before, the studies are oriented to decrease cooling loads and energy consumption and increase thermal comfort. Thus, the main indicators are kWh/m^2 year and Energy Use Intensity (EUI).

Other variables need to be studied to achieve those objectives. Such is the case of heat transfer analysis through the building envelope to evaluate a building's thermal behavior, given its influence on energy performance. Based on these studies, decision-makers can take action accordingly. An important indicator is the Overall Thermal Transfer Variable (OTTV) employed in [51].

3.2.4. Retrofit Decision Approaches

Economic analyses are widely applied, given their ability to infer whether an alternative is profitable. LCC is an indicator implemented to evaluate each strategy available and find the most economically feasible. This includes the capital cost and the cost generated by operation, maintenance, and transportation; in other words, all the costs related to the project from its initial to its end.

LCC was analyzed in [48] and helped choose the correct option (type of window), where the most energetically viable was not selected because of its high cost. This action resulted in a 27% cost savings. LCC was also studied in the investigation by M.J. Abden [61], and the option selected (a combination of EPS and FSPCM) had the highest cost. Nevertheless, the payback period was appropriate (2.2 years), which is why, under this consideration, this option was chosen. Moreover, CO_2 savings were also considered in Australia because of the carbon pricing that must be paid. In this case, the strategy chosen represented 6.17 USD/m^2 in environmental savings and 380.5 kg/m^2 in the saving in $CO_{2\text{-eq}}$.

In [66], a cost analysis was performed focusing on cost–energy effectiveness through the Dynamic Generation Cost (DGC) indicator, which takes into account the cost and the energy-saving strategy in a respective year (USD/kWh). Thus, it was possible to recognize the most feasible strategy. This tool helped to identify that window and roof replacement resulted in cost-effectiveness that was greater than appliance replacement. A similar approach was developed in [67]. However, they considered energy generation through the Levelized Cost of Energy (LCOE), which was studied to define energy generation costs. The LCOE considers capital expenditure, operating maintenance costs, fuel costs, tax costs, and energy generation per year. They achieve 0.125 USD/kWh as the LCOE value, the lowest energy tariff established in Ghana. The payback period is an important indicator that refers to the time it takes to recover the investment. In the case of Ghana, it was 6 to 10 years. Based on the outcomes, it was found that installing PV systems was feasible.

Economic analysis is important since there are different ways to develop an energy-saving strategy. For example, a low-cost house can reduce its energy consumption, cooling demand, or internal heat gains with a low-cost strategy. In the case of [64], implementing a roof cover as a low-cost technology improved thermal comfort (acceptability of 80%).

Based on this literature review, there is a decision-making process in most cases to find the correct energy-saving method. An important comparison between every criterion must be made to carry out the decision-making process. A technique often used is parametric analysis. To perform such an analysis, different simulation software is required. One is the Grasshopper plugin in Rhinoceros software (Rhinoceros 7), which uses some plug-ins, such as Ladybug and Honeybee. The first deals with meteorological data and the second relates to different software and develops parametric analysis. Another often-used simulation tool is DesignBuilder simulation software (DesignBuilder v6.1.0.006), an EnergyPlus interface where parametric analysis is possible. In addition to performing a parametric analysis, a study of phase change material is also recommended [60].

There is an additional decision-making tool that requires a more involved procedure. This is multi-criteria decision-making, which corresponds to a method that evaluates more than two criteria, and the most significant variable is given priority in the decision process. Based on the results, the best energy-saving strategy is selected accordingly. This methodology was carried out in [48], where a survey was applied to define the priority order of the considered criteria, assigning a specific weight to each one. In this case, some professionals were considered for the task. The authors identified four factors to be evaluated as follows: GWP, which refers to the energy stored by a specific gas for long periods; LCC, which refers to embodied energy; and operational energy. According to the results, GWP took the first position, followed by LCC, and the operational and embodied energy were in third and fourth place. Finally, after considering all the criteria, the double-glazed windows performed the best. It is worth noting that, concerning the authors, academic professionals prioritize environmental aspects, and the economic field has greater importance for construction professionals.

Multi-criteria decision-making has a subcategory called multi-objective decision-making; the latter was considered in [75] to reduce energy consumption, meet an optimal thermal comfort level, and decrease energy and refurbishment costs. In that study, a sensitivity analysis was performed to identify the variables strongly associated with those objectives. After the analysis, the occupancy hours, external wall construction, and cooling set point temperature were identified as the most correlated variables. Those variables were studied and modified to reach the objectives.

An alternative to a decision-making process was used in [66], where a metamodel was developed that did not require a complex simulation. For this, a Bayesian model based on the Gaussian Process (GP) was performed. A Bayesian model and the GP work together because the latter generates a specific output, which usually depends on variables that can integrate an unknown value, which the Bayesian model will estimate. In this case, the GP was used to identify the energy consumption before and after a retrofit, which depended on air temperature and other variables that were considered. The Bayesian model was required to define the unknown variables and develop the calibration model, for instance, the infiltration rate. A cost-energy effectiveness analysis was integrated into this model to establish a saving strategy ranking given in USD/kWh. The lowest value generated in that analysis corresponded to the best strategy. In this case, the appliance was replaced over the roof, and the window and wall were substituted.

4. Discussion

This study aimed to focus on the tropical climate. According to the Köppen climate classification, this research revealed that Af (Tropical Rainforest Climate) and Aw (Tropical Savanna Climate) were the most prevalent categories. For Af climates, the primary focus of the reviewed studies was office buildings, followed by residential structures, with educational buildings being the least studied, as shown in Figure 8. In Aw climates, residential buildings were the primary focus, followed by office facilities and educational structures, which were also analyzed. This is important since different tropical climates exist, leading to the implementation of different strategies that may not be interchangeable. For instance, there are regions with high humidity with low temperatures and high humidity with high temperatures with the same solar radiation intensity. Thus, highlighting the climate type can help to choose better strategies.

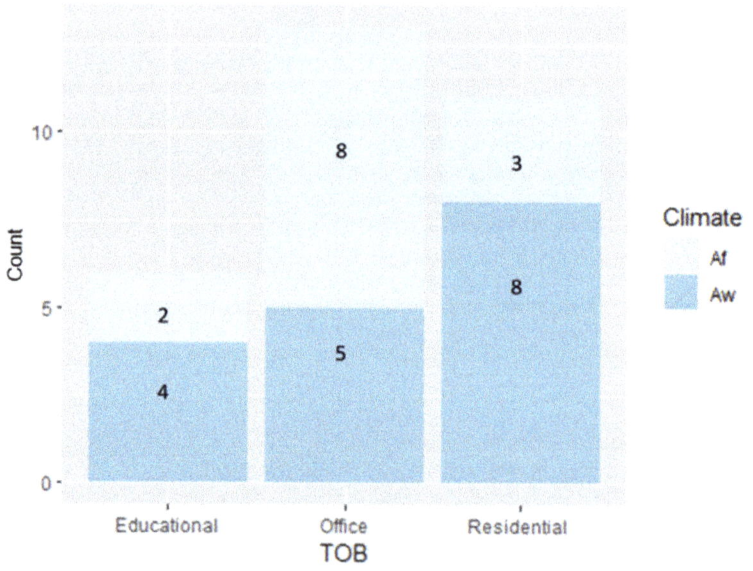

Figure 8. Building types presented in the studies found.

This literature review focused on nations characterized by tropical and humid climates. Figure 9 illustrates the studies that discussed retrofit strategies in each region. The figure illustrates the frequency at which each study was presented in various countries. The leading region is the Asia continent (India, Singapore, Malaysia, and Indonesia), followed by Latin America (Brazil, Ecuador, and Panama), the African continent (Mauritius, Madagascar, and Ghana), and the Oceania continent (Australia). This helps us understand the efforts currently being made towards studying energy-based building retrofits.

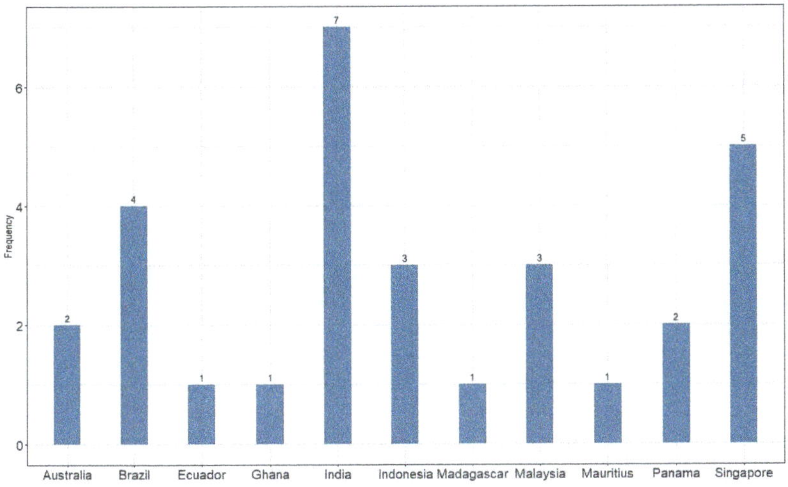

Figure 9. Countries represented in the studies found.

In this study, it was determined that the country within the tropical sector with the most studies was India, specifically the cities of Bangalore, Chennai, and Mumbai. Next was Singapore. In the Americas region, the tropical countries found were Brazil, Panama, and Ecuador (Figure 9).

Different purposes were found for carrying out a retrofit analysis, and it was determined that the main reason was to achieve energy savings, followed by savings directly in cooling demand since it is widely used given climatic conditions. Some studies considered thermal comfort and achieving a net-zero energy building (NZEB) (Figure 10).

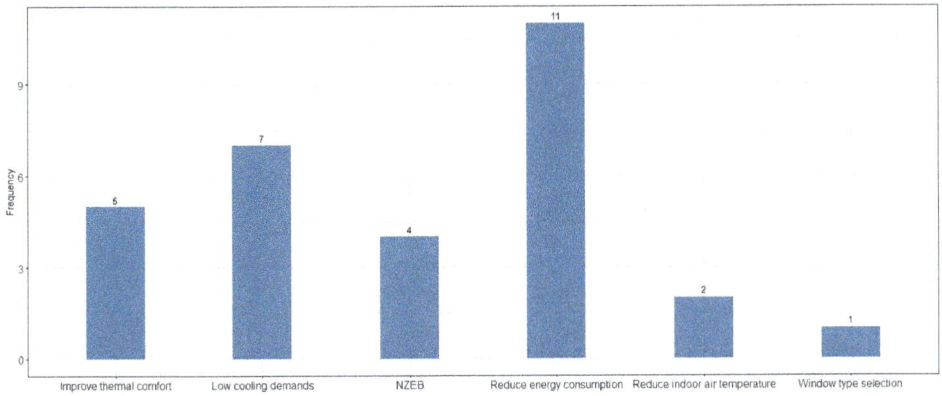

Figure 10. Purposes of energy retrofits.

Moreover, passive strategies remained crucial in office buildings within both climate types. Residential buildings implemented a mix of passive and active strategies, while educational buildings mostly relied on passive strategies and predominantly implemented passive approaches.

The reviewed research utilized three classification approaches including simulation, experimental, and in situ measures. Most studies utilized simulation approaches, especially for passive strategies. Some studies incorporated in situ measures, such as implementing changes to enhance building performance without structural alterations, while only one study used experimental approaches (Table A1).

Although the simulation approach was efficient in experimental validation and achieving better results, an emerging trend showed some studies applied in situ measures. This area presents a potential for further exploration, with resource availability influencing a shift towards more in situ approaches than experimental ones because it provides insights directly relevant to real-life situations and circumstances. Such insights are invaluable for guiding decision-making across industries, governments, and other practical domains and the impact of scientific research.

Various retrofit strategies were applied in residential and office buildings, including phase change materials, roof cover, daylight control, and WWR. Educational buildings did not implement phase change material strategies; instead, educational buildings focused on HVAC system replacement, shading devices, and shading with trees, while office buildings predominantly used wall insulation and photovoltaic strategies. Similarly, several strategies were applied in office buildings, including glazing enhancements, insulation additions, solar protection, shading devices, and advanced window technologies. Several authors [77] proposed solutions to reduce material consumption in residential buildings. However, there is a notable absence of retrofit strategies for office and educational buildings compared with other types of structures, indicating a potential area for application. Authors such as [78] advocate for an active strategy involving PV (photovoltaic), defined as bidirectional reflectance PV (BRPV), to address this gap. They evaluated its performance in a school building, which saw an increase in efficiency from 34.1% to 65.8%. Additionally, Historical or Heritage Buildings and institutional structures should be studied more.

Within the strategies with the best performance, the following were found: air conditioning (AC) set point, appliance replacement with more efficient ones, cooling operating schedule daylighting control and solar protection for windows, double and triple glazing, envelope insulation (which involves wall insulation, roof insulation), HVAC system improvement, natural ventilation, phase change material (PCMs), photovoltaic (PV) systems, and Window-to-Wall ratio improvement, keeping it in the optimal range (20–40%) according to the literature [46]. According to the outcomes, envelope insulation (for windows, walls, or roofs) corresponds to the most common strategy with the best result in terms of energy saving, as shown in Figure 11. Natural ventilation was barely studied; however, the conclusion of the few investigations suggests that natural ventilation is not enough to provide optimal thermal comfort by itself. Despite passive strategies such as envelope insulation or ventilation, some authors pointed out that excessive envelope insulation combined with insufficient ventilation are primary factors contributing to fungal growth in energy-efficient buildings [79].

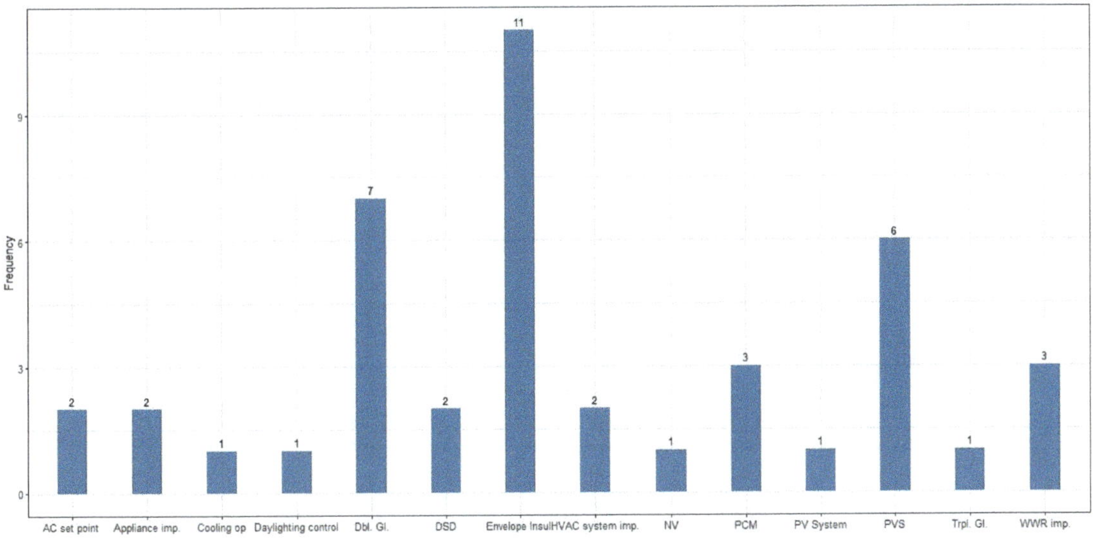

Figure 11. Strategies with the best performance within the studies found.

Most studies did not present specific cost analyses for retrofit strategies, with a few studies offering global value ranges, indicating a need for baseline values according to climate types. Higher upfront costs were highlighted as a challenge in retrofit techniques [77]. Additionally, the cost-optimal methodology should undergo cost-optimal calculations every five years to validate and subsequently revise existing national requirements. The examination of Member States' advancements in implementing this methodology suggests an overall positive trend in development [80].

Passive strategies have been widely used, yet it is evident that these strategies are not universally applicable across different building types. This highlights the opportunity to categorize and implement measures specific to each building type, thus improving strategies customized for specific building types. Educational and office buildings have yet to be studied, thus representing further research opportunities. Combined measures could enhance the cost-effectiveness of building energy retrofits. A study provided a practical framework for decision-making on energy retrofits, emphasizing the importance of comprehensive guidelines for city renewal, particularly in institutional buildings [81].

Compared with a wider range of retrofit studies and experiences in other climates, the literature indicated insufficient guidance and information on existing housing stock in the U.K. to support realistic plans for reducing carbon emissions [82]. Retrofit measures primarily focused on building envelopes, HVAC systems, lighting, and photovoltaic systems. Financial barriers, lack of standards, and regulatory support were key challenges in evaluating retrofit measures [83].

Other technical approaches to energy-efficient building retrofits included law regulation, financial incentives, and practical considerations (performance-based architectural design) [84]. The current mandatory building energy regulations in certain regions (e.g., the hot summer–cold winter region of China) were deemed insufficient to achieve significantly lower carbon emissions, suggesting the need for more ambitious regulations [85]. Opportunities for achieving nearly zero-energy buildings (nZEB) and enhancing energy efficiency in tropical climates were highlighted as part of the retrofit opportunities [86].

5. Conclusions

This research extensively reviewed and synthesized the existing literature on retrofitting strategies for buildings in tropical and humid climates. The main aim was to outline efficient approaches to improving energy efficiency, thermal comfort, and the overall performance of buildings in these geographic zones.

This study examined tropical climates, emphasizing Af (Tropical Rainforest Climate) and Aw (Tropical Savanna Climate) as the primary categories according to the Köppen climate classification. Across these classifications, this research highlighted varying focuses, revealing a bias towards office buildings in Af climates, whereas Aw climates strongly emphasized residential structures.

In office buildings in both climate types, passive strategies retained paramount importance. Conversely, a mix of passive and active approaches was employed in residential buildings, while the approaches in educational structures leaned towards passive strategies. This study reviewed three classification methodologies including simulation, experimental, and in situ measures, with simulation predominantly used, particularly for passive strategies. An intriguing emerging trend showed the budding application of in situ measures, hinting at potential exploration avenues.

Retrofit strategies varied significantly across building types, with offices leading in strategy implementation and residences and educational buildings, where strategies were relatively scant. Notably, educational structures lacked retrofit strategies compared with other building types, illuminating an area for extensive study and potential application. The absence of specific cost values for retrofit strategies underscored the need for baseline values, with high initial costs identified as a significant challenge in retrofit techniques.

In conclusion, research into retrofitting buildings in tropical and humid climates has primarily emphasized passive strategies. However, there is a noticeable gap in integrating active methods or blending both approaches, particularly in educational buildings. This highlights promising opportunities for further exploration and advancement. Utilizing a combination of measures could significantly enhance the cost-effectiveness of building energy retrofits, emphasizing the importance of comprehensive guidelines, especially in institutional buildings. The literature reveals significant shortcomings in guidance for reducing carbon emissions, indicating a pressing need for more ambitious regulations. This underscores the potential for achieving nearly zero-energy buildings and improving energy efficiency in tropical climates through retrofit opportunities.

Supplementary Materials: The following supporting information can be downloaded at: https://www.mdpi.com/article/10.3390/buildings14061633/s1, The PRISMA checklist.

Author Contributions: Original concept, formal analysis, data curation, and writing—review and editing, K.C.-C., J.G. and M.C.A. Investigation, methodology, and writing of most of this manuscript, K.C.-C. and J.G. Project administration, M.C.A. and D.M. Supervision and funding acquisition, M.C.A., C.C., D.M. and N.A. All authors have read and agreed to the published version of the manuscript.

Funding: This research was funded by the Panamanian Institution Secretaría Nacional de Ciencia, Tecnología e Innovación SENACYT (https://www.senacyt.gob.pa/, accessed on 30 December 2023), under the project code IDDS22-30, and supported by the Sistema Nacional de Investigación (SNI).

Data Availability Statement: All data supporting the reported results are included in this paper.

Acknowledgments: The authors would like to thank the Technological University of Panama and the Faculty of Mechanical Engineering (https://fim.utp.ac.pa/, accessed on 30 December 2023) for their collaboration, along with the Research Group ECEB. Special thanks is also given to the Department of Mechanical, Energy and Management Engineering (DIMEG, https://www2.unical.it/portale/strutture/dipartimenti_240/dimeg/, accessed on 30 December 2023), University of Calabria (UNICAL, https://www2.unical.it/portale/, accessed on 30 December 2023).

Conflicts of Interest: The authors declare no conflicts of interest. The funders had no role in the design of this study; in the collection, analyses, or interpretation of data; in the writing of this manuscript; or in the decision to publish the results.

Abbreviations

Acronym	Definition
AC	air conditioning
Af	Tropical Rainforest Climate
Aw	Tropical Savanna Climate
BIPV	building-integrated photovoltaic
BRPV	bidirectional reflectance PV
CAPEX	capital expenditure
CBA	Cost–Benefit Analysis
DGC	Dynamic Generation Cost
EG	energy generation
EPBD	Energy Performance of Buildings Directive
EPS	expanded polystyrene
EUI	Energy Use Intensity
FC	fuel cost
FSPCM	form stable PCM
GHG	greenhouse gas
GP	Gaussian Process
GWP	global warming potential
HATCS	hybrid air treatment cooling system
HVAC	heating, ventilation, and air conditioning.
IEA	International Energy Agency
LCC	life cycle cost
LCOE	Levelized Cost of Energy
Mtoe	megatonne oil equivalent
NZEB	net-zero energy building
nZEB	nearly zero energy building
OM	operating maintenance
OTTV	overall thermal transfer value
PCMs	phase change materials
PMV	predicted mean vote
PV	photovoltaic
TC	tax cost
TCT	thermal comfort time
VSA	vertical shadow angle (VSA)
WOEA	window opening effective area
WWR	Window-to-Wall Ratio

Appendix A

Table A1 displays information on countries, climates, project types, building types, costs, energy generation, and U-values, particularly focusing on studies that applied retrofit strategies.

Table A1. Type of project and economics studies in tropical climate.

City, Country	Climate Type	Type of Strat-egy	Building Typology	Type of Project	Cost	Energy Generation kWh/m²y	U (W m^{-2} K^{-1}) Conventional or Original					U (W m^{-2} K^{-1}) Improved					Year	Ref.
							Walls	Windows	Titles	Roof	Glazing	Walls	Windows	Titles	Roof	Glazing		
Bamako, Mali; Bangalore, India; Kolkata, India; Dar es Salaam, Tanzania	Aw	Passive	Residential	Simulation	-	-	-	-	-	-	-	-	-	-	-	-	2019	[59]
Bangalore, India	Aw	Passive	Office	Simulation	-	108.02	0.46	-	-	-		0.46	-	-	-	-	2023	[69]
Chennai, India	Aw	Passive	Educational	Simulation	-	-	2.13	5.77	3.20	-	-	0.85	3	0.20	-	-	2023	[44]
Chennai, India	Aw	Passive	Office	Simulation	64 USD/m²	-	-	-	-	-	-	-	-	-	-	-	2022	[46]
Darwin, Australia	Aw	Passive	Residential	Simulation	106 USD/m²	-	-	-	-	-	-	-	-	-	-	-	2022	[61]
Darwin, Australia	Aw	Passive	Residential	Simulation	-	-	-	-	-	-	-	-	-	-	-	-	2022	[63]
Ecuador	Aw	Passive	Educational	Simulation	USD 3200	61.65	-	-	-	-	-	-	-	-	-	-	2021	[56]
Ghana	Aw	Passive	Residential	Simulation	USD 6,484.53	68.4-78.43	-	-	-	-	-	-	-	-	-	-	2022	[67]
Indonesia	Af	Passive	Educational	Simulation	-	-	3.56	5.6				-	-	-	-	-	2023	[51]
Indonesia	Af	Passive	Educational		-	-	-	-	-	-	-	-	-	-	-	-	2020	[55]
Madagascar	Af	Passive	Office	Simulation	-	-	2.62			2.01		-	-	-	-	-	2020	[62]
Malaysia	Af	Passive	Residential	In situ measures	3.6 USD/m²	-	-	-	-	-	-	-	-	-	-	-	2020	[64]
Malaysia	Af	Passive	Residential	Simulation			1.3.				6.12	-	-	-	-	-	2021	[60]
Malaysia	Af	Passive	Office	Simulation	79.70 USD/m²	-		7.3				-	2.1	-	-	-	2022	[48]
Mauritius	Aw	Passive	Office	Simulation						2.2.04		-	-	-	-	-	2023	[70]
Mumbai, India	Aw	Passive	Educational	Simulation	-	-	2.85	2.36		2.86		-	-	-	-	-	2022	[67]
Mumbai, India	Aw	Passive	Educational	Simulation	-	-	2.85	2.36		2.85		0.1	0.5	0.1		-	2022	[58]
Panamá		Passive	Residential	Simulation		330.87	-	-	-	-	-	-	-	-	-	-	2022	[68]
Rio de Janeiro, Brazil	Aw	Passive	Office	Simulation	-	-	-	-	-	-	-	-	-	-	-	-	2021	[45]
Sao Paulo, Brazil	Af	Active	Office	Simulation	-	-	2.38	5.8	-	-	-	2.8	2.8	-	-	-	2020	[49]
Singapore	Af	Passive	Office	Simulation			-	4.8	-	-	-	-	-	-	4.1	-	2019	[78]
Singapore	Af	Passive	Office	In situ measures	-	-	-	4.96	-	-	-	-	4.1	-	-	-	2020	[52]
Singapore	Af	Active	Office	Simulation and in situ measures	-	-	-	5.649	-	-	-	-	1.998	-	-	-	2020	[50]
Singapore	Af	Active	Residential	Simulation	-	-	-	-	-	-	-	-	-	-	-	-	2019	[66]
Singapore; Miami, USA; Darwin, Australia	Af	Active	Office	Experimental and simulation	-	-	-	-	-	-	-	-	-	-	-	-	2019	[65]

References

1. Aste, N.; Del Pero, C.; Leonforte, F. Toward Building Sector Energy Transition. In *Handbook of Energy Transitions*; CRC Press: Boca Raton, FL, USA, 2022; pp. 127–150.
2. Azhgaliyeva, D.; Rahut, D.B. *Promoting Green Buildings: Barriers, Solutions, and Policies*; ADBI Working Paper 1331; Asian Development Bank Institute: Tokyo, Japan, 2022.
3. Anuja, N.; Akalya, B.; Karthika, R.; Venkateshwari, P. Controlling of CO_2 emission in buildings: An overview. *Int. J. Civ. Eng. Constr.* **2022**, *1*, 1–5. [CrossRef]
4. Alkhatib, F.; Alawag, A.M. Building Information Modelling (BIM) and Energy Performance of Building—A Review. *J. Appl. Artif. Intell.* **2022**, *2*, 22–31. [CrossRef]
5. Das, A.K.; Sharma, A. Chapter 1—Climate change and the energy sector. In *Advancement in Oxygenated Fuels for Sustainable Development*; Kumar, N., Mathiyazhagan, K., Sreedharan, V.R., Kalam, A., Eds.; Elsevier: Amsterdam, The Netherlands, 2023; pp. 1–6.
6. United Nations Statistics Division, *The Sustainable Development Goals Report*; United Nations Statistics Division: New York, NY, USA, 2023; Volume 07988.
7. Goubran, S.; Cucuzzella, C. Integrating the Sustainable Development Goals in Building Projects. *J. Sustain. Res.* **2019**, *1*, e190010. [CrossRef]
8. Goubran, S. On the Role of Construction in Achieving the SDGs. *J. Sustain. Res.* **2019**, *1*, e190020. [CrossRef]
9. European Commission. Energy Performance of Buildings Directive. Available online: https://energy.ec.europa.eu/topics/energy-efficiency/energy-efficient-buildings/energy-performance-buildings-directive_en (accessed on 10 May 2024).
10. Lee, J.; Shepley, M.M.; Choi, J. Exploring the effects of a building retrofit to improve energy performance and sustainability: A case study of Korean public buildings. *J. Build. Eng.* **2019**, *25*, 100822. [CrossRef]
11. Regnier, C.; Sun, K.; Hong, T.; Piette, M.A. Quantifying the benefits of a building retrofit using an integrated system approach: A case study. *Energy Build.* **2018**, *159*, 332–345. [CrossRef]
12. Liu, G.; Tan, Y.; Li, X. China's policies of building green retrofit: A state-of-the-art overview. *Build. Environ.* **2020**, *169*, 106554. [CrossRef]
13. Benzar, B.-E.; Park, M.; Lee, H.-S.; Yoon, I.S.; Cho, J. Determining retrofit technologies for building energy performance. *J. Asian Arch. Build. Eng.* **2020**, *19*, 367–383. [CrossRef]
14. Geraldi, M.S.; Ghisi, E. Building-level and stock-level in contrast: A literature review of the energy performance of buildings during the operational stage. *Energy Build.* **2020**, *211*, 109810. [CrossRef]
15. Deb, C.; Schlueter, A. Review of data-driven energy modelling techniques for building retrofit. *Renew. Sustain. Energy Rev.* **2021**, *144*, 110990. [CrossRef]
16. Liu, Y.; Liu, T.; Ye, S.; Liu, Y. Cost-benefit analysis for Energy Efficiency Retrofit of existing buildings: A case study in China. *J. Clean. Prod.* **2018**, *177*, 493–506. [CrossRef]
17. United States Department of Energy. *Stepping Up to the Challenge Together—Better Buildings Progress Report 2022*; United States Department of Energy: Washington, DC, USA, 2022.
18. Ministry of Energy of Thailand. Thailand's 20-Year Energy Efficiency Development Plan. 2011. Available online: http://www.enconfund.go.th/pdf/index/EEDP_Eng.pdf (accessed on 10 May 2023).
19. Socialist Republic of Vietnam, Ministry of Construction in Vietnam (MOC). *National Technical Regulation on Energy Efficiency Buildings Hanoi 2017*; Ministry of Construction in Vietnam: Hanoi, Vietnam, 2017.
20. Directive (eu) 2023/1791 of the European Parliament and of the Council of 13 September 2023 on Energy Efficiency and Amending Regulation (eu) 2023/955 (recast) (Text with eea Relevance). Official Journal of the European Union. 2018. Available online: https://eur-lex.europa.eu/eli/dir/2023/1791/oj (accessed on 20 December 2023).
21. Bureau of Energy Efficiency Ministry of Power. *Guidelines for Financing Energy Efficiency Projects in India*; Bureau of Energy Efficiency Ministry of Power: New Delhi, India, 2017.
22. UNEP Finance Initiative; United Nations Global Compact. Delivering Net Zero Emissions in Japan. Available online: https://dwtyzx6upklss.cloudfront.net/Uploads/t/e/i/pri_netzerobriefing2021japan_583956.pdf (accessed on 10 May 2023).
23. *Energy Star Canada*; Government of Canada: Ottawa, ON, Canada, 2021.
24. Economidou, M.; Todeschi, V.; Bertoldi, P. *Accelerating Energy Renovation Investments in Buildings—Financial and Fiscal Instruments Across the EU*; Publications Office of the European Union: Luxembourg, 2019.
25. The Economist Intelligence Unit. *Achieving Scale in Energy-Efficient Buildings in China*; The Economist Intelligence Unit: London, UK, 2013.
26. Building and Construction Authority. *Singapore Green Building Masterplan Public Engagement Report*; Building and Construction Authority: Singapore, 2021; pp. 1–25.
27. Svendsen, A.; Schultz, P.C. *Roadmap for an Energy Efficient, Low-Carbon Buildings and Construction Sector in Indonesia*; Danish Energy Agency: Copenhagen, Denmark, 2022; pp. 1–95.
28. Verdote, N.; Oliver, T.; du Pont, P.; Velasco, L.; Prijyanonda, C. The Philippines Green Buildings program: Developing a market niche for energy efficiency. *ACEEE Summer Study*, June–July. 2000; Asilomar, CA, USA; Volume 4, 4377–4390.

29. United Nations Development Programme. Energy Efficiency in Buildings: Accelerating Low-carbon Development in Cambodia Policy Brief & In-country Case Studies. Available online: https://www.undp.org/sites/g/files/zskgke326/files/migration/kh/UNDP2020_Energy-Efficiency-in-Building-Policy-Brief-Cambodia_ENG_Small.pdf (accessed on 10 May 2023).
30. Sustainable and Renewable Energy Development Authority; Power Division Ministry of Power, Energy and Mineral Resources Government of the People's Republic of Bangladesh. *Energy Efficiency and Conservation Master Plan up to 2030*; Ministry of Power Energy and Mineral Resources Government of the People's Republic of Bangladesh: Dhaka, Bangladesh, 2015.
31. Ranawaka, I.; Mallawaarachchi, H. A risk-responsive framework for green retrofit projects in Sri Lanka. *Built Environ. Proj. Asset Manag.* **2018**, *8*, 477–490. [CrossRef]
32. Shari, Z.; Mohamad, N.L.; Dahlan, N.D. Building Envelope Retrofit for Energy Savings in Malaysian Government High-Rise Offices: A Calibrated Energy Simulation. *J. Teknol.* **2023**, *85*, 1–15. [CrossRef]
33. Asadi, S.; Mostavi, E.; Boussaa, D.; Indraganti, M. Building energy model calibration using automated optimization-based algorithm. *Energy Build.* **2019**, *198*, 106–114. [CrossRef]
34. Royapoor, M.; Roskilly, T. Building model calibration using energy and environmental data. *Energy Build.* **2015**, *94*, 109–120. [CrossRef]
35. Saffari, M.; de Gracia, A.; Ushak, S.; Cabeza, L.F. Passive cooling of buildings with phase change materials using whole-building energy simulation tools: A review. *Renew. Sustain. Energy Rev.* **2017**, *80*, 1239–1255. [CrossRef]
36. Asadi, E.; da Silva, M.G.; Antunes, C.H.; Dias, L. A multi-objective optimization model for building retrofit strategies using TRNSYS simulations, GenOpt and MATLAB. *J. Affect. Disord.* **2012**, *56*, 370–378. [CrossRef]
37. Wang, B.; Xia, X.; Zhang, J. A multi-objective optimization model for the life-cycle cost analysis and retrofitting planning of buildings. *Energy Build.* **2014**, *77*, 227–235. [CrossRef]
38. Jafari, A.; Valentin, V. Selection of optimization objectives for decision-making in building energy retrofits. *J. Affect. Disord.* **2018**, *130*, 94–103. [CrossRef]
39. Mejjaouli, S.; Alzahrani, M. Decision-making model for optimum energy retrofitting strategies in residential buildings. *Sustain. Prod. Consum.* **2020**, *24*, 211–218. [CrossRef]
40. Feng, W.; Zhang, Q.; Ji, H.; Wang, R.; Zhou, N.; Ye, Q.; Hao, B.; Li, Y.; Luo, D.; Lau, S.S.Y. A review of net zero energy buildings in hot and humid climates: Experience learned from 34 case study buildings. *Renew. Sustain. Energy Rev.* **2019**, *114*, 109303. [CrossRef]
41. Kwong, Q.J.; Adam, N.M.; Sahari, B. Thermal comfort assessment and potential for energy efficiency enhancement in modern tropical buildings: A review. *Energy Build.* **2014**, *68*, 547–557. [CrossRef]
42. Azima, M.; Seyis, S. Science mapping the knowledge domain of energy performance research in the AEC industry: A scientometric analysis. *Energy* **2023**, *264*, 125938. [CrossRef]
43. Van Eck, N.J.; Waltman, L. Software survey: VOSviewer, a computer program for bibliometric mapping. *Scientometrics* **2010**, *84*, 523–538. [CrossRef]
44. Surendran, V.M.; Irulappan, C.; Jeyasingh, V.; Ramalingam, V. Thermal Performance Assessment of Envelope Retrofits for Existing School Buildings in a Hot–Humid Climate: A Case Study in Chennai, India. *Buildings* **2023**, *13*, 1103. [CrossRef]
45. Fontenelle, M.R.; Bastos, L.E.G.; Lorente, S. Natural ventilation for office building retrofit in dense urban context under hot and humid climate. *Ambient. Constr.* **2021**, *21*, 67–87. [CrossRef]
46. Chandrasekaran, C.; Sasidhar, K.; Madhumathi, A. Energy-efficient retrofitting with exterior shading device in hot and humid climate—Case studies from fully glazed multi-storied office buildings in Chennai, India. *J. Asian Arch. Build. Eng.* **2022**, *22*, 2209–2223. [CrossRef]
47. Shandilya, A.; Hauer, M.; Streicher, W. Optimization of Thermal Behavior and Energy Efficiency of a Residential House Using Energy Retrofitting in Different Climates. *Civ. Eng. Arch.* **2022**, *8*, 335–349. [CrossRef]
48. Balasbaneh, A.T.; Yeoh, D.; Ramli, M.Z.; Valdi, M.H.T. Different alternative retrofit to improving the sustainability of building in tropical climate: Multi-criteria decision-making. *Environ. Sci. Pollut. Res.* **2022**, *29*, 41669–41683. [CrossRef]
49. Gomes, V.; Loche, I.; Saade, M.R.; Pulgrossi, L.; Franceschini, P.B.; Rodrigues, L.L.; Pimenta, R.G.; Neves, L.O.; Kowaltowski, D.C.C.K. Operational and embodied impact assessment as retrofit decision-making support in a changing climate. In Proceedings of the 11th Windsor Conference on Thermal Comfort, Windsor, UK, 16–20 April 2020; pp. 936–948. Available online: https://www.researchgate.net/publication/341194240_Operational_and_embodied_impact_assessment_as_retrofit_decision-making_support_in_a_changing_climate (accessed on 10 May 2023).
50. Koh, W.S.; Liu, H.; Somasundaram, S.; Thangavelu, S.R.; Chong, A.; Pillai, K.; Kojima, H.; Mori, Y. Evaluation of glazing retrofitting solution for the tropics. *Energy Build.* **2020**, *223*, 110190. [CrossRef]
51. Sebayang, S.; Alkadri, M.F.; Chairunnisa, I.; Dewi, O.C. Retrofit design strategies for educational building through shading and glazing modification. *BIO Web Conf.* **2023**, *62*, 05001. [CrossRef]
52. Somasundaram, S.; Thangavelu, S.R.; Chong, A. Effect of Existing Façade's Construction and Orientation on the Performance of Low-E-Based Retrofit Double Glazing in Tropical Climate. *Energies* **2020**, *13*, 2016. [CrossRef]
53. Somasundaram, S.; Chong, A.; Wei, Z.; Thangavelu, S.R. Energy saving potential of low-e coating based retrofit double glazing for tropical climate. *Energy Build.* **2020**, *206*, 109570. [CrossRef]

4. Alwi, N.M.; Flor, J.-F.; Anuar, N.H.; Mohamad, J.; Hanafi, N.N.H.; Muhammad, N.H.; Zain, M.H.K.M.; Nasir, M.R.M. Retrofitting measures for climate resilience: Enhancing the solar performance of Malaysian school buildings with passive design concepts. *IOP Conf. Ser. Earth Environ. Sci.* **2022**, *1102*, 012014. [CrossRef]
5. Khidmat, R.P.; Ulum, M.S.; Lestari, A.D.E. Façade Components Optimization of Naturally Ventilated Building in Tropical Climates through Generative Processes. Case study: Sumatera Institute of Technology (ITERA), Lampung, Indonesia. *IOP Conf. Ser. Earth Environ. Sci.* **2020**, *537*, 012015. [CrossRef]
6. Litardo, J.; Palme, M.; Hidalgo-León, R.; Amoroso, F.; Soriano, G. Energy Saving Strategies and On-Site Power Generation in a University Building from a Tropical Climate. *Appl. Sci.* **2021**, *11*, 542. [CrossRef]
7. Bakker, A.A.M.A.; Taskeen, A.Z.F.; Priya, A.K.I. Energy-Efficient Retrofitting with Kinetic Shading Device in Tropical Climate. *Eng. Technol.* **2023**, *10*, 948–958.
8. Gupta, V.; Deb, C. Energy retrofit analysis for an educational building in Mumbai. *Sustain. Futur.* **2022**, *4*, 100096. [CrossRef]
9. Bimaganbetova, M.; Memon, S.A.; Sheriyev, A. Performance evaluation of phase change materials suitable for cities representing the whole tropical savanna climate region. *Renew. Energy* **2020**, *148*, 402–416. [CrossRef]
0. Al-Absi, Z.A.; Hafizal, M.I.M.; Ismail, M.; Ghazali, A. Towards Sustainable Development: Building's Retrofitting with PCMs to Enhance the Indoor Thermal Comfort in Tropical Climate, Malaysia. *Sustainability* **2021**, *13*, 3614. [CrossRef]
1. Abden, J.; Tao, Z.; Alim, M.A.; Pan, Z.; George, L.; Wuhrer, R. Combined use of phase change material and thermal insulation to improve energy efficiency of residential buildings. *J. Energy Storage* **2022**, *56*, 105880. [CrossRef]
2. Nematchoua, M.K.; Noelson, J.C.V.; Saadi, I.; Kenfack, H.; Andrianaharinjaka, A.-Z.F.R.; Ngoumdoum, D.F.; Sela, J.B.; Reiter, S. Application of phase change materials, thermal insulation, and external shading for thermal comfort improvement and cooling energy demand reduction in an office building under different coastal tropical climates. *Sol. Energy* **2020**, *207*, 458–470. [CrossRef]
3. Sangwan, P.; Mehdizadeh-Rad, H.; Ng, A.W.M.; Tariq, M.A.U.R.; Nnachi, R.C. Performance Evaluation of Phase Change Materials to Reduce the Cooling Load of Buildings in a Tropical Climate. *Sustainability* **2022**, *14*, 3171. [CrossRef]
4. Tuck, N.W.; Zaki, S.A.; Hagishima, A.; Rijal, H.B.; Yakub, F. Affordable retrofitting methods to achieve thermal comfort for a terrace house in Malaysia with a hot–humid climate. *Energy Build.* **2020**, *223*, 110072. [CrossRef]
5. Cui, X.; Islam, M.; Chua, K. Experimental study and energy saving potential analysis of a hybrid air treatment cooling system in tropical climates. *Energy* **2019**, *172*, 1016–1026. [CrossRef]
6. Yuan, J.; Nian, V.; Su, B. Evaluation of cost-effective building retrofit strategies through soft-linking a metamodel-based Bayesian method and a life cycle cost assessment method. *Appl. Energy* **2019**, *253*, 113573. [CrossRef]
7. Ohene, E.; Hsu, S.-C.; Chan, A.P. Feasibility and retrofit guidelines towards net-zero energy buildings in tropical climates: A case of Ghana. *Energy Build.* **2022**, *269*, 112252. [CrossRef]
8. Austin, M.C.; Carpino, C.; Mora, D.; Arcuri, N. A Methodology to identify appropriate refurbishment strategies towards zero energy buildings in a hot and humid climate. *J. Phys. Conf. Ser.* **2022**, *2385*, 012020. [CrossRef]
9. Azeem, A.; Thomas, A. Net-zero Energy Retrofit of an Existing Commercial Building in Temperate Climate Zone of India. In Proceedings of the 6th International Conference on Modeling and Simulation in Civil Engineering, Kollam, India, 1–3 December 2022.
0. Jhumka, H.; Yang, S.; Gorse, C.; Wilkinson, S.; Yang, R.; He, B.-J.; Prasad, D.; Fiorito, F. Assessing heat transfer characteristics of building envelope deployed BIPV and resultant building energy consumption in a tropical climate. *Energy Build.* **2023**, *298*, 113540. [CrossRef]
1. Kaewpraek, C.; Ali, L.; Rahman, A.; Shakeri, M.; Chowdhury, M.S.; Jamal, M.S.; Mia, S.; Pasupuleti, J.; Dong, L.K.; Techato, K. The Effect of Plants on the Energy Output of Green Roof Photovoltaic Systems in Tropical Climates. *Sustainability* **2021**, *13*, 4505. [CrossRef]
2. Arenandan, V.; Wong, J.K.; Ahmed, A.N.; Chow, M.F. Efficiency enhancement in energy production of photovoltaic modules through green roof installation under tropical climates. *Ain Shams Eng. J.* **2022**, *13*, 101741. [CrossRef]
3. Costa, J.F.W.; Amorim, C.N.D.; Silva, J.C.R. Retrofit guidelines towards the achievement of net zero energy buildings for office buildings in Brasilia. *J. Build. Eng.* **2020**, *32*, 101680. [CrossRef]
4. Ardiani, N.; Sharples, S.; Mohammadpourkarbasi, H. Multi-Objective Optimisation OF Energy Retrofit in Hot-Humid Climates' Office Building. In Proceedings of the 10th International Conference on Architecture and Built Environment, S.ARCH, Berlin, Germany, 4–6 April 2023; p. 298.
5. Chacón, L.; Austin, M.C.; Castaño, C. A Multiobjective Optimization Approach for Retrofitting Decision-Making towards Achieving Net-Zero Energy Districts: A Numerical Case Study in a Tropical Climate. *Smart Cities* **2022**, *5*, 405–432. [CrossRef]
6. Naves, A.X.; Esteller, L.J.; Haddad, A.N.; Boer, D. Targeting Energy Efficiency through Air Conditioning Operational Modes for Residential Buildings in Tropical Climates, Assisted by Solar Energy and Thermal Energy Storage. Case Study Brazil. *Sustainability* **2021**, *13*, 12831. [CrossRef]
77. Pal, N. A Critical Review of Energy Retrofitting Techniques in Building. *Res. Rev. J. Archit. Des.* **2023**, *5*, 1–9. [CrossRef]
78. Lee, K.H.; Song, Y.-H. Analysis of Energy Reduction and Energy Self-Sufficiency Improvement Effects by Applying a Bidirectional Reflectance PV Array with Integrated External Shading at a School Building. *Buildings* **2023**, *13*, 2915. [CrossRef]
79. Carpino, C.; Loukou, E.; Austin, M.C.; Andersen, B.; Mora, D.; Arcuri, N. Risk of Fungal Growth in Nearly Zero-Energy Buildings (nZEB). *Buildings* **2023**, *13*, 1600. [CrossRef]

80. Zangheri, P.; D'agostino, D.; Armani, R.; Bertoldi, P. Review of the Cost-Optimal Methodology Implementation in Member States in Compliance with the Energy Performance of Buildings Directive. *Buildings* **2022**, *12*, 1482. [CrossRef]
81. Lu, Y.; Li, P.; Lee, Y.P.; Song, X. An integrated decision-making framework for existing building retrofits based on energy simulation and cost-benefit analysis. *J. Build. Eng.* **2021**, *43*, 103200. [CrossRef]
82. Alabid, J.; Bennadji, A.; Seddiki, M. A review on the energy retrofit policies and improvements of the UK existing buildings, challenges and benefits. *Renew. Sustain. Energy Rev.* **2022**, *159*, 112161. [CrossRef]
83. Hong, Y.; Ezeh, C.I.; Deng, W.; Hong, S.-H.; Peng, Z. Building Energy Retrofit Measures in Hot-Summer–Cold-Winter Climates: A Case Study in Shanghai. *Energies* **2019**, *12*, 3393. [CrossRef]
84. Fernandes, J.; Santos, M.C.; Castro, R. Introductory Review of Energy Efficiency in Buildings Retrofits. *Energies* **2021**, *14*, 8100. [CrossRef]
85. Liu, C.; Sharples, S.; Mohammadpourkarbasi, H. A Review of Building Energy Retrofit Measures, Passive Design Strategies and Building Regulation for the Low Carbon Development of Existing Dwellings in the Hot Summer–Cold Winter Region of China. *Energies* **2023**, *16*, 4115. [CrossRef]
86. Chung-Camargo, K.; González, J.; Solano, T.; Yuil, O.; Velarde, V.; Austin, M.C. *Energy-Efficiency Measures to Achieve Zero Energy Buildings in Tropical and Humid Climates*; IntechOpen: London, UK, 2023. [CrossRef]

buildings

MDPI

Article

Optimisation of Nearly Zero Energy Building Envelope for Passive Thermal Comfort in Southern Europe

Jaime Resende and Helena Corvacho *

CONSTRUCT (LFC), Faculty of Engineering (FEUP), University of Porto, 4200-465 Porto, Portugal; jaime.resende@fe.up.pt
* Correspondence: corvacho@fe.up.pt

Abstract: The pursuit of sustainable and energy-efficient construction is vital to mitigate climate change and reduce carbon emissions. The application of the concept of nearly Zero Energy Building (nZEB) is now a reality for new buildings in the European Union, helping to achieve those goals. However, there is significant complexity in achieving acceptable thermal comfort levels in warmer climates such as the one in Southern Europe. This study carried out a multi-objective optimisation of the nZEB envelope using current construction solutions and nZEB regulations currently in force in different climate zones in this region, aiming to reduce thermal discomfort according to EN 16798-1. The results indicate that passive measures induced by regulatory requirements can significantly reduce discomfort at an affordable cost. However, great caution must be taken in relation to regulatory requirements, mainly for the cooling season, aiming to avoid summer overheating of dwellings and guaranteeing that nZEB's buildings are sustainable and comfortable in the Mediterranean climate regions. In addition, designers should be aware that increasing the insulation layer beyond regulatory requirements does not necessarily imply an increase in passive thermal comfort. Often, this implies, in addition to an increase in construction costs, an increase in discomfort, particularly during the cooling season.

Keywords: nZEB/NZEB; envelope optimisation; thermal comfort; regulatory requirements; Southern Europe

1. Introduction

1.1. Introduction and Background

The quest for sustainable and energy-efficient construction practices has become increasingly indispensable in the context of global climate change and the pressing need to reduce carbon emissions. Buildings represent 40% of final energy consumption and 36% of energy-related greenhouse gas emissions in the European Union, while around 75% of its buildings are still energy inefficient [1].

In this context, Europe adopted the nZEB concept as an obligation for all new buildings constructed in its Member States (MS) [2,3]. Each MS has the autonomy to transpose the Directives according to its local particularities in order to achieve, in its own way, the common goals defined by the Directives. In addition to the European-wide definition of the concept, there are other applicable definitions worldwide. The same occurs with the concept of Net Zero Energy Buildings (NZEBs), and due to different contexts and realities around the world, there is no common understanding regarding this kind of concept [4]; thus, a common definition may never be reached [5]. However, two main strategies are always involved in nZEB/NZEB construction: reducing energy needs through the implementation of passive strategies and using energy from renewable sources, preferably local, to meet those already-reduced needs [6].

Besides the first nZEB, most basic approaches are already included in MS national regulations; further approaches are more comprehensive and consider the whole life cycle of the building, considering the embodied energy in each of its materials and components [7] and the embodied carbon [8]. In this scope, the new European Directive that was published very recently goes further in this direction than the last transposed version.

However, although the nZEB/NZEB simplest "recipe" seems simplistic, it is not quite so. Its design requires a holistic approach, and its objective can be achieved with the best combination of envelope, systems, and energy sources under technical and financial constraints that change with context and time. Social engagement and the attitudes of key stakeholders in the industry are also important to achieve the goal in practice [9]. The use of multi-objective optimisation tools can certainly help in this complex relationship between the inputs and outputs of nZEB/NZEB construction [10,11].

Regardless of the concept or metric considered, minimising energy use must be a fundamental design requirement and the highest priority of nZEB/NZEB. Energy efficiency measures are typically the most cost-effective approach with the highest return on investment. Prior to developing local renewable energy strategies, optimising efficiency opportunities would reduce the number and cost of renewable energy sources [6].

Building envelopes, including walls, windows, floors and roofs, have a substantial impact on energy needs, so these construction elements must be selected with awareness [12,13]. It is known that the construction of nZEB/NZEBs is becoming more feasible owing to the advancements in building technology, renewable energy systems and academic research [6]. Given that improving energy performance has become a priority for current buildings, it is possible to ensure that nZEB/NZEB construction objectives can be achieved through current practices. However, a fundamental factor, namely thermal comfort, has been neglected in the most recent studies [14].

1.2. Literature Review

The theoretical basis of adaptive thermal comfort models has matured, and, therefore, several models have already been integrated into regulations worldwide, which is fundamental for their acceptance and implementation [15]. In Europe, the EN 16798-1 [16] standard establishes limits for indoor conditions to ensure occupant comfort. A critical interplay exists between thermal comfort parameters and energy demand, particularly in regions with warmer climates, as in Southern European countries [17]. That interplay is critical and difficult to predict due to various factors, from the designer's choices to the users' preferences and behaviour. This underscores the need for tailored solutions that consider regional climate, socioeconomic conditions and occupant behaviour. A study [18] proposed that for existing homes, the most effective initial reduction in energy use should come from changing occupant behaviour, followed by improving existing building envelopes. The adoption of a realistic adaptive thermal comfort model can help in reducing energy demand while ensuring occupant comfort, especially in Mediterranean regions [19].

There is complexity in achieving nearly or net-zero energy objectives while maintaining indoor thermal comfort, which requires a careful balance between energy efficiency measures and occupant comfort requirements [14]. In [20], the authors shed light on the importance of considering occupants' cultural practices and traditional adaptive behaviours in achieving thermal comfort in vernacular dwellings.

As we strive to move towards a future of sustainable building practices, it is essential to recognise the challenges and opportunities. The most advanced technologies are available, from the production of building components to the automation of a building's operational phase, making it a so-called smart building. However, a comprehensive assessment is needed to ensure that their application corresponds to a truly sustainable improvement because, sometimes, current simple solutions may be the most sustainable. For a nearly zero-energy building design, De Masi et al. [21] analysed the environmental impact of increasing envelope insulation and the integration of photovoltaic panels in different climates, and they concluded that the most energy-efficient solutions do not always correspond to the optimal environmental result, showing the need for a case-by-case assessment. Furthermore, the articulation of energy-saving potentials, as discussed in [22], underscores the significance of simple adaptive strategies, such as natural ventilation and adaptive setpoint temperatures, in mitigating energy expenditure.

Adaptive comfort is also crucial in relation to overheating in dwellings, a global concern that is increasing due to global warming and more frequent and extreme heat waves. Forecasts indicate that cooling will be increasingly needed in buildings [23]. A study presented in [24] revealed that apartments built under current Spanish energy regulations do not show a significant reduction in hours of indoor overheating compared to those built before any energy regulations came into force. This demonstrates there is an increasing need to adapt current building energy regulations to face warmer conditions, which the world is already facing and can be expected to get worse.

An extensive literature review presented in [25] on strategies for achieving net-zero energy and net-zero carbon buildings identified studies that made a critical analysis of the existing governance guidelines. In a recent article [26], Capeluto calls attention to the "unsustainable direction of Green Building Codes", proposing an Occupation Correction Factor that could alter the energy ratings of buildings, taking into account the occupancy and actual usage patterns since these factors are crucial for achieving broader environmental and social sustainability goals. All the raised questions show how complex the challenge can be of transforming sustainable building concepts into real-life practices.

The trend of European regulations towards a constant increase in the required level of thermal insulation of the building envelope raises justified concerns in Southern European countries due to an increased risk of overheating.

Considering all the mentioned aspects, from the scarcity of studies that consider adaptive thermal comfort to the importance of multi-objective optimisation and focusing on the particularities of the Southern European climate, the present envelope optimisation study aims at minimising passive thermal discomfort and initial construction cost, allowing at the same time the discussion of some aspects of regulatory requirements.

By doing so, this study tries to respond to the following questions:

- To what extent do current regulatory requirements prevent passive discomfort?
- Given that winter and summer passive comfort point mostly to opposite strategies, what would be the most balanced envelope options for achieving nZEB/NZEB status?
- Is the trend of an increasing regulatory requirement for the level of envelope thermal insulation recommendable for Southern Europe?
- What would be the consequences for the choice of heating and cooling systems and renewable energy sources of the optimised envelope options?

2. Materials and Methods

2.1. Climate Zoning

Climate zoning in Portugal is defined by dividing the territory into three winter climate zones (I1, I2 and I3) and three summer climate zones (V1, V2 and V3). Zones numbered with one have milder climates, which get more severe from one to three. Winter climate zones are defined according to the number of heating degree-days (HDD) for a base temperature of 18 °C, corresponding to the heating season. Summer climate zones are defined based on the average outdoor temperature for the conventional cooling season ($\theta_{ext,v}$) [27].

To calculate HDD and $\theta_{ext,v}$, reference values are indicated for each region of the country, and an adjustment for each specific location must be made depending on its altitude [28,29]. Figure 1 shows the geographic representation of winter and summer climate zoning. After overlaying these maps and checking the criteria for Portuguese municipalities through their altitude, eight possible combinations were identified in mainland Portugal (there is no location in mainland Portugal that corresponds to the combination I1V1) [27,30]. Table 1 shows the municipalities chosen to represent each of these climate zone combinations, their altitude, and the criteria for defining the zones. In Figure 1, the geographic locations of the representative municipalities are also identified.

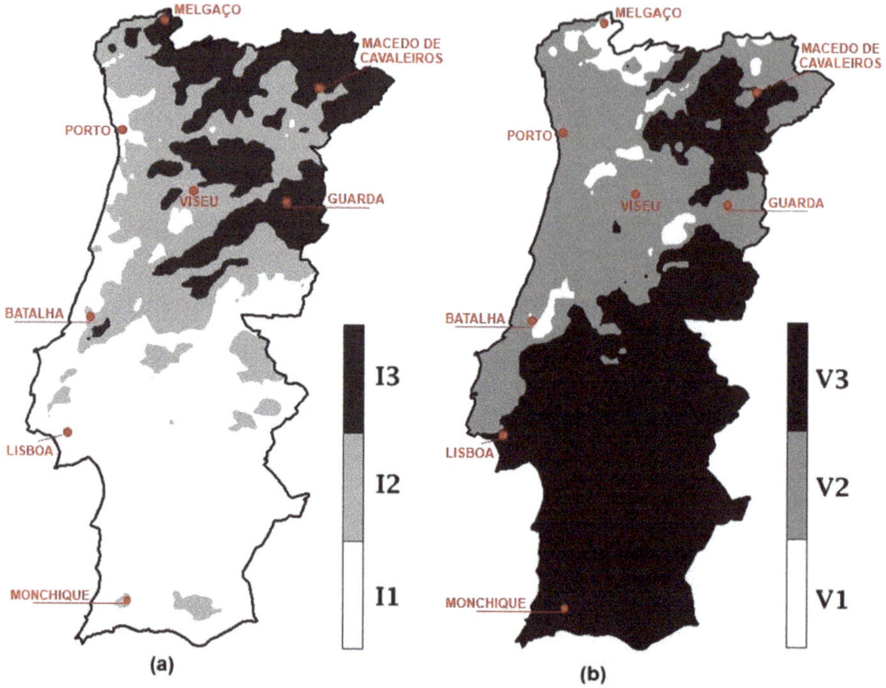

Figure 1. Graphical representation of the (**a**) winter climate zones and (**b**) summer climate zones and locations of the studied municipalities.

Table 1. Combinations of climate zones, representative municipality, location altitude and criteria for defining the Portuguese climate zones.

Climate Zones	Municipality		Altitude
I1V2	Porto		83 m
I1V3	Lisboa		4 m
I2V1	Batalha		247 m
I2V2	Viseu		475 m
I2V3	Monchique		446 m
I3V1	Melgaço		678 m
I3V2	Guarda		1017 m
I3V3	Macedo de Cavaleiros		572 m
Winter climate zone	I1	I2	I3
Winter criteria	HDD $\leq$ 1300	1300 < HDD $\leq$ 1800	HDD > 1800
Summer climate zone	V1	V2	V3
Summer criteria	$\theta_{ext,v} \leq 20$ °C	20 °C < $\theta_{ext,v} \leq 22$ °C	$\theta_{ext,v} > 22$ °C

2.2. Standard Dwelling

For the purposes of the study, a standard dwelling (Figure 2), a single house, is considered based on the data available for the floor area of new single houses at the Building Energy Certification System statistics since nZEB regulatory requirements came into force in July 2021 [31].

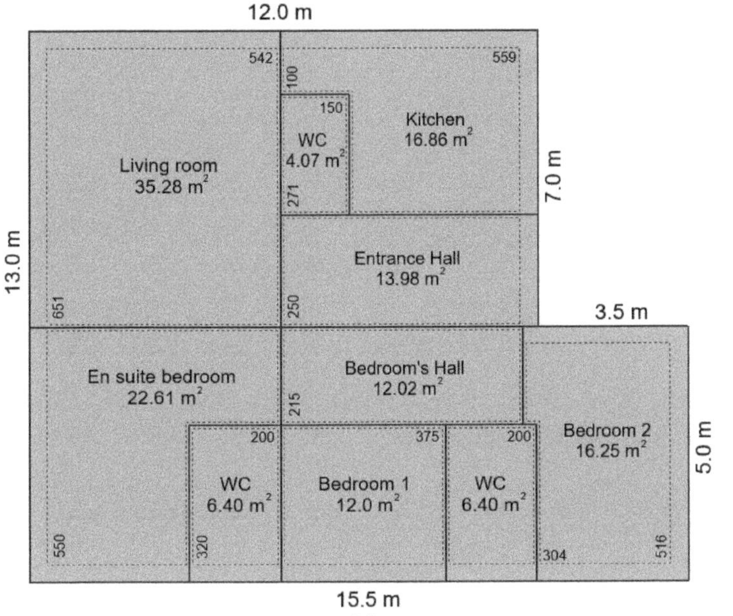

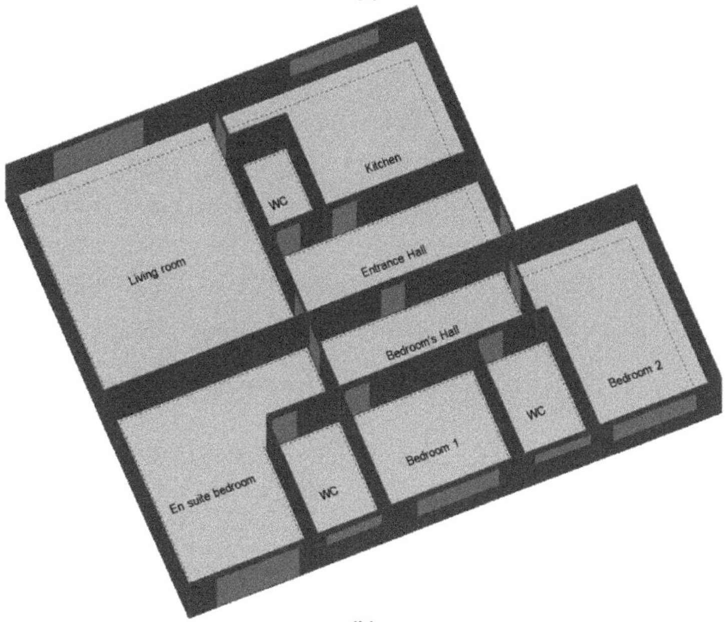

Figure 2. (**a**) Floor plan and (**b**) axonometric projection of the considered standard dwelling.

Table 2 presents the reference U-values defined by the Portuguese regulations for each element of the building envelope [27,32].

Table 2. Regulatory reference U-values for the envelope components.

Envelope Component	Winter Climate Zone	Regulatory Reference U-Value [W/m^2.°C]
External Wall	I1 I2 I3	0.50 0.40 0.35
Roof	I1 I2 I3	0.40 0.35 0.30
Ground Floor	I1, I2 and I3	0.50
Glazing	I1 I2 I3	2.80 2.40 2.20

For the standard single house, various combinations of envelope solutions were considered, for which U-values and costs were estimated.

Table 3 presents the considered envelope solutions and their characteristics. For each of the studied locations, current construction options with a U-value equal to the regulatory reference U-value were defined, as well as solutions with slightly less demanding U-values and with slightly more demanding U-values. According to the regulations in force, U-values above the regulatory reference value are permitted as long as the building meets the requirement of maximum allowable nominal useful energy needs for heating and cooling [32–34], with a maximum of 0.90 W/m^2.°C for opaque components and 3.00 W/m^2.°C for glazing.

Table 3. Studied envelope solutions.

Envelope Component	Winter Climate Zone	U-Value [W/m^2.°C]	Solution	Cost [€/m^2] [35]
External Wall	I1	0.40	Double wall [1]	145.50
		0.40	Single wall [2]	135.81
		0.50	Double wall [1]	142.88
		0.50	Single wall [2]	133.18
		0.60	Double wall [1]	141.07
		0.60	Single wall [2]	131.38
	I2	0.30	Double wall [1]	149.77
		0.30	Single wall [2]	140.24
		0.40	Double wall [1]	145.50
		0.40	Single wall [2]	135.81
		0.50	Double wall [1]	142.88
		0.50	Single wall [2]	133.18
	I3	0.25	Double wall [1]	153.37
		0.25	Single wall [2]	143.68
		0.35	Double wall [1]	147.31
		0.35	Single wall [2]	137.61
		0.45	Double wall [1]	144.03
		0.45	Single wall [2]	134.33

Table 3. *Cont.*

Envelope Component	Winter Climate Zone	U-Value [W/m^2.°C]	Solution	Cost [€/m^2] [35]
Roof	I1	0.30	Flat roof (α = 0.5) [3]	104.49
		0.30	Flat roof (α = 0.4) [3]	107.04
		0.40	Flat roof (α = 0.5) [3]	100.06
		0.40	Flat roof (α = 0.4) [3]	102.61
		0.50	Flat roof (α = 0.5) [3]	97.44
		0.50	Flat roof (α = 0.4) [3]	99.99
	I2	0.25	Flat roof (α = 0.5) [3]	107.94
		0.25	Flat roof (α = 0.4) [3]	110.49
		0.35	Flat roof (α = 0.5) [3]	102.03
		0.35	Flat roof (α = 0.4) [3]	104.58
		0.45	Flat roof (α = 0.5) [3]	98.75
		0.45	Flat roof (α = 0.4) [3]	101.30
	I3	0.20	Flat roof (α = 0.5) [3]	113.18
		0.20	Flat roof (α = 0.4) [3]	115.73
		0.30	Flat roof (α = 0.5) [3]	104.49
		0.30	Flat roof (α = 0.4) [3]	107.04
		0.40	Flat roof (α = 0.5) [3]	100.06
		0.40	Flat roof (α = 0.4) [3]	102.61
Ground Floor	I1, I2 and I3	0.40	Ground floor with insulation [4]	81.77
		0.50	Ground floor with insulation [4]	75.68
		0.60	Ground floor [4]	72.03
Glazing	I1	2.40	Double glass	467.91
		2.80	Double glass	406.19
		3.20	Double glass	308.54
	I2	2.00	Low-e double glass	497.68
		2.40	Low-e double glass	463.54
		2.80	Low-e double glass	438.50
	I3	1.80	Low-e double glass	537.36
		2.20	Low-e double glass	475.82
		2.60	Low-e double glass	451.10

[1] Thermal insulation between walls; [2] external thermal insulation; [3] flat roof with external thermal insulation, where α is the solar absorptance; [4] reinforced ground floor with light concrete.

The maximum allowable values of the annual nominal needs of useful heating energy (N_i) and the annual nominal needs of useful cooling energy (N_v) are defined by the regulations based on reference values of the relevant parameters applied to the geometry of the actual building. The values of the annual nominal needs of useful heating energy (N_{ic}) and cooling energy (N_{vc}) of the actual building are determined using the real thermal characteristics of the building under analysis and considering the heat transfer by transmission through the building envelope, the heat transfer by ventilation and the useful thermal gains resulting from solar and internal gains [27]. The latter are not considered for the calculation of the maximum allowable values, Ni and Nv.

For winter climate zones, the regulations establish the following requirements: for I1, N_{ic}/N_i must be less than or equal to 75%; for I2, N_{ic}/N_i must be less than or equal to 85%; for I3, N_{ic}/N_i must be less than or equal to 90% [33].

Table 4 shows the dimensions and the glazed area of windows and glazed doors, as can be seen in Figure 2b. All glazed areas have external roller shutters closed at night

(7 p.m. to 7 a.m.) throughout the year. In the conventional cooling season (June, July, August and September), they are also partially closed in the daytime (7 a.m. to 9 p.m.) for solar protection, as follows: on the north façade, it is not necessary to close in the daytime; on the façades that face east, west and south, the shutters are configured as closed 60% of daytime. For the analysis, the orientation of the building was rotated, with the main façade successively facing the four cardinal points, which is the one with the most significant number and area of glazing (the façade of the bedrooms). The air change rate is equal to 0.6 air changes per hour.

Table 4. Dimensions and area of windows and glazed doors.

Indoor Space	Dimensions	Glazing Area
Living room	2.60 m × 2.20 m	5.72 m^2
Kitchen	2.60 m × 1.20 m	3.12 m^2
En suite bedroom	2.40 m × 2.20 m	5.28 m^2
WCs (with windows)	1.60 m × 0.70 m	1.12 m^2
Bedroom 1	2.40 m × 1.20 m	2.88 m^2
Bedroom 2	2.40 m × 1.20 m	2.88 m^2

Later, when analysing overheating in summer, additional tests were carried out by positioning the shutters on the indoor side of the windows to explore this type of option that is sometimes chosen by designers for new buildings. The risk of overheating in summer was also analysed for the case in which the users do not use the shutters in the daytime.

Natural ventilation is used under the following conditions: whenever the indoor temperature is higher than the outdoor temperature, the outside temperature is at least 18 °C and less than 25 °C and the indoor temperature is above 24 °C, with a limit of four air changes per hour (a constant air change rate of 0.6 is considered, but when the conditions defined for natural ventilation to work are met, this rate can reach a maximum of 4.0). The occupancy was configured to represent a family of four people in the house daily, from 6 p.m. to 8 a.m. No HVAC system was used, nor were lighting and equipment considered.

2.3. *Discomfort Assessment*

The discomfort analysis is carried out using the concept of Degree Hours of Discomfort combined with the EN 16798-1 Standard [16]. The optimal operative temperature (Θc) is given by (1), which considers the weighted average outdoor temperature (Equation (2)). For this study, firstly, Category II was considered, which defines the upper limit at 3 °C more and the lower limit at 4 °C less than the optimal operative temperature. When the operative temperature of each thermal zone of the building is outside the upper or lower limit, the difference is accounted for. A calculation is carried out weighted by the area of each thermal zone, and a sum of all hours analysed is made:

$$\Theta_c = 0.33\ \Theta_{rm} + 18.8 \quad (1)$$

$$\Theta_{rm} = (\Theta_{ed\text{-}1} + 0.8\ \Theta_{ed\text{-}2} + 0.6\ \Theta_{ed\text{-}3} + 0.5\ \Theta_{ed\text{-}4} + 0.4\ \Theta_{ed\text{-}5} + 0.3\ \Theta_{ed\text{-}6} + 0.2\ \Theta_{ed\text{-}7})/3.8 \quad (2)$$

where Θ_{rm} = Outdoor Running mean temperature for the considered day; $\Theta_{ed\text{-}1}$, $\Theta_{ed\text{-}2}\ldots\ \Theta_{ed\text{-}i}$ = daily mean outdoor air temperature for the i-th previous day.

After this first step, to analyse summer overheating, an optimisation was performed for the four months of the cooling season with the upper limit following Category II criteria, with an adaptation to the recommended design values of operative temperature for residential buildings design, as presented in the Standard, with an upper limit not exceeding 26 °C [16]. This adaptation is explained further on.

2.4. Optimisation Settings

DesignBuilder version 7.2 software was used in this study to create models and run dynamic simulations and optimisations. DesignBuilder is an EnergyPlus-based software tool for assessing and controlling energy, carbon emissions, lighting and comfort. It is widely used to create and assess building design projects. However, due to the large volume of simulations and the lengthy duration of each optimisation process, the jEPlus software version 2.1.0 was also used to carry out all possible combinations more quickly, using the EnergyPlus IDF files generated by DesignBuilder.

The objectives of the optimisations are to find the optimal cases by minimising the initial construction cost (considering only the sum of the envelope components costs presented in Table 3) and minimising the amount of Degree Hours of Discomfort. The variables are shown in Table 3, in which, for each climate zone, six external wall options, six roof options, three ground floor options and three glazing options are considered. Initially, optimisations are performed for the 8760 h of the year. The main façade faces each of the four cardinal points successively, leading to four optimisation processes.

The Degree Hours of Discomfort analysis is not among the objectives available by default in the DesignBuilder optimisation options. However, starting with version 7 of the software, creating Scripts to perform calculations according to specific objectives is possible. Therefore, a Python Script was developed to calculate this optimisation objective.

To analyse summer overheating, the Script was adapted to encompass just the 2928 h of the cooling season and to consider that the upper limit follows the Category II criteria of Standard EN 16798-1 until reaching the value of 26 °C fixed from then on. The temperature of 26 °C was chosen following the maximum operative temperature for cooling in residential buildings with mechanical cooling systems, established by Standard EN 16798-1.

2.5. Checking nZEB Regulatory Requirements

In the optimisation process, there was a possibility that some solutions containing U-values greater than the maximum regulatory reference values would prove to be optimal. Therefore, additional verification of their compliance with the requirements of nominal useful energy needs for heating and cooling was necessary. Therefore, the spreadsheet developed and made available by the Platform for Energy Efficiency in Buildings (P3E) of the Institute for Research and Technological Development in Construction, Energy, Environment and Sustainability (Itecons) [36] was used to verify compliance with nZEB regulatory standards.

3. Results

3.1. Annual Optimisation

Figure 3 shows the result of optimising the standard dwelling envelope in Porto, with the main façade facing south. This graph was chosen as an example in order to explain how other results will be shown throughout this study. The Y axis presents the construction cost, considering only the envelope components (presented in Table 3), and the X axis gives the number of Degree Hours of Discomfort, calculated using EN 16798-1 Category II criteria.

Solutions with lower cost and discomfort combinations form a Pareto front of optimal solutions along the lower left edge of the "cloud" of data points. The optimal solutions are highlighted in orange and joined by a line. All points in the cloud that are not considered optimal solutions are greyed out. Among the solutions of the Pareto front, a point is marked in green, representing the solution immediately before the first significant increase in discomfort for just a slight or zero reduction of cost, when the Pareto front is analysed from the highest to the lowest cost. This solution was considered as the ideal solution.

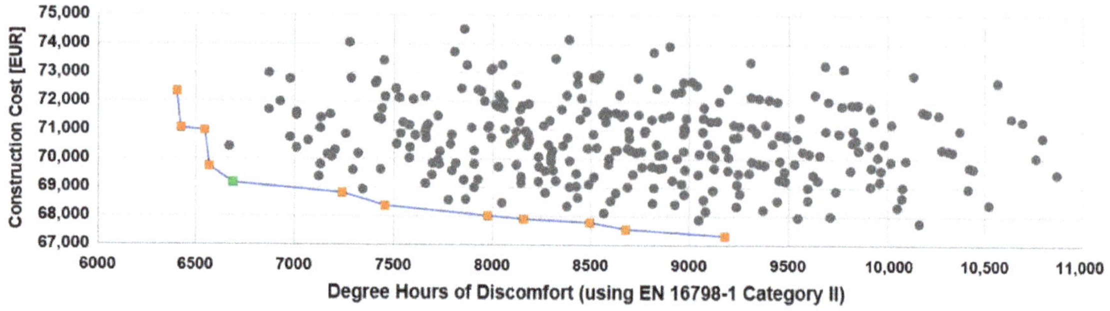

Figure 3. Annual optimisation result for the standard dwelling with the main façade facing south, located in Porto.

The results in several of the following figures will show more than one optimisation process in the same graph, which is why the corresponding data clouds are not represented, showing only the Pareto front with the optimal solutions of each optimisation. The point that best represents the balance between both objectives is also highlighted on the curve in green.

Table 5 shows the composition of the solutions that form the Pareto front of the optimisation of the standard dwelling facing south and located in Porto, detailing each of the four variables. The solution highlighted in green in Figure 3 is highlighted in bold in the table as the best balance between the two variables. It is interesting to note that the ground floor solution, in all optimal solutions, is the less demanding tested floor option in terms of U-value. The ideal solution, in bold, presents the following composition and the following kind of demand: a single external wall with the most demanding option (lowest U-value), a flat roof with the most demanding option (lowest U-value with a solar absorptance of 0.50), glazing with the less demanding option (highest U-value) and ground floor with the less demanding option (highest U-value).

Table 5. Characterisation of the variables that make up the Pareto front solutions for the annual optimisation of standard dwelling facing south, located in Porto.

Objective	Objective	Variables			
Degree Hours of Discomfort (EN 16798-1 Standard—Cat. II)	Total Construction Cost [EUR]	External Wall [W/m^2.°C]	Flat Roof [W/m^2.°C]	Glazing [W/m^2.°C]	Ground Floor Construction [W/m^2.°C]
6401.93	72,348.70	DW U = 0.40	U = 0.30	U = 2.40	U = 0.60
6422.74	71,089.20	SW U = 0.40	U = 0.30	U = 2.40	U = 0.60
6544.56	70,986.84	DW U = 0.40	U = 0.30	U = 2.80	U = 0.60
6567.63	69,727.34	SW U = 0.40	U = 0.30	U = 2.80	U = 0.60
6687.49	**69,161.38**	**SW U = 0.40**	**U = 0.30**	**U = 3.20**	**U = 0.60**
7237.85	68,819.56	SW U = 0.50	U = 0.30	U = 3.20	U = 0.60
7455.38	68,377.34	SW U = 0.40	U = 0.40	U = 3.20	U = 0.60
7974.40	68,035.52	SW U = 0.50	U = 0.40	U = 3.20	U = 0.60
8158.52	67,913.50	SW U = 0.40	U = 0.50	U = 3.20	U = 0.60
8491.26	67,801.62	SW U = 0.60	U = 0.40	U = 3.20	U = 0.60
8678.19	67,571.68	SW U = 0.50	U = 0.50	U = 3.20	U = 0.60
9180.67	67,337.78	SW U = 0.60	U = 0.50	U = 3.20	U = 0.60

DW = Double wall; SW = Single wall.

Figure 4 presents the optimisation results for each orientation and the two analysed locations in winter climate zone I1, Porto and Lisboa.

For both graphs presented in Figure 4, the point that would be the ideal solution in each of these optimisation processes is highlighted in green. This ideal solution has the same composition for all eight Pareto fronts generated for winter climate zone I1, which corresponds to the same kind of demand already described in Figure 3.

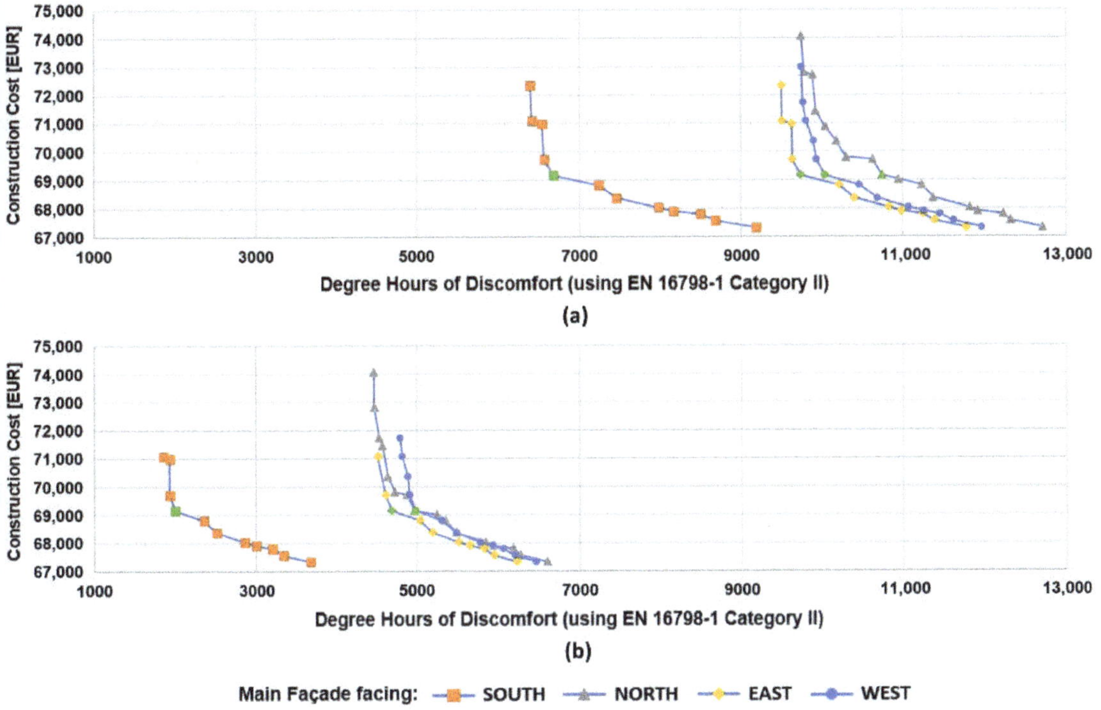

Figure 4. Annual optimisation results (Pareto fronts) for each orientation and two I1 locations: (**a**) Porto and (**b**) Lisboa.

Figure 5 presents the optimisation results for each orientation and the three analysed locations in winter climate zone I2: Batalha, Viseu and Monchique. In all the curves presented in Figure 5, the point that would be the ideal solution is highlighted in green. As shown in the results in Figure 4, this ideal solution has the same composition in all Pareto fronts for south, east and west orientations generated for the winter climate zone I2, which corresponds to the same kind of demand previously described but now concerning zone I2 defined U-values. However, the Pareto fronts for the north orientation in the three I2 locations present an ideal solution with a more demanding ground floor U-value that is equal to the regulatory reference U-value.

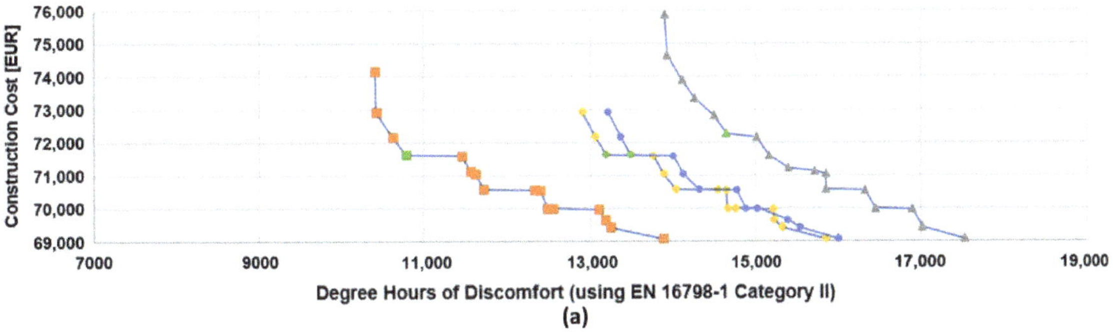

Figure 5. *Cont.*

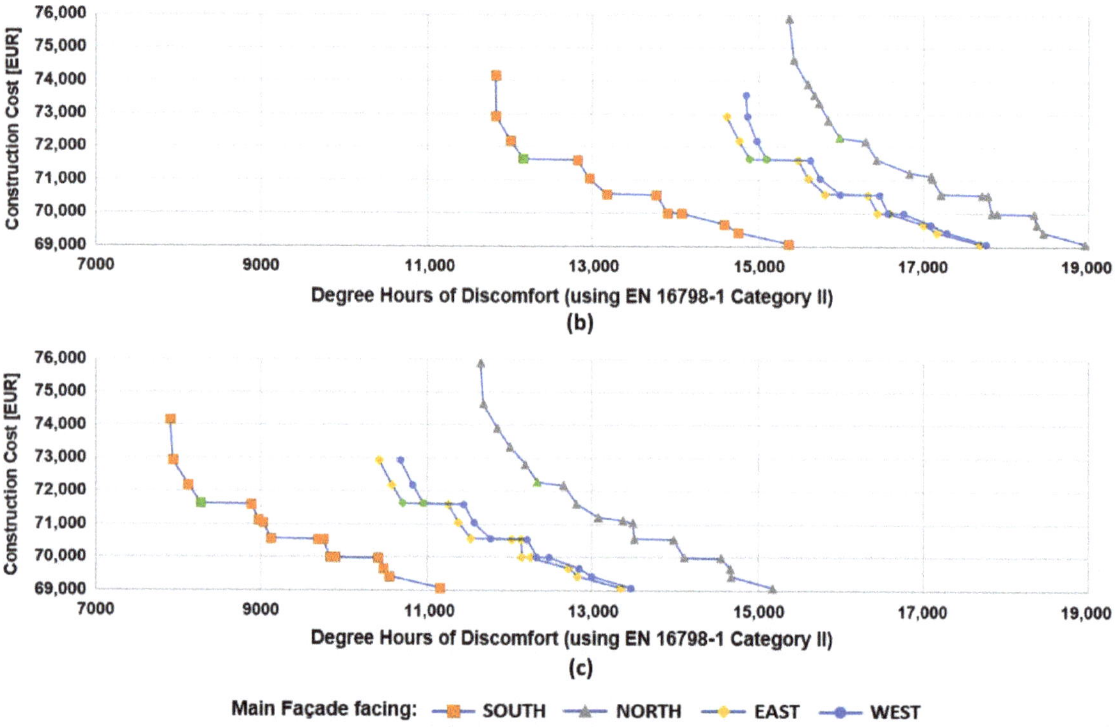

Figure 5. Annual optimisation results (Pareto fronts) for each orientation and three I2 locations: (**a**) Batalha, (**b**) Viseu and (**c**) Monchique.

Figure 6 presents the optimisation results for each orientation and the three analysed locations in winter climate zone I3: Melgaço, Guarda and Macedo de Cavaleiros. Again, in all the curves presented in Figure 6, the point that would be the ideal solution is highlighted in green. As shown in the results in Figure 5, this ideal solution has the same composition in all Pareto fronts for south, east and west orientations generated for the winter climate zone I3, which corresponds to the same kind of demand previously described but now concerning zone I3 defined U-values. However, the Pareto fronts for the north orientation in the three I3 locations present an ideal solution with a more demanding ground floor U-value equal to the regulatory reference U-value.

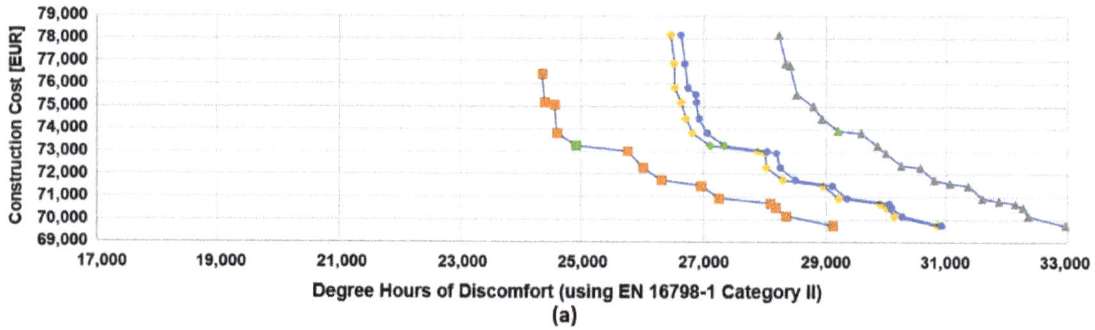

Figure 6. *Cont.*

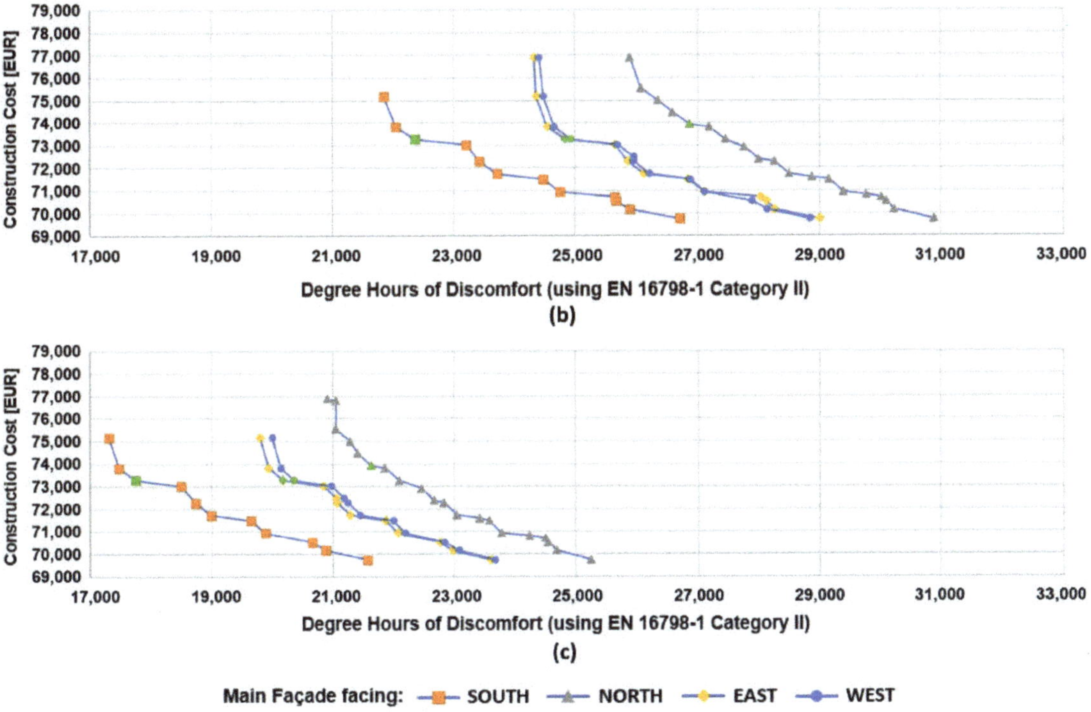

Figure 6. Annual optimisation results (Pareto fronts) for each orientation and three I3 locations: (**a**) Melgaço, (**b**) Guarda and (**c**) Macedo de Cavaleiros.

3.2. Checking nZEB Regulatory Requirements of Ideal Solutions

The glazing of all the ideal solutions presents U-values above the regulatory reference values, and the ground floor of the great majority of the ideal solutions also presents U-values above the regulatory reference values. Therefore, it was necessary to verify whether this variable combination of the ideal solution meets the regulatory requirement for nominal useful energy needs for heating and cooling [32–34]. Table 6 presents the results of applying Portuguese regulations to verify this issue [36]. It can be noted that regulatory requirements ($N_{vc}/N_v < 100\%$ for all summer climate zones; $N_{ic}/N_i < 75\%$ for I1 locations, $N_{ic}/N_i < 85\%$ for I2 locations, and $N_{ic}/N_i < 90\%$ for I3 locations) are met for each orientation and for all eight locations.

Table 6. Results of nominal useful energy needs for heating and cooling, in kWh/m^2.year, of the identified ideal solutions according to Portuguese regulations.

	Orientation South						Orientation North					
	N_i	N_{ic}	N_{ic}/N_i	N_v	N_{vc}	N_{vc}/N_v	N_i	N_{ic}	N_{ic}/N_i	N_v	N_{vc}	N_{vc}/N_v
Porto	58.45	34.15	**58.43%**	9.13	3.86	**42.28%**	58.45	39.76	**68.02%**	9.13	4.10	**44.91%**
Lisboa	39.01	18.41	**47.19%**	16.50	11.69	**70.85%**	39.01	22.75	**58.32%**	16.50	12.25	**74.24%**
Batalha	70.42	42.68	**60.61%**	6.55	2.22	**33.89%**	70.42	45.91	**65.19%**	6.55	2.67	**40.76%**
Viseu	75.10	45.34	**60.37%**	10.32	4.87	**47.19%**	75.10	48.75	**64.91%**	10.32	5.62	**54.46%**
Monchique	74.40	52.41	**70.44%**	18.97	13.92	**73.38%**	74.40	54.84	**73.71%**	18.97	15.02	**79.18%**
Melgaço	104.25	74.21	**71.18%**	4.24	1.25	**29.48%**	104.25	77.04	**73.90%**	4.24	1.52	**35.85%**
Guarda	103.24	72.62	**70.34%**	7.32	2.64	**36.07%**	103.24	75.65	**73.28%**	7.32	3.13	**42.76%**
Macedo de Cavaleiros	83.67	55.55	**66.39%**	13.67	8.51	**62.25%**	83.67	58.39	**69.79%**	13.67	9.44	**69.06%**

Table 6. *Cont.*

	Orientation East						Orientation West					
	N_i	N_{ic}	N_{ic}/N_i	N_v	N_{vc}	N_{vc}/N_v	N_i	N_{ic}	N_{ic}/N_i	N_v	N_{vc}	N_{vc}/N_v
Porto	58.45	40.59	**69.44%**	9.13	4.67	**51.15%**	58.45	40.59	**69.44%**	9.13	4.70	**51.48%**
Lisboa	39.01	23.42	**60.04%**	16.50	13.49	**81.76%**	39.01	23.42	**60.04%**	16.50	13.55	**82.12%**
Batalha	70.42	50.39	**71.56%**	6.55	2.87	**43.82%**	70.42	50.39	**71.56%**	6.55	2.89	**44.12%**
Viseu	75.10	53.52	**71.26%**	10.32	6.02	**58.33%**	75.10	53.52	**71.26%**	10.32	6.07	**58.82%**
Monchique	74.40	59.27	**79.66%**	18.97	16.29	**85.87%**	74.40	59.27	**79.66%**	18.97	16.35	**86.19%**
Melgaço	104.25	83.39	**79.99%**	4.24	1.52	**35.85%**	104.25	83.39	**79.99%**	4.24	1.54	**36.32%**
Guarda	103.24	81.97	**79.40%**	7.32	3.35	**45.77%**	103.24	81.97	**79.40%**	7.32	3.38	**46.17%**
Macedo de Cavaleiros	83.67	63.66	**76.08%**	13.67	9.69	**70.89%**	83.67	63.66	**76.08%**	13.67	9.76	**71.40%**

3.3. *Cooling Season Optimisation*

The following results refer to the conventional four months of the cooling season (from 1st June to 30th September). Initially, the optimisation was carried out strictly following the criteria established by Standard EN 16798-1—Category II, and the results of the optimal solutions for all simulations showed no discomfort. Therefore, the combination with the lowest absolute cost was, logically, considered the optimal case. In fact, the methodology of the Standard allows very high indoor temperatures if the average outdoor temperature is also very high. Despite warm summers, the use of cooling systems is not common in Portuguese dwellings. However, because of global warming, higher temperatures are expected. Thus, if these kinds of analyses are carried out during the design phase of a new building, admitting an excessive tolerance towards high indoor temperatures instead of imposing a demanding upper limit will certainly lead to an increase in the use of those systems.

Therefore, for a more demanding assessment of summer overheating risk, in the scope of the present study, the Standard EN 16798-1—Category II criteria were subjected to an adaptation concerning the recommended design values of indoor temperature for residential buildings, introducing a limitation for its upper limit definition. The application of EN 16798-1 formulae for the calculation of that upper limit must not lead to a temperature exceeding 26 °C.

Figure 7 shows the limits considered for the evaluation of summer overheating, where it is possible to notice the adaptation of the upper limit: when the Operative Temperature (Θ_O) reaches 26 °C, the upper limit is fixed and no longer follows Equation (1), represented by the dotted line.

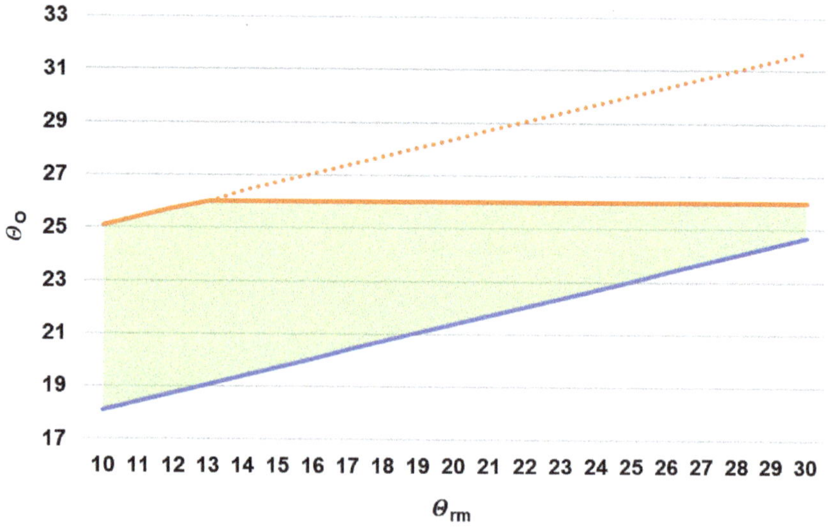

Figure 7. Adaptation of EN 16798-1—Category II criterium for the Degree Hours of Discomfort analysis in the cooling season.

To simplify the visualisation of the results, only two orientations will be presented for each analysed location: south and west. This is because, due to the balanced distribution of the glazed area on opposite façades, the results for the other two orientations are very similar.

All the presented results for annual optimisations correspond to the use, as solar protection, of a roller shutter placed on the outdoor side of the windows according to a schedule previously described. For the specific cooling season analysis, the use of a shutter placed on the indoor side of the windows was also tested with the same schedule, as well as the non-use of any shutter by the occupants in the daytime.

The two locations in the V1 summer climate zone, Batalha and Melgaço, did not present significant discomfort in any of the three shading configurations, even with the upper limit adaptation explained above, for the summer overheating analysis. Therefore, graphical results for locations in this climate zone are not displayed for this specific analysis.

Figure 8 presents the optimisation results for the cooling season of the three summer climate zone V2 locations: Porto, Viseu and Guarda. As shown in the previous figures, the points considered the ideal solution in each graph are highlighted in green. In this figure, the comparison is made between three locations that present different combination possibilities since they are located in different winter climate zones for which the regulatory reference values were established, and it is possible to see that the composition of the ideal solution and its cost is different for each location. However, in relative terms, the ideal solution, coloured in green, has the same kind of demand for the 18 Pareto fronts, which is as follows: a single external wall with the less demanding option, a roof with the less demanding option (but with a solar absorptance of 0.40), glazing with the less demanding option and ground floor with the less demanding option.

Figure 9 presents the optimisation results for the cooling season of the three summer climate zone V3 locations: Lisboa, Monchique and Macedo de Cavaleiros. Each ideal solution is highlighted in green, which, in relative terms, presents the same kind of demand described for the V2 zone.

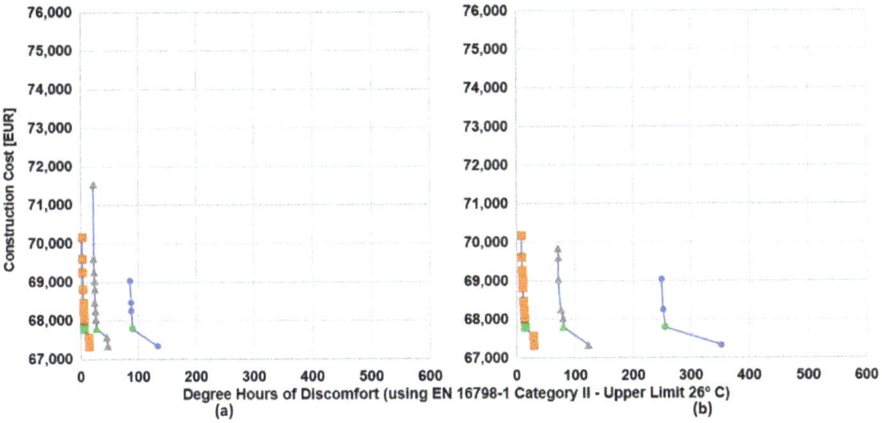

Figure 8. *Cont.*

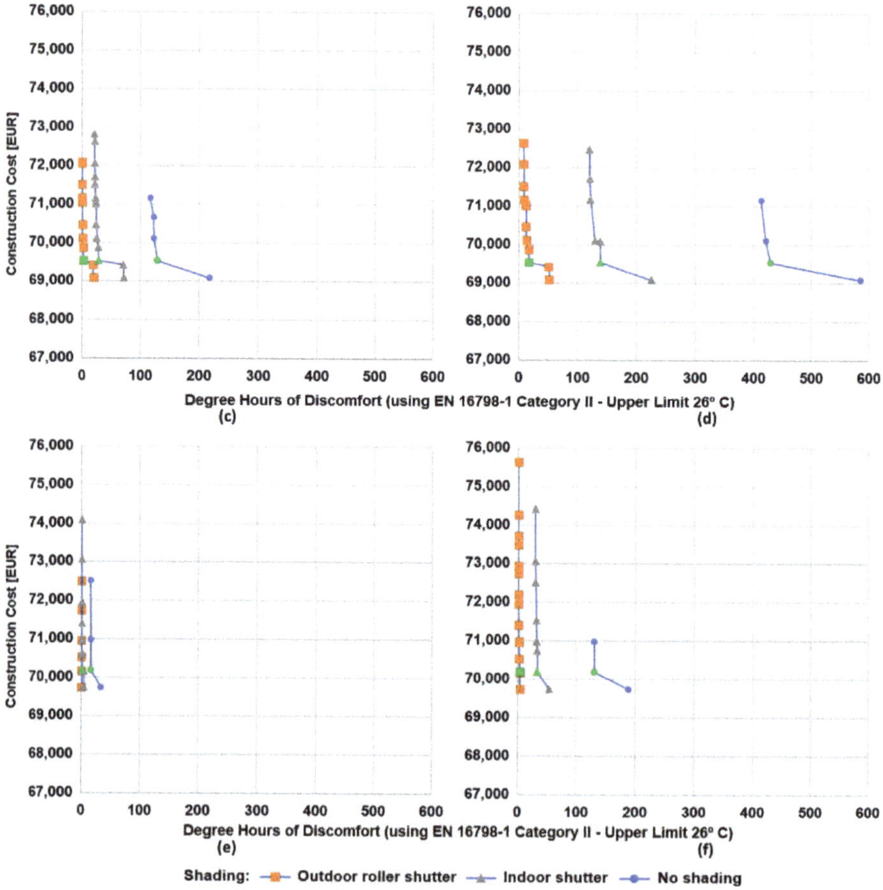

Figure 8. Optimisation results for the cooling season for V2 locations: Porto (**a**) south and (**b**) west orientations; Viseu (**c**) south and (**d**) west orientations; and Guarda (**e**) south and (**f**) west orientations.

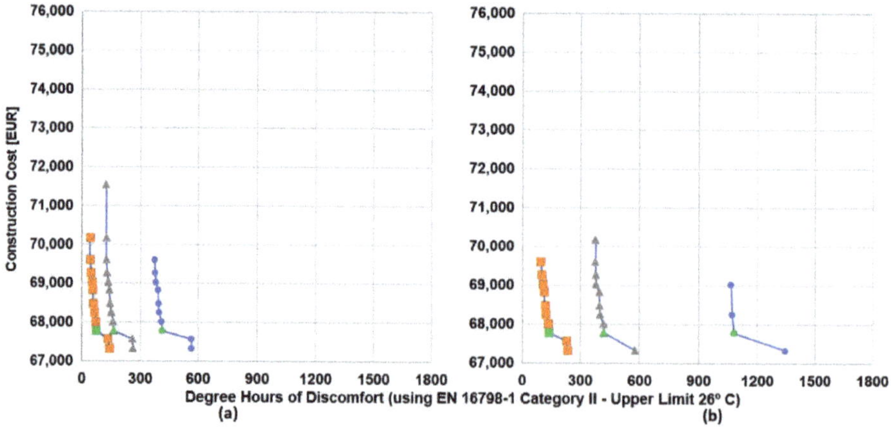

Figure 9. *Cont.*

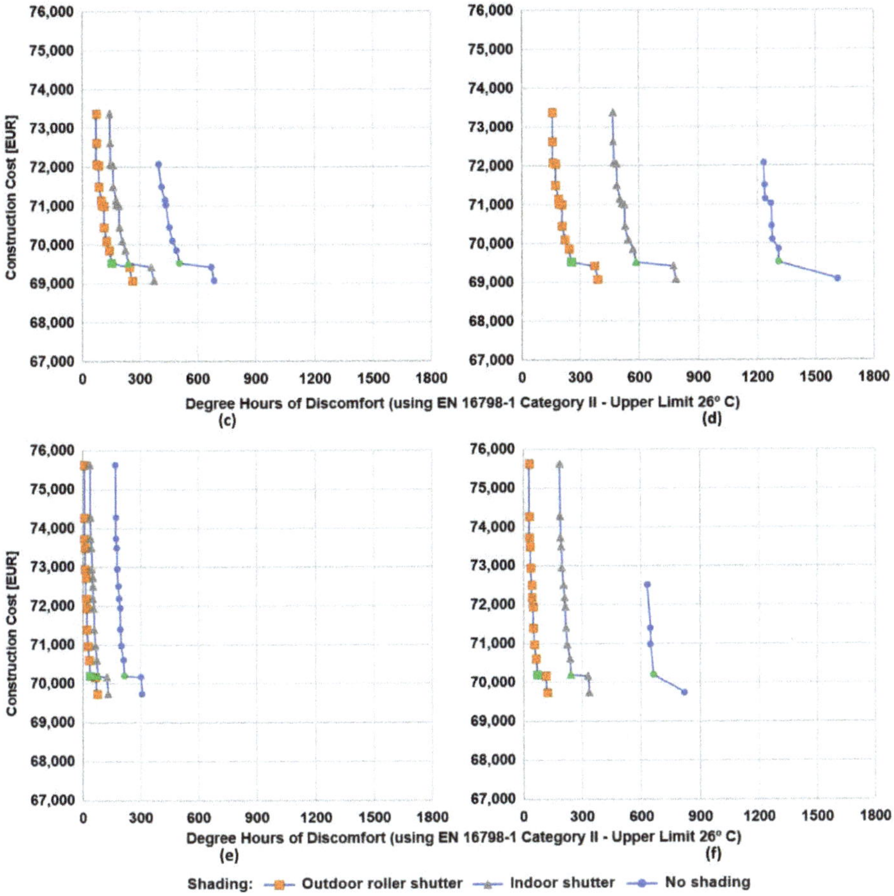

Figure 9. Optimisation results for the cooling season for V3 locations: Lisboa (**a**) south and (**b**) west orientations; Monchique (**c**) south and (**d**) west orientations; and Macedo de Cavaleiros (**e**) south and (**f**) west orientations.

3.4. Additional Analysis on Summer Overheating

The analyses presented in the results of Figures 8 and 9 were carried out according to the base case configurations defined for this study. However, this base case follows what we can consider as good practices, namely having high thermal inertia, an excellent window-to-floor ratio and adequate operation of natural ventilation and the shutters. Unfortunately, these practices may not be followed by designers regarding the inertia and glazing ratio and by occupants in relation to the operation of natural ventilation and shutters. With this in mind, it is necessary to make a quick analysis of each of these situations when they do not follow the practices determined for the base case.

To carry out this analysis, the house with the main façade facing West, located in Lisboa, was chosen, which, in the base case (Figure 8b), already has some discomfort during the cooling season. To analyse the situation of low thermal inertia, the insulation layer was positioned on the indoor side of the external walls, roof and floor (when applicable). In the analysis of a larger glazing area, the percentage of the window-to-floor ratio was doubled (in the base case, it was around 15%; in this new analysis, it is around 30%). In relation to natural ventilation, for it to be running, it no longer depends on the difference between the outdoor and indoor temperatures but rather only on the user's routine; the user will open the windows every morning and close them only in the late afternoon, regardless of outdoor and indoor temperatures, maintaining the limit of four air changes per hour.

Aiming to bring more elements for further discussion, instead of presenting only the Pareto fronts with the optimal solutions of each optimisation, the complete cloud with all possible combinations of each case will be presented, highlighting the optimal solutions that form the Pareto curves. However, to bring information with this presentation, it was necessary to leave aside the adoption of the same scale for the Degree Hours of Discomfort. Therefore, when analysing the graphs in Figures 10–12 it is necessary to pay particular attention to the different scales of the X axes.

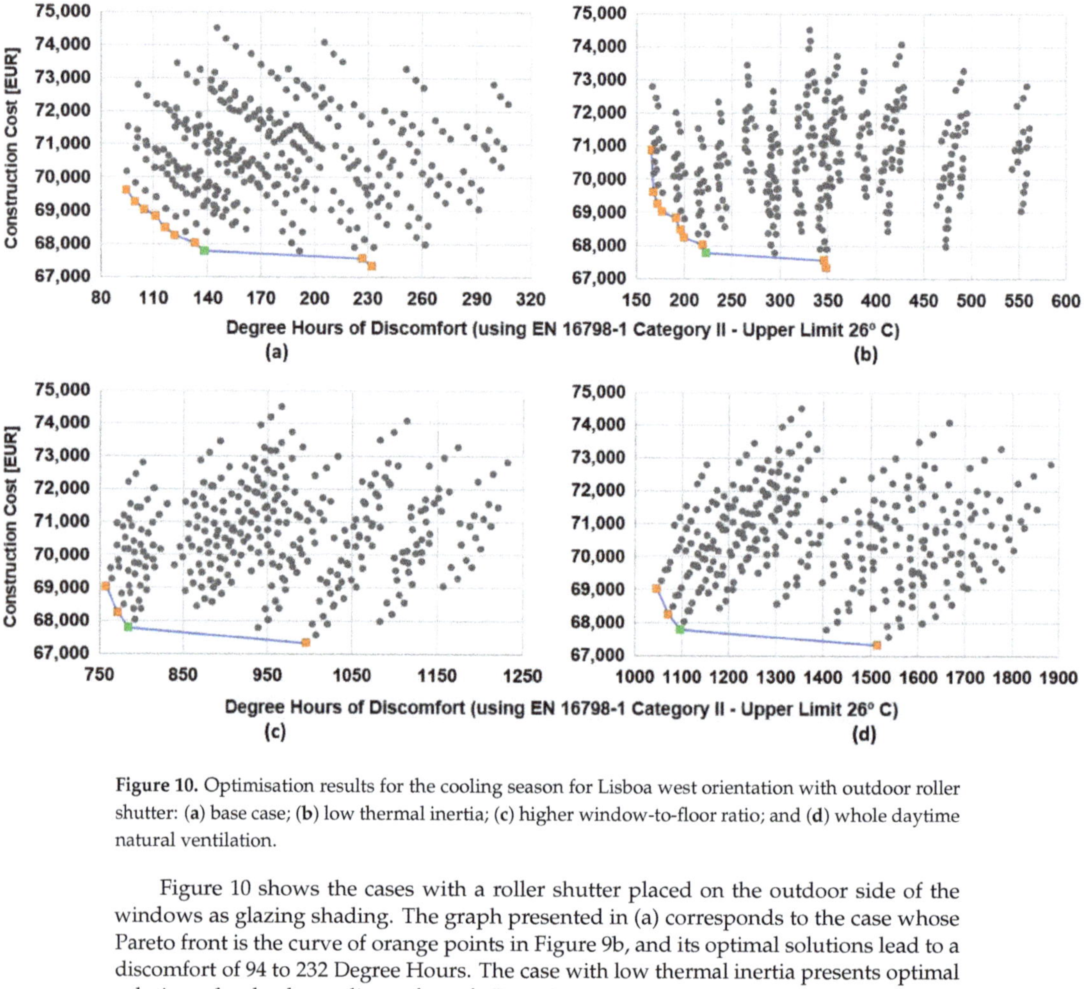

Figure 10. Optimisation results for the cooling season for Lisboa west orientation with outdoor roller shutter: (**a**) base case; (**b**) low thermal inertia; (**c**) higher window-to-floor ratio; and (**d**) whole daytime natural ventilation.

Figure 10 shows the cases with a roller shutter placed on the outdoor side of the windows as glazing shading. The graph presented in (a) corresponds to the case whose Pareto front is the curve of orange points in Figure 9b, and its optimal solutions lead to a discomfort of 94 to 232 Degree Hours. The case with low thermal inertia presents optimal solutions that lead to a discomfort of 171 to 348 Degree Hours. The case with a higher window-to-floor ratio presents optimal solutions corresponding to 757 to 996 Degree Hours of Discomfort. Finally, the case with whole daytime natural ventilation presents optimal solutions that show from 1045 to 1513 Degree Hours of Discomfort. The ideal solution, highlighted in green, for all the scenarios follows the same configuration previously presented in the base case: a single external wall with the least demanding option, a roof with the least demanding option (with a solar absorptance of 0.40), glazing with the least demanding option and ground floor with the least demanding option.

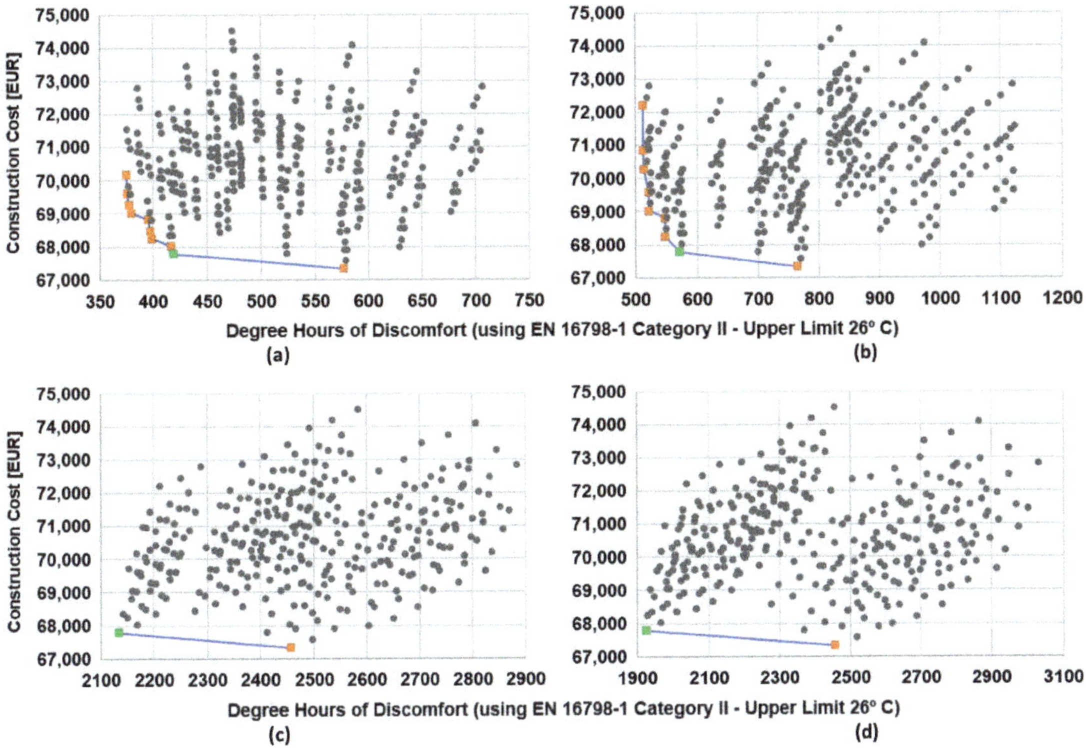

Figure 11. Optimisation results for the cooling season for Lisboa west orientation with indoor shutter: (**a**) base case; (**b**) low thermal inertia; (**c**) higher window-to-floor ratio; and (**d**) whole daytime natural ventilation.

Figure 11 shows the cases with an indoor shutter as unique solar protection of the window. The graph presented in (a) corresponds to the case whose Pareto front is the curve of grey points in Figure 9b, and its optimal solutions lead to a discomfort from 374 to 577 Degree Hours. The case with low thermal inertia presents optimal solutions that lead to a discomfort from 512 to 764 Degree Hours. The case with a higher window-to-floor ratio presents optimal solutions with 2134 to 2455 Degree Hours of Discomfort. Finally, the case with whole daytime natural ventilation presents optimal solutions corresponding to a discomfort of 1926 to 2456 Degree Hours. The ideal solution, highlighted in green, follows the same configuration previously presented in the base case.

Figure 12 shows the cases with no glazing shading. The graph presented in (a) corresponds to the case whose Pareto front is the curve of blue points in Figure 9b, and its optimal solutions lead to a discomfort between 1064 and 1345 Degree Hours. The case with low thermal inertia presents optimal solutions that lead to a discomfort from 1302 to 1637 Degree Hours. The case with the higher window-to-floor ratio presents optimal solutions with 4718 to 5175 Degree Hours of Discomfort. Finally, the case with whole daytime natural ventilation presents optimal solutions corresponding to a discomfort of 3826 to 4499 Degree Hours of Discomfort. The ideal solution, highlighted in green, follows the same configuration previously presented in the base case.

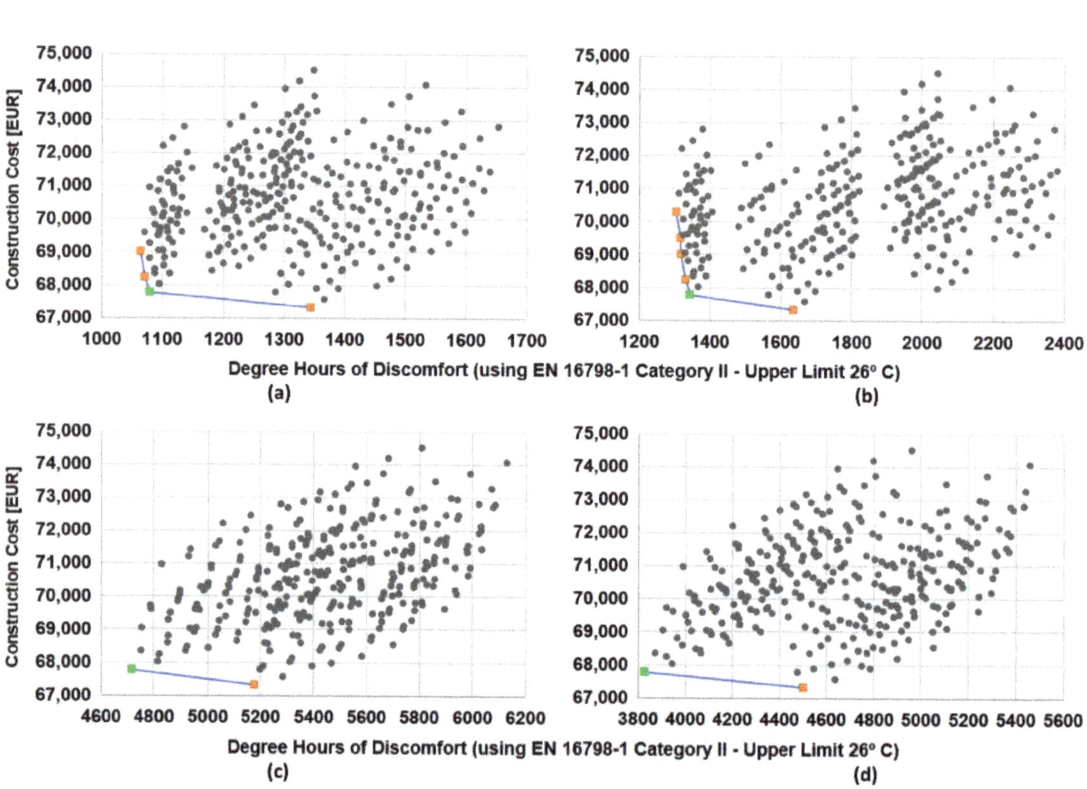

Figure 12. Optimisation results for the cooling season for Lisboa west orientation with no glazing shading: (**a**) base case; (**b**) low thermal inertia; (**c**) higher window-to-floor ratio; and (**d**) whole daytime natural ventilation.

4. Discussion

To begin the discussion of the results, it is essential to highlight some choices made in preparing this study. Firstly, the main focus is on obtaining passive comfort through the optimisation of the building envelope. Therefore, the HVAC system was not used in this investigation, aiming to get the best comfort levels without energy consumption through best construction practices. This choice meets the most common behaviour of the Southern European population in a Mediterranean climate, where people do not have the habit of maintaining their indoor environment at a highly demanding level of comfort through active acclimatisation. In contrast to Central and Northern Europe, this cultural behaviour was, in a certain way, shaped both by the milder climate and by the average financial conditions of the populations of these regions. Considering these factors, Category II of Standard EN 16798-1 was used in the annual comfort analysis in the investigated locations.

Although the exact definitions of nZEB/NZEB concepts vary greatly between organisations and countries, there is a "rule" that is universal to all definitions and has been very well established in the literature for a long time: "tackle demand first, then supply" [37,38]. Portuguese regulations, as mentioned before, allow the criteria related to passive comfort to be separated from those related to renewable energy supply. In this sense, it was possible and necessary to verify compliance with the nZEB requirements of the passive part of the construction. This also explains the choice of the values for the variation of the U coefficient, which were taken around the regulatory reference U-values within a narrow range. Logically, the nZEB/NZEB concepts require, at their very core, a supply of renewable energy. However, this step is outside the scope of the present paper.

It was also an option of the authors to perform the annual optimisation on a base case that complies with good practices related to other characteristics of the building and their use, namely, a recommendable window-to-floor ratio (15%), a high thermal inertia, the sensible use of external roller shutters and an energy-conscious natural ventilation schedule.

According to the Pareto fronts, the optimal annual results are presented in Figures 4–6. It is noticed that the results clearly agree with a principle that is already very well established as a good practice for energy efficiency in buildings, which is that the façade with the largest glazing area should be facing south for locations in the northern hemisphere. In the eight studied locations, it is possible to notice that the south orientation curve notably presents the lowest annual thermal discomfort.

It is interesting to note that in all 24 Pareto fronts related to the south, east and west orientations, the so-called ideal solution, highlighted in green in all of them, has the same kind of demand in relative terms (obviously respecting the different requirements in different climate zones). This ideal solution presents the best theoretical balance between the two optimisation objectives: minimisation of the cost of the analysed envelope components and minimisation of thermal discomfort based on the concept of Degree Hours of Discomfort, calculated using EN 16798-1—Category II. For each climate zone, the ideal solution consists of a single external wall with external thermal insulation, with a U-value lower than its regulatory reference value, a roof with external thermal insulation also with a U-value lower than its regulatory reference, glazing with a U-value higher than its reference value and a ground floor also with a U-value higher than its reference value.

Concerning the Pareto fronts for the north orientation, which is the ideal solution for zones I2 and I3, they present practically the same previous kind of demand, with the exception of the ground floor, which must now be a little more demanding, with a U-value equal to the regulatory reference.

These results can be explained by some factors, considering, of course, the best balance between the optimisation objectives. The external wall and roof variables, which have a larger exposed area and consequently have a more significant influence on heat exchange, are the ones that presented the most demanding options in the composition of the ideal solution. Meanwhile, the ground floor and glazing, in their less demanding options, provide sufficient conditions within the analysed objectives. It is essential to highlight that even though the ground floor presents, in most cases, the least demanding option, with a U-value above the maximum regulatory reference values, it continues to be a variable option with an acceptable and effective U-value.

Glazing, as the envelope component with the highest cost per m^2 but without a too-large area that could allow high heat losses in winter, seems to be the component where the cost is the most determinant. However, the least demanding option of the glazing for climate zone I1 (3.2 $W/m^2.°C$), which makes up the optimal solutions, cannot be used in current construction practice because the regulations limit the maximum U-value for glazing up to 3.0 $W/m^2.°C$ [32]. However, this option was considered, mainly to follow the same kind of variation as the other climate zone options. If this restriction did not exist, the combination would pass all other regulatory requirements, as shown in the results in Table 6. Therefore, this issue deserves greater attention in the subsequent revisions of the regulations.

It is also essential to mention the importance of an informed and reasoned choice within the possible existing variables. As an additional example of the importance of choosing an optimised solution for the users' comfort, let us assume that the designer had opted for the further-to-the-right point case in the cloud of the graph in Figure 3 (a single external wall with the least demanding option, a roof with the least demanding option (with a solar absorptance of 0.40), glazing with the least demanding option and ground floor with the most demanding option). The nZEB regulatory requirements related to passive options would also be met (Nic/ Ni < 75% and Nvc/Nv < 100%), but the total cost of this choice would be slightly higher than that of the ideal solution (€69,513 versus €69,161) and, more significantly, it would lead to around a 62% higher number of Degree

Hours of Discomfort (10,867 against 6687). Figure 13 compares the optimised solution (the one identified as ideal) and this possible uninformed choice (non-optimised solution but still complying with the regulations). In the coldest week of the year, it is possible to notice that the indoor hourly operative temperature is more than one degree higher in the ideal solution than in this other possible solution. This comparison translates into a smaller difference from the lower limit of the EN 16798-1 Standard; therefore, lower Degree Hours of Discomfort in a situation without an HVAC system, which would consequently translate into lower energy consumption for heating due to a passive choice made thoughtfully.

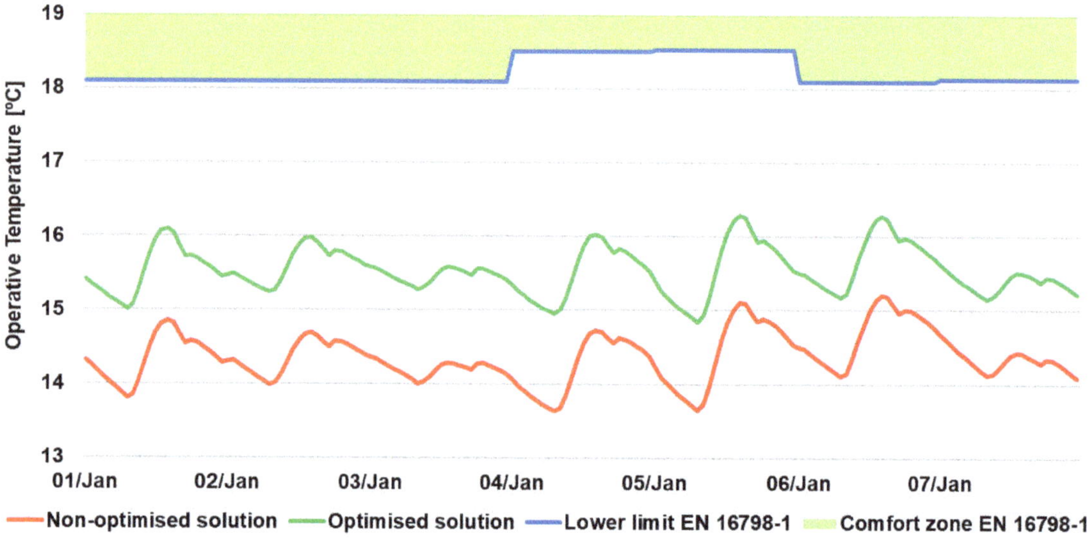

Figure 13. Comparison between the optimised solution and a non-optimised solution in the coldest week of the year for the standard dwelling with the main façade facing south, located in Porto.

As explained before, with the aim of analysing summer overheating risk, an adaptation was made to Standard EN 16798-1, with the upper limit fixed at 26 °C. The results presented in Figures 8 and 9 show that the ideal solution would combine all less-demanding variables.

In the base case, the values of Degree Hours of Discomfort of the optimal solutions for the cooling season are quite low when compared to the annual results, showing that without HVAC, discomfort in winter is much higher. However, in this case, the assessment was made for ideal conditions.

Thus, some additional analyses were carried out, and the results are presented in Figures 10–12 to expand the analysis of summer overheating risk. For this analysis, it was essential to present the entire cloud of results in each graph because they show us, in many graphs, the shape of the cloud with a tilt to the right, which means that a higher cost would translate into a greater number of hours of discomfort. This fact reaffirms that excessive insulation is harmful for summer comfort in this region. In the most recent years, excessive insulation has often been the choice of designers who go beyond regulatory requirements, believing they are making a better choice.

In addition to these issues, several others can be noted in the analyses of the cooling season. It is worth highlighting the importance of paying attention to solar gains in this season. It is clear that when shading is placed on the outdoor side of the glazing, there is less discomfort in all situations, a factor that designers must consider. Furthermore, the behaviour of users is also fundamental in this matter. When they do not follow good practices, the consequence is a very significant increase in discomfort. When no shading is used and, for example, there is a large glazing area (Figure 12c), the Degree Hours of

Discomfort can reach 6000 or more. If this value is compared to the values of the annual Pareto front for Lisboa west orientation (Figure 4b), it can be seen that in certain conditions, discomfort in summer may turn out to be the most critical. This is most likely to occur in regions of a particularly mild winter (I1) and a severe summer (V3), like Lisboa.

From the additional analyses, it is also observed that the option for low thermal inertia or a higher window-to-floor ratio translates into greater discomfort in summer for all analysed cases, two factors for which designers have a responsibility and must be carefully chosen. Again, on the part of users' behaviour, the use of natural ventilation in a prudent manner can be translated into a reduction in discomfort.

Another point that is important to highlight is an absorptance value of 0.40 for the ideal roof solutions in the cooling season analyses, while in the annual analyses, the optimal cases (not just the ideal solutions) present a solar absorptance of 0.50. The choice must be made through a sensible balance between annual analysis and seasonal analysis, taking into account all the factors involved in a case-by-case assessment. However, in Southern Europe, where solar irradiation can be very intense, light colours are preferable for thermal comfort in summer but also, in large measure, for the higher durability of the outer material layers. In winter, the possible increase in solar gains through an opaque element by increasing its solar absorptance is normally negligible when the envelope component in question has a low U-value.

Furthermore, reference year climate data were used in this study, which is the most common procedure in scientific studies. However, in recent years, more constant and longer-lasting heat waves have occurred, and they may tend to get worse in the future. This fact is still not reflected in the reference year climate data. Researchers are aware of the limitations of using these data. However, increased attention must be paid to this issue, especially when the comfort analysis is carried out for regions of the world that have a greater tendency to overheat in summer, as is the case.

Finally, the key characteristic of the optimisation results is to highlight the optimal cases that form the Pareto front, but it is necessary to always keep in mind when carrying out the analyses that these cases are the optimal ones for the combination of the two selected objectives. The practical choice depends, obviously, on the relative importance of each objective.

5. Conclusions

As stated before, focusing on Southern European climate and considering a standard single house complying with nZEB requirements, the present study aimed to perform an optimisation of the envelope by minimising passive thermal discomfort and initial construction cost, allowing at the same time the discussion of some aspects of regulatory requirements.

The main conclusions of this paper can be summarised by answering the questions raised at the beginning of it:

- To what extent do current regulatory requirements prevent passive discomfort?

With current construction solutions that comply with the regulatory requirements for nZEB, it is possible to find a set of optimised solutions that do not totally prevent discomfort but lead to a considerably reduced value of annual Degree Hours of Discomfort at an affordable cost. However, other design options and the pattern of usage have a great influence on comfort conditions, and even with envelope solutions complying with the reference values for the nZEB envelope, the discomfort may result in being much higher than one of the optimised solutions.

- Given that winter and summer passive comfort point mostly to opposite strategies, what would be the most balanced envelope options for achieving nZEB/NZEB status?

Winter passive comfort relies mainly on the thermal insulation level of the envelope and its U-value, as well as solar heat gains and the capacity of the building to store them.

Summer passive comfort relies mainly on protection from solar heat gains and the potential for night heat losses through the envelope and by ventilation.

Winter passive strategies are not very dependent on the user, and the designer can easily optimise them.

On the contrary, several very influential factors on summer comfort were identified, and some of them, namely the ones that depend on user behaviour, are obviously not controllable. Some design options that have an influence on summer passive comfort are also free from any requirements or just have indirect requirements, and designers are less aware of their influence. This scenario, associated with climate change predictions of higher temperatures and longer and more intense heat waves, stresses the need for increased attention to summer requirements.

For the moment, as the present study shows, the annual optimisation process is highly influenced by the heating season since it is the longer one, and the differences between outdoor and indoor comfort temperatures are undeniably greater than those differences in the cooling season. However, this scenario tends to change with time.

It can be concluded that envelope U-values around the current reference U-values allow for optimised solutions. But there is a need for a compromise between the contradictory optimal solutions for winter and summer, especially for the external walls and roof, the larger and more exposed components of the envelope: in winter, the ideal solutions recommend a U-value slightly lower than the reference; and in summer, the ideal solutions propose a U-value slightly higher than the reference. Since, in both seasons, other passive strategies can (much easier in winter than in summer, though) compensate for the fewer good options for U-values, as a general rule, for a balanced envelope solution, U-values should be as close as possible to the reference, and designers should optimise solar heat gains in winter and provide the means for allowing night ventilation of the building in summer. Regulations should reinforce the requirements for summer by prescribing the external shading of the windows and/or defining a maximum window-to-floor ratio.

Concerning solar absorptance, summer optimisation should be prevalent.

- Is the trend of an increasing regulatory requirement for the level of envelope thermal insulation recommendable for Southern Europe?

As can be implied by the answer to the former question and confirmed by the presented results, a regulatory requirement imposing a higher level of thermal insulation for the building envelope in Southern Europe is not recommendable. For the moment, the current requirements seem adequate.

- What would be the consequences for the choice of heating and cooling systems and renewable energy sources of the optimised envelope options?

As a known general principle, heating and cooling systems must be as efficient as possible, and for an nZEB, renewable energy sources must cover at least a regulated percentage of the energy needs of the building. An optimised envelope, by reducing the discomfort, also reduces heating and cooling energy needs, obviously. Thus, the demand put on the systems and on renewable energy sources will be lower. Even if changes to the optimised solutions must be made to reach a compromise between winter and summer comfort, other passive strategies, as said before, will compensate for those changes.

Author Contributions: Conceptualisation, J.R. and H.C.; Formal analysis, J.R. and H.C.; Funding acquisition, J.R. and H.C.; Investigation, J.R.; Methodology, J.R. and H.C.; Resources, H.C.; Software, J.R.; Supervision, H.C.; Validation, H.C.; Visualisation, J.R.; Writing—original draft, J.R.; Writing—review & editing, H.C. All authors have read and agreed to the published version of the manuscript.

Funding: This work was financially supported by Base Funding—UIDB/04708/2020 with DOI 10.54499/UIDB/04708/2020 (https://doi.org/10.54499/UIDB/04708/2020) of the CONSTRUCT—Instituto de I&D em Estruturas e Construções—funded by national funds through the FCT/MCTES (PIDDAC). J.R. would like to acknowledge the financial support of Fundação para a Ciência e a Tecnologia (FCT) by the doctoral grants with DOI 10.54499/2021.05343.BD (https://doi.org/10.54499/2021.05343.BD).

Data Availability Statement: The original contributions presented in the study are included in the article, further inquiries can be directed to the corresponding author.

Conflicts of Interest: The authors declare no conflicts of interest.

References

1. The European Parliament; The Council of the European Union. *Directive (EU) 2024/1275 of the European Parliament and of the Council of 24 April 2024 on the Energy Performance of Buildings (Recast)*; European Union: Maastricht, The Netherlands, 2024.
2. The European Parliament; The Council of the European Union. *Directive 2010/31/EU*; European Union: Maastricht, The Netherlands, 2010; Volume L153, pp. 13–35.
3. The European Parliament; The Council of the European Union. *Directive 2018/844/EU Energy Performance of Buildings*; European Union: Maastricht, The Netherlands, 2018.
4. IEA. *IEA SHC Task 40/EBC Annex 52 towards Net Zero Energy Solar Buildings: A Review of 30 Net ZEBs Case Studies*; International Energy Agency: Paris, France, 2014; Volume 130.
5. Marszal, A.J.; Heiselberg, P.; Bourrelle, J.S.; Musall, E.; Voss, K.; Sartori, I.; Napolitano, A. Zero Energy Building—A Review of Definitions and Calculation Methodologies. *Energy Build.* **2011**, *43*, 971–979. [CrossRef]
6. Jaysawal, R.K.; Chakraborty, S.; Elangovan, D.; Padmanaban, S. Concept of Net Zero Energy Buildings (NZEB)—A Literature Review. *Clean. Eng. Technol.* **2022**, *11*, 100582. [CrossRef]
7. Tagliabue, L.C.; Di Giuda, G.M.; Villa, V.; De Angelis, E.; Ciribini, A.L.C. Techno-Economical Analysis Based on a Parametric Computational Evaluation for Decision Process on Envelope Technologies and Configurations. *Energy Build.* **2018**, *158*, 736–749. [CrossRef]
8. Chastas, P.; Theodosiou, T.; Kontoleon, K.; Bikas, D. The Effect of Embodied Impact on the Cost-Optimal Levels of Nearly Zero Energy Buildings: A Case Study of a Residential Building in Thessaloniki, Greece. *Energies* **2017**, *10*, 740. [CrossRef]
9. Alam, S.; Airaksinen, M.; Lahdelma, R. Attitudes and Approaches of Finnish Retrofit Industry Stakeholders toward Achieving Nearly Zero-Energy Buildings. *Sustainability* **2021**, *13*, 7359. [CrossRef]
10. Fabrizio, E. Zero Energy Buildings: A Reached Target or a Starting Point? *Appl. Sci.* **2020**, *10*, 512. [CrossRef]
11. Santos-Herrero, J.M.; Lopez-Guede, J.M.; Flores-Abascal, I. Modeling, Simulation and Control Tools for NZEB: A State-of-the-Art Review. *Renew. Sustain. Energy Rev.* **2021**, *142*, 110851. [CrossRef]
12. Yang, L.; Yan, H.; Lam, J.C. Thermal Comfort and Building Energy Consumption Implications—A Review. *Appl. Energy* **2014**, *115*, 164–173. [CrossRef]
13. Moon, J.W.; Han, S.-H. Thermostat Strategies Impact on Energy Consumption in Residential Buildings. *Energy Build.* **2011**, *43*, 338–346. [CrossRef]
14. Tang, F.; Chen, J.; Li, J.; Rodriguez, D. Energy Saving Actions toward NZEBs with Multiple-Criteria Optimization in Current Residential Buildings. *Energy Rep.* **2020**, *6*, 3008–3022. [CrossRef]
15. Carlucci, S.; Bai, L.; de Dear, R.; Yang, L. Review of Adaptive Thermal Comfort Models in Built Environmental Regulatory Documents. *Build. Environ.* **2018**, *137*, 73–89. [CrossRef]
16. *EN 16798-1*; Energy Performance of Buildings-Ventilation for Buildings-Part 1: Indoor Environmental Input Parameters for Design and Assessment of Energy Performance of Buildings Addressing Indoor Air Quality, Thermal Environment, Lighting and Acoustics-Module M1-6. European Committee for Standardization: Brussels, Belgium, 2019.
17. Guillén-Lambea, S.; Rodríguez-Soria, B.; Marín, J.M. Comfort Settings and Energy Demand for Residential NZEB in Warm Climates. *Appl. Energy* **2017**, *202*, 471–486. [CrossRef]
18. Davidson, R.K. Cultural Impacts on Occupant Behaviour and Energy Efficiency. *Int. J. Energy Prod. Manag.* **2017**, *2*, 186–195. [CrossRef]
19. Bienvenido-Huertas, D.; Pulido-Arcas, J.A.; Rubio-Bellido, C.; Pérez-Fargallo, A. Feasibility of Adaptive Thermal Comfort for Energy Savings in Cooling and Heating: A Study on Europe and the Mediterranean Basin. *Urban Clim.* **2021**, *36*, 100807. [CrossRef]
20. Costa-Carrapiço, I.; González, J.N.; Raslan, R.; Sánchez-Guevara, C.; Redondas Marrero, M.D. Understanding Thermal Comfort in Vernacular Dwellings in Alentejo, Portugal: A Mixed-Methods Adaptive Comfort Approach. *Build. Environ.* **2022**, *217*, 109084. [CrossRef]
21. De Masi, R.F.; Gigante, A.; Vanoli, G.P. Are NZEB Design Solutions Environmental Sustainable? Sensitive Analysis for Building Envelope Configurations and Photovoltaic Integration in Different Climates. *J. Build. Eng.* **2021**, *39*, 102292. [CrossRef]
22. Bienvenido-Huertas, D.; Rubio-Bellido, C.; Pérez-Fargallo, A.; Pulido-Arcas, J.A. Energy Saving Potential in Current and Future World Built Environments Based on the Adaptive Comfort Approach. *J. Clean. Prod.* **2020**, *249*, 119306. [CrossRef]
23. Zhai, Z.J.; Helman, J.M. Implications of Climate Changes to Building Energy and Design. *Sustain. Cities Soc.* **2019**, *44*, 511–519. [CrossRef]
24. Arriazu-Ramos, A.; Bes-Rastrollo, M.; Sánchez-Ostiz Gutiérrez, A.; Monge-Barrio, A. Building Parameters That Influence Overheating of Apartment Buildings in a Temperate Climate in Southern Europe. *Build. Environ.* **2023**, *228*, 109899. [CrossRef]

25. Lou, H.-L.; Hsieh, S.-H. Towards Zero: A Review on Strategies in Achieving Net-Zero-Energy and Net-Zero-Carbon Buildings. *Sustainability* **2024**, *16*, 4735. [CrossRef]
26. Capeluto, I.G. The Unsustainable Direction of Green Building Codes: A Critical Look at the Future of Green Architecture. *Buildings* **2022**, *12*, 773. [CrossRef]
27. Ambiente e Ação Climática. *Despacho n.o 6476-H/2021*; Diário da República: Porto, Portugal, 2021; pp. 330-(66)–330-(316).
28. DB City. Available online: www.db-city.com (accessed on 31 January 2024).
29. Yamazaki, D.; Ikeshima, D.; Tawatari, R.; Yamaguchi, T.; O'Loughlin, F.; Neal, J.C.; Sampson, C.C.; Kanae, S.; Bates, P.D. A High-Accuracy Map of Global Terrain Elevations. *Geophys. Res. Lett.* **2017**, *44*, 5844–5853. [CrossRef]
30. *Despacho no 15793-F Zonamento Climático*; 2.a Série; Diário da República: Porto, Portugal, 2013; pp. 26–31.
31. ADENE. Estatística do Sistema de Certificação Energética dos Edifícios. Available online: https://www.sce.pt/estatisticas/ (accessed on 31 January 2024).
32. *Ambiente e Ação Climática e Infraestruturas e Habitação*; Portaria n.o 138-I/2021 de 1 de Julho; Diário da República: Porto, Portugal, 2021; pp. 128-(12)–128-(53).
33. *Ambiente e Ação Climática Despacho n.o 6476-E/2021*; Diário da República: Porto, Portugal, 2021; pp. 330-(30)–300-(32).
34. *Presidência do Conselho de Ministros*; Decreto-Lei n.o 101-D/2020 de 7 de Dezembro; Diário da República: Porto, Portugal, 2020; pp 7-(21)–7-(45).
35. CYPE Ingenieros, S.A. Gerador de Preços Para Construção Civil—Portugal. Available online: http://www.geradordeprecos.info/ (accessed on 7 March 2024).
36. P3E. Itecons Ferramentas de Cálculo DL 101-D/2020. Available online: https://www.itecons.uc.pt/p3e/index.php (accessed on 31 January 2024).
37. Attia, S. Evolution of Definitions and Approaches. In *Net Zero Energy Buildings (NZEB)*; Elsevier: Amsterdam, The Netherlands, 2018; pp. 21–51.
38. Crawley, D.; Pless, S.; Torcellini, P. Getting to Net Zero. *ASHRAE J.* **2009**, *51*, 18–25.

buildings

MDPI

rticle

Retrofit Measures for Achieving NZE Single-Family Houses in a ropical Climate via Multi-Objective Optimization

ristina Carpino [1], Miguel Chen Austin [2,3,4,*], Dafni Mora [2,3,4] and Natale Arcuri [1]

1 Department of Mechanical, Energy and Management Engineering, University of Calabria, Via Pietro Bucci 46C, 87036 Rende, Italy; cristina.carpino@unical.it (C.C.); natale.arcuri@unical.it (N.A.)
2 Faculty of Mechanical Engineering, Universidad Tecnológica de Panamá, Panama City 0819-07289, Panama; dafni.mora@utp.ac.pa
3 Centro de Estudios Multidisciplinarios en Ciencias, Ingeniería y Tecnología (CEMCIT-AIP), Panama City 0819-07289, Panama
4 Sistema Nacional de Investigación (SNI), Clayton 0816-02852, Panama
* Correspondence: miguel.chen@utp.ac.pa

Abstract: To achieve sustainable cities and communities, it is necessary to decarbonize existing buildings. Actions need to be taken to reduce the buildings' energy demand and ensure that the low remaining demand is met by energy produced from renewable sources. This leads to Net Zero Energy Buildings (NZEBs), whose impact on energy consumption is zero or positive, meaning that they are able to produce more energy than they require. The "zero" objective may be difficult to reach in hot and humid climates, where the cooling demand is prevalent. In this case, a combination of active and passive measures, together with appropriate interaction with users, is a viable way to obtain NZEBs. The present study aims to explore technological solutions for renovating existing buildings to NZEBs in a tropical climate. The analysis is developed through a parametric analysis, a sensitivity analysis, and an optimization directed at minimizing the site's net energy and hours of discomfort. Evaluations are conducted for a case study consisting of a single-family house located in Panama City. The results showed that photovoltaic size, cooling operation schedule, and cooling set-point temperature are the most influential variables for the attainment of NZEBs in a hot climate. Regarding the building envelope, the outcomes suggest the low insulation of dispersing structures and local solar shading of windows as recommended measures.

Keywords: building renovation; net zero energy buildings; optimization; parametric analysis; renewable energy; sensitivity analysis; thermal comfort; tropical climate

1. Introduction

Building retrofit or building refurbishment is necessary to improve the building's energy performance to achieve an objective imposed by local energy regulations, second-life purposes, or users' requirements [1]. Due to recent the global goals regarding energy and emissions, local buildings' regulations objectives are becoming stricter [2]. Among such objectives is the improving the efficiency of all existing buildings to achieve nearly Zero Energy Buildings (nZEB) or even Net Zero Energy Buildings (NZEB). Buildings classified as NZEB are considered highly performant, managing to equally balance their annual energy consumption with on-site renewable energy generation, e.g., the electricity and thermal energy produced from solar energy at the building site. This balance is calculated annually per building's total surface area in terms of primary energy (kWh/m^2 year). An nZEB is a building which uses the minimum requirements to achieve a high performance. These requirements depend on the country, where the annual energy balance in primary energy terms is "close" to 0 kWh/m^2 year. This "close" value varies significantly per country [3].

The renovation of existing buildings is, in fact, one of the obstacles encountered in achieving the objectives of climate neutralization [4]. Government support is essential to encourage the renovation process [5] and the amount of incentives required increases when trying to reach nZEB levels and to ensure internal comfort conditions [6]. New indexes can be considered the basis for defining incentives, like the one proposed by [7], which takes into account CO_2-neutral exported electrical energy.

In this context, many countries have initiated retrofit actions in the building sector, which include improvements at the envelope level, at cooling or heating at the plant level, and also can go as far as focusing on the occupants' behavior level, while assuring the thermal comfort of the latter. In the literature, different approaches can be found, referring to simulation and experimental applications to optimize various aspects of the renovation measures. These include the four categories of retrofit [8], i.e., active strategies, passive strategies, occupants' interaction, and renewable systems. For example, Sami et al. [9] used TRNSYS (Thermal Energy System Specialists, LLC, Madison, WI, USA) to evaluate a cogeneration system designed for a multi-story building in Iran, while Aste et al. [10] simulated a four-story residential building in Italy using EnergyPlus (Department of Energy, USA), and TRNSYS (Thermal Energy System Specialists, LLC, Madison, WI, USA). Gas and electricity consumption and internal parameters (temperature, relative humidity, and CO_2) were monitored for one year by García-Ballano et al. [11] in a residential complex in Zaragoza (Spain) after the renovation according to the nZEB target in order to verify the percentage of improvement in energy efficiency.

Three steps are generally used to achieve an NZEB [12]. These include the use of passive strategies, the use of energy-efficient technologies (HVAC, lighting, and appliances), and the integration of renewable sources. Sometimes a smart control can be added to obtain "Smart Sustainable Buildings" [13]. The level of performance varies based on the climate zone, the state of the buildings (new or existing), and the intended use (residential and non-residential). For example, a study conducted on the Member States of the European Union [14] shows that values for non-renewable primary energy for NZEBs in Europe vary from 20 to 150 kWh/m^2 year for new buildings, and from 20 to 157 kWh/m^2 year for existing buildings. In addition to energy savings, the high level of perceived comfort and satisfaction of occupants living in nZEB buildings should be highlighted [15].

The technical solutions for the building envelope mainly include the addition of thermal insulation and high-efficiency windows. Reference values report a thermal transmittance value for the external walls of 0.15–0.20 W/m^2 K for new buildings and 0.10–0.20 W/m^2 K for existing buildings. For the roof, 0.10–0.25 W/m^2 K was reported for new buildings and 0.10–0.20 for existing buildings. For window glass, 0.85–1.0 W/m^2 K was reported for new buildings and 0.9–1.15 W/m^2 K for existing buildings [14]. For a non-residential NZEB building in a Mediterranean climate, [16] obtained optimal transmittance values of 0.23 W/m^2 K for the external walls and roof, and 0.9 W/m^2 K for the windows. For a case study in Morocco, the optimal thermal transmittance values were 0.496 W/m^2 K for the external walls and 0.685 W/m^2 K for the roof. For the windows, double glazing with argon was used [17]. In a hot and humid climate, 6 mm single clear glass (U-value: 5.78 W/m^2 K) performs better than double and triple glass [18]. Passive strategies result in the highest percentage of energy savings in tropical climates [19]. Passive cooling solutions include the use of solar shading, natural ventilation and natural lighting, thermal mass, and night cooling.

The percentage of renewable energy sources used in NZEBs varies from 20 to 50% [14]. Photovoltaic is the most used renewable source and the heat pump is the most widespread generation system for heating and domestic hot water [14]. The variability in the size of the photovoltaic system significantly affects the ratio of renewable energy for heating, cooling, and domestic hot water [20]. Izadi et al. [21] explored the application of a hybrid renewable system for a medium three-story office in Iran. The hybrid system includes photovoltaic, wind and hydrogen storage. Between 35% and 49% of the electricity required by the building can be generated by photovoltaics and wind turbines and 70% to 88% can be covered by the combination of renewable sources and hydrogen storage. A total of 45% of the building's energy load is covered by renewable energy in a residential building in Morocco [17]. A newly built hotel in Croatia showed that 53.94% of the energy coverage was obtained from renewable sources [22]. The integration of renewable sources (photovoltaic and wind turbines) reduces the interaction with the grid [23]. The production of "distributed" renewable energy and the different forms of storage (heat, cold, and electricity) are crucial in the creation of nearly zero energy communities [24]. Huang and Sun [25] highlighted the advantages of grouping nZEB buildings to allow for them to share their energy generation and energy storage, and collaborate in load management, minimizing their interaction with the grid. Time-of-use energy management strategies help shift periods of peak power [26]. Storage solutions are also essential to obtain zero energy buildings in the case of "stand-alone" buildings [27]. For buildings connected to the grid, demand-side management can help optimize the load by decreasing the pressure on the network during peak hours [28]. However, this is not enough; in order to create the conditions for the large-scale diffusion of NZEB buildings, the flexible interaction of buildings and the grid must be improved, promoting the advancement of grid-friendliness indicators [29]. The development of advanced controls provides an additional opportunity to improve the behavior of each individual technical component, the environmental efficiency and energy-efficiency of buildings, and the interface with the energy grid [30]. In hot climates, solar energy can be optimized for the production of cooling energy through solar cooling systems [31]. However, the unpredictability of renewable sources and the variability of user loads can lead to a mismatch between energy supply and availability [32]. Furthermore, for multistory buildings, only the roof area is available to install photovoltaics on site, making it difficult to achieve energy self-sufficiency. D'Agostino et al. [33] have suggested that, as a preliminary design in a Mediterranean climate, square-plan buildings can achieve the NZEB target for a building height of up to seven stories. As buildings with a rectangular or L-shaped plan have a higher rate of energy consumption, energy self-sufficiency can be achieved for up to a maximum of six or five floors, respectively [33].

Few renovation studies conducted in buildings in a tropical climate can be found in the literature. Little attention is paid to the application of NZEBs in hot climates [34,35]. Renovation measures that include the improvement of the envelope and the use of active systems contribute to the decarbonization process. However, the objectives are still focused on reducing heating and DHW consumption, even in countries where the demand for cooling is important, such as those in Southern Europe. No specific indicators are used to limit cooling demand. This aspect should be regulated in view of climate changes that potentially increase cooling needs [36]. As pointed out by Cerezo-Narváez et al. [37], adaptation to the new standards produces significant energy savings (from 69% to 127%), which can be achieved with the restructuring of the envelope, and the installation of thermal and photovoltaic solar systems. However, these measures are effective for reducing heating demand, but do not produce significant effects mitigating cooling needs. The authors suggest that regulations should include proposals to alleviate climate consequences in locations with significant summer weather severity. An overview of the concept of zero energy buildings in Latin America was provided by Austin et al. [38]. Retrofit guidelines to obtain zero energy buildings in hot climates are defined by Costa et al. [34]. Kitagawa et al. [39] suggest the combination of ventilative cooling and radiant cooling to improve the cooling performance of buildings in hot and humid climates and improve the thermal comfort of the occupants. A combination of the variable thermal transmittance of the envelope (including phase-change materials), integrated photovoltaics, solar thermal systems, and solar greenhouses was explored by Buonomano et al. [16] for a non-residential building in the Mediterranean area. The integration of PCM into the envelope of a school in Catalonia has also been addressed by Berardi et al. [40]. A saving of 46% for the total final energy consumption was achieved for a four-story office in Brasilia [34]. To reduce energy consumption for air conditioning, a reduced WWR, a glass with reduced SHCG, the addition of solar shading, and natural ventilation are among the measures that were found to be most effective for reaching the NZEB goal. According to a study conducted by Abdou et al. [17], 40% of the total cooling energy can be saved by combining passive measures and renewable sources in Morocco. In hot and humid climates, passive strategies based on natural ventilation, solar shading, and the airtightness of the envelope allow for greater energy savings than thermal insulation [41]. In any case, occupant behavior and actual building use should be used as more realistic input parameters to avoid deviations between the design and the actual building performance [42]. In fact, occupant behavior can lead to discrepancies between expected consumption and actual consumption. Buildings designed to be nZEBs can show significantly higher consumption (up to double) under real operating conditions [15]. Nevertheless, some studies show a full correspondence between the calculated savings and those actually obtained after the renovation of buildings according to nZEB standards [11].

Several studies have addressed the problem of the optimization of zero-energy buildings, focusing on different objectives and adopting different methods. Energy conservation, carbon emissions, costs, and comfort were considered by Wu et al. [43] through an intelligent optimization process applied to a university building in Wuhan, China. Ecology, energy and economy were integrated into the multi-criteria optimization algorithm for zero-energy buildings developed by Zhu et al. [44]. The method aims to optimize the Life Cycle Energy and Life Cycle Cost and includes indices of different perspectives, such as the effectiveness of solar energy application, sociability, energy savings, and equipment economy. Energy savings, Life Cycle Cost, and thermal comfort were the subject of the optimization conducted by Abdou et al. [17] in the climatic zones of Morocco. A method for the integrated optimization of the envelope and the HVAC system for the nZEB refurbishment of a hotel in Croatia is proposed by Delač et al. [45]. The optimal solutions are achieved when the system meets the design load with adequate thermal insulation. Excessive thermal insulation and oversized systems have a negative impact on the cost.

Regarding investigations carried out in a hot climate, the reduction in the U-value of the external walls, the introduction of Energy Recovery Ventilation (ERV), the control of the ventilation rate, and the use of an HVAC system with a high SEER have been found to be significant factors that reduce the energy intensity of buildings and achieve NZEBs in tropical climate [46]. An optimization of the external Venetian blinds was conducted by Huo et al. [47] with reference to the inclination angle of the slats, the orientation, and the WWR ratio in different locations in China. The results demonstrated that solar shadings allow for the cooling load to be reduced and are useful for obtaining nZEBs. Ascione et al. [48] combined model predictive control with comfort models to optimize space cooling. Optimal control strategies allow the cooling load to be predicted and the hourly values of the set-point temperature to be optimized based on the weather forecast. This allows for savings of 28% to be obtained compared to the case in which a fixed set-point temperature is assumed. The economic optimization conducted by Usman et al. [49] for a zero-energy solar house in a subtropical climate showed that the Net Present Value of the solar house becomes lower than that of a conventional house in eight years. Li and Wang [50] highlighted the importance of coordinating the design process of the envelope and HVAC systems to optimize the performance of buildings in cooling-dominated subtropical regions.

Objective

The objective of this study is to propose an optimization procedure for the identification of the most effective building renovation measures to obtain Net Zero Energy Buildings in a tropical climate. The majority of optimization studies that can be found in the literature focus on objective functions that concern the minimization of energy demand, the minimization of emissions, the maximization of renewable energy, and the minimization of costs [51]. Few studies have addressed the problem of optimization from the perspective of comfort. Those that are available generally use PMV, and they are not applied in tropical climates. The current study aims to fill this gap by analyzing a case study set in Panama, characterized by a hot and humid climate. An optimization framework is proposed, aimed at reducing energy demand and, differently from other studies available in the literature, this work addresses the optimization of comfort not using the PMV index but the minimization of hours of discomfort inside the building.

2. Methodology

A case study is considered for the application of the methodology. First, possible energy renovation options are identified. Analysis variables corresponding to each measure are modeled. A parametric analysis is conducted for each variable in order to evaluate the impact, in terms of the extent of the achievable reduction in consumption and the effect produced. Subsequently, the sensitivity analysis allows for the identification of the variables with the greatest impact, in which it is convenient to invest to facilitate the achievement of the Zero Energy objective. Finally, an optimization study allows the optimal solution to be identified based on the minimization of Net Site Energy and discomfort hours, evaluated according to the ASHRAE with 55–80% acceptability [52]. Figure 1 schematizes the methodology's workflow.

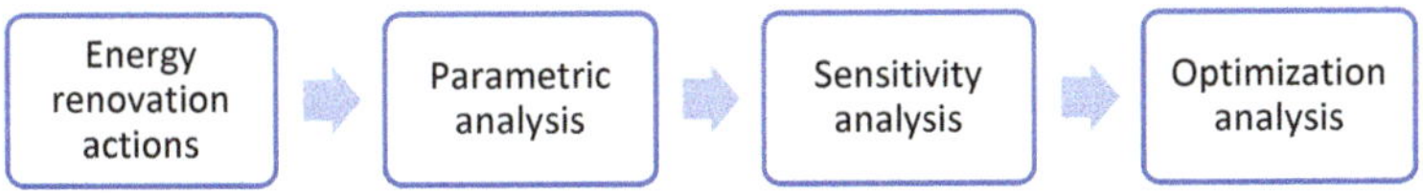

Figure 1. Workflow scheme.

2.1. Case Study Description

The case study building is a single-family house located in Panama City. The climate is identified as Aw according to the Köppen–Geiger classification (tropical wet and dry/savanna climate), with a pronounced dry season in the low-sun months, no cold season, and a wet season in the high-sun months [53]. According to historical data obtained from CLIMdata online platform from Solargis© (Bratislava, Slovakia), the climate of the case study is characterized by a dominant wind direction coming from the northwest, with a most frequent speed range between 1 and 3 m/s (annual average of 2.5 m/s) at 10 m. Two seasons are observed: dry (December to April) and rainy (May and November), where March is the hottest month and October is the rainiest month, with an annual average global radiation of 4.96 kWh/m^2 day and an annual rainfall of 580 kg/m^2. Moreover, the maximum temperature in a day is reached at 3:00 p.m. and the lowest at 6:00 a.m., with an annual average outdoor air temperature range between 22.2 °C and 35.6 °C near the ground. The annual average outdoor relative humidity ranges from 33% to 100%, with an average of 78%.

The software Designbuilder (Version 7.0.2.6, DesignBuilder Software Ltd., Stroud, Gloucs, UK) was used as the simulation tool. It is based on the EnergyPlus software (Version 9.4, Department of Energy, USA) as a Calculation Engine developed by the Department of Energy of the United States Government. This software has additional integrated modules for the easy-run calculation of a parametric analysis, a sensitivity analysis, and both single-objective and multiple-objective optimization analyses.

To perform thermal and energy calculations, the hourly weather data were implemented as site data to define the boundary conditions, which are normally included in the software or provided by the user. The following considerations are accounted for in the modeling approaches used by the software for each aspect of the model:

- The natural ventilation was set up as "calculated" in the software, which refers to the airflows at each opening, estimated via the airflow network method, which are already integrated in the software. This is implemented as the default by the software.
- Solar distribution modelling was set as the complete exterior.
- The Conduction Transfer Function (CTF) was left as default as the heat transfer solution algorithm with constant thermal properties, with no consideration of humidity movement.

- A constant infiltration rate of 0.7 ach was considered always available, considering the good crack state.
- Thermal bridges were not considered.
- The ground thermal behavior modeling was set by default because this study is interested in the comparison of designs where the ground modelling remains the same, among other characteristics. Any differences that are found are a result of the renovations only.

The building is on one floor, with a net surface area of 65.80 m^2. The house includes three bedrooms, a dining room, a living room, a kitchen, two bathrooms, and a storage room. The floor plan of the house is shown in Figure 2.

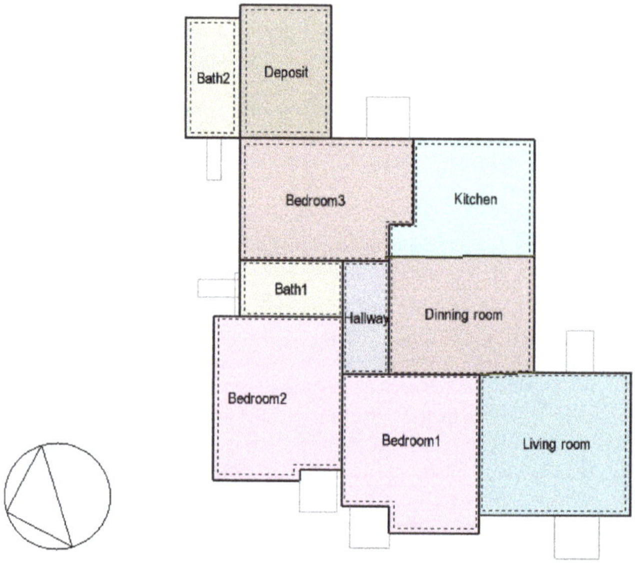

Figure 2. Floor plan of the analyzed case study building.

The roof is pitched, with a north–south orientation and an unconditioned space underneath. The house is considered representative of this category of buildings, based on a survey previously carried out in the research area [54].

Regarding the construction characteristics, the building envelope is not insulated. Table 1 reports the thermal transmittance values of the opaque elements of the construction.

Table 1. Thermal transmittance of the opaque elements of the building envelope.

Building Element	Thermal Transmittance Value [W/m^2 K]
External walls	4.02
Roof	7.14
Dispersing ceiling	4.54
Ground floor	2.30
Windows	5.80

Only the three bedrooms are equipped with a cooling system, consisting of split units with a COP of 3. The cooling system is used in bedroom 1 from 7 p.m. to midnight, with a set point temperature of 23 °C. In bedrooms 2 and 3 it is used from 11 p.m. to 7 a.m. with a set point temperature of 23 °C. Occupation varies in the various rooms of the house, based on the typical profiles defined throughout the study [54]. The use of equipment and lighting was also defined based on the typical behavior identified in the same study. The hourly climate file used in the simulations was obtained based on real data recorded in the study area.

Regarding the comfort assessments, the following metabolic rate value was assumed in the environments: 90 W/person in the bedrooms, 110 W/person in the dining and living rooms, 120 W/person in the bathroom, and 160 W/person in the kitchen.

A clothing of 0.50 was assumed throughout the year, as there is no seasonal difference in the considered climate; rather, it is summer throughout the year.

2.2. Retrofitting Actions

The renovation interventions are primarily designed with reference to the minimum requirements defined by the standard [55]. This establishes the performance characteristics required for different construction elements to achieve an estimated level of energy saving. The rule actually refers to newly constructed buildings. However, as specific regulations on existing buildings are not available in Panama, it was deemed appropriate to refer to this source. Furthermore, in the hypothesis where an existing building is transformed into a Net Zero Energy Building, an extensive level of renovation is expected, which can potentially be assimilated into newly constructed buildings in terms of regulation. Table 2 summarizes the requirements for residential buildings, referring to "design option 3", provided by Regulation [55], which allows for energy savings of 16%.

Table 2. Target values for residential buildings (design variant 3) [55].

External walls U-value [W/m^2 K]	0.8
Windows U-value [W/m^2 K] g-value (SHGC)	 5.80 0.60
Exterior window shade Projection factor	0.30 South and West
Window-to-wall ratio	40%
Roof U-value [W/m^2 K]	1.04
HVAC COP	3.0

The present study intends to go beyond the performance required by law, identifying further renovation measures capable of bringing the building to have a zero or positive energy balance. In summary, the renovation actions include passive measures, active measures, renewable sources, and interaction with occupants.

Passive measures concern the improvement of the thermal performance of the building envelope. This is achieved through the thermal insulation of the opaque dispersing elements and the use of highly energy-efficient windows, with reduced thermal transmittance. The choice of the characteristics of the transparent surfaces, particularly the solar gain of the glass, is also one of the passive measures, as well as the use of solar shadings. The window-to-wall ratio can also be classified as a passive measure. However, this specific parameter was not analyzed in the present study, as it is difficult to change in the case of existing buildings. The size of the windows, in fact, is linked to the internal lighting requirements and the minimum ventilation required for the healthiness of the rooms. Furthermore, in the case of existing buildings, this could be prevented from modifying the extension and shape of the windows to avoid altering the façades. However, in the building under consideration, the WWR does not comply with the minimum imposed by the standard, as it is lower than 40% for all exposures.

The active measures concern the use of the cooling system, which is expected to improve its efficiency.

The efficiency of the building in relation to the interaction with the occupants is examined considering two parameters: the cooling set point temperature and the operation schedule of the cooling system.

Finally, the installation of a photovoltaic system on the roof of the building for the production of electricity on-site was planned. The system occupies the pitch of the roof facing south.

2.3. Modelling of the Design Variables

In relation to each analyzed aspect, a set of variables was defined in order to perform a parametric analysis to evaluate the impact of each single variable on the energy balance of the building. The variation ranges were defined with reference to the different technical solutions that can be adopted for the analyzed building. Uniform distributions were assumed for the different variation options. Table 3 shows the list of variables used and the minimum and maximum values containing the variation ranges. Discrete increments were assumed between the limit values, with variable steps. For photovoltaics, three options were analyzed, corresponding to 3 kW_p, 6 kW_p, and 9 kW_p. For the cooling set-point, the options are supported by the reported results from surveys found in [54,56]. For the cooling schedule, six options were defined, which are detailed in Table 4.

Table 3. Definition of the design variables.

Design Variable	Regulation Target Value (with Variant n°3)	Min	Max	Step
U external walls [W/m^2 K]	0.8	0.2	4	0.2
U windows [W/m^2 K]	5.8	1.8	5.8	0.2
U roof [W/m^2 K]	1.04	0.2	7	0.2
U ground floor [W/m^2 K]	-	0.2	4	0.2
U ceiling [W/m^2 K]	-	0.2	4.6	0.2
g-value [-]	0.6	0.35	0.6	0.025
Solar shading [%] of lit area	0.3	0	1	-
Cooling COP [-]	3	3	4	0.1
Photovoltaic [kW_p]	-	3	9	3
Cooling schedule	-	Off/24	On/24	-
Cooling set-point [°C]	[23.5–28.5]	16	28	1

Table 4. Cooling operation schedule options included in the parametric analysis.

Cooling Operation Schedule	Operation Hours	Description
C1	Always "Off"	Always "Off" 24/7
C2	"On" from 23:00 to 7:00	"On" eight hours during the night
C3	"On" from 8:00 to 17:00	"On" nine hours during the day
C4	"On" from 19:00 to 24:00	"On" five hours during the night
C5	"On" from 19:00 to 6:00	"On" eleven hours during the night
C6	Always "On"	Always "On" 24/7

3. Results

3.1. Energy Consumption at Current State

The baseline model was calibrated according to the ASHRAE Guidelines 14 [57]. The Mean Bias Error (MBE %) and Coefficient of Variation of Root Mean Square Error (CVRMSE %) indices were calculated considering the monthly metered electricity data, obtaining deviations lower than the limits allowed by the standard (±5% for the MBE and 15% for the CVRMSE).

The house, at present, has an energy demand of 75.55 kWh/m^2 year, which does not follow the base line proposed in [55] for Panama (60 kWh/m^2 year, although this proposed base line is only for new buildings). The graph in Figure 3 shows the monthly trend in electricity demand. There are no renewable systems, so all the necessary energy is taken from the grid. In the current state, the reference case shows 2431.17 discomfort hours, evaluated according to the ASHRAE 55 considering the 80% acceptability limit [52].

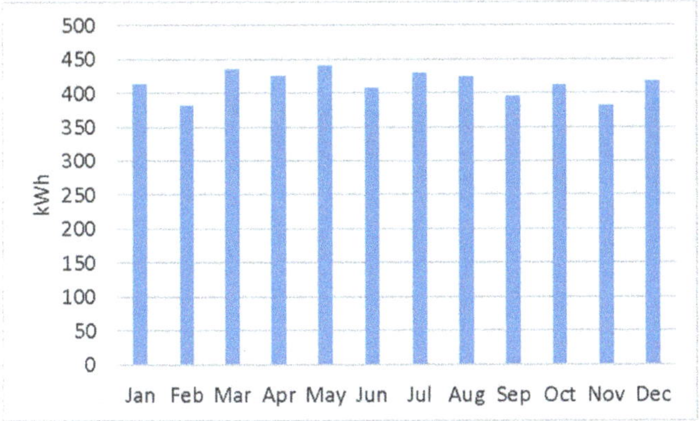

Figure 3. Monthly electricity consumption for the analyzed building in its current state.

Electricity consumption includes, in addition to cooling, electricity for the operation of household appliances and lighting, deriving from the modeling of the typical occupancy profiles of the considered area. On average, cooling consumption takes up approximately 50% of the total electricity consumption assessed for the home.

With the aim of finding the best combinations of measures capable of obtaining Net Zero Energy Buildings in a tropical climate, the net energy at the site was considered as the output of the analysis. This is calculated as the difference between the electricity demand and the renewable electricity generated on site by the photovoltaic system. The balance is calculated at the site. The lower the net site energy, the more energy-efficient and independent the building. Negative net site energy values indicate that the generated energy is greater than the demanded energy, with the surplus resulting in a positive energy building. Surplus electricity is assumed to be fed into the grid. There are no storage systems.

3.2. Parametric Analysis

The results of the parametric analysis allow for first idea to be obtained of intervention methods that are suitable for pursuing the goal of Zero Energy. Regarding the thermal insulation of the opaque envelope, the observed trend is that the addition of a layer of thermal insulation and the consequent reduction in thermal transmittance leads to a decrease in cooling energy demand. This is true for the external walls, for which a more significant decrease in energy needs is recorded as the thermal insulation thickness increases, going from 4973.27 kWh/year to 4216.47 kWh/year, and also for the ceiling and roof, which produce less significant reductions. The opposite trend is instead observed for the ground floor slab. In this case, the positive effect is registered as the thermal insulation thickness decreases. The higher the thermal transmittance of the floor in contact with the ground, the lower the net site energy. This behavior is explained by the fact that, in a hot climate, such as the tropical climate of Panama, heat losses towards the ground favor the passive cooling of internal environments, with a consequent reduction in the required energy. The graphs in Figures 4 and 5 show the results of the parametric analysis conducted for the net site energy as a function of the thermal transmittance of the external walls and the ground floor, respectively.

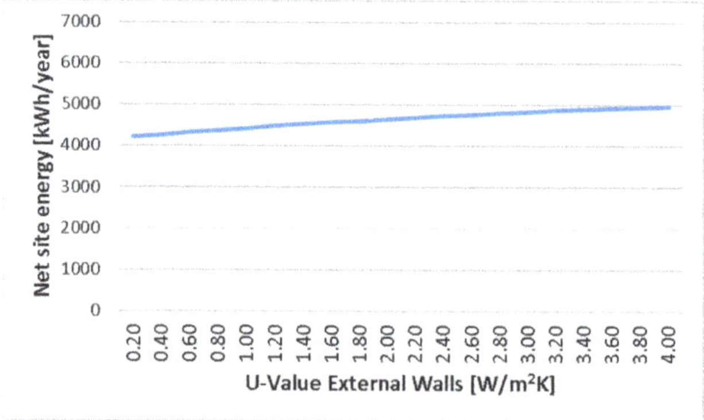

Figure 4. Parametric analysis of the net site energy during variations in the thermal transmittance of the external walls.

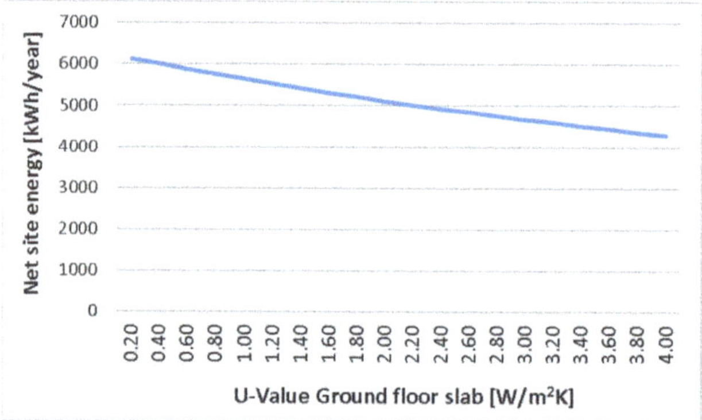

Figure 5. Parametric analysis of the net site energy during variations in the thermal transmittance of the ground floor slab.

Regarding windows, the variation in thermal transmittance does not produce significant variations in energy demand. The net site energy remains almost constant using windows with different thermal transmittance values. This is due to the limited extension of the openings compared to opaque surfaces. Furthermore, the solar gain factor of the glass and the different solar shading levels that are analyzed do not appreciably change the energy requirements. This is assumed because, for the specific case in question, the windows are already in the shade due to the geometry of the construction itself, so the impact of solar radiation is already reduced, even in the current configuration.

Regarding the efficiency of the cooling system, the energy demand decreases as the cooling efficiency increases. The graph in Figure 6 shows the trend of the net site energy upon variations in the cooling COP.

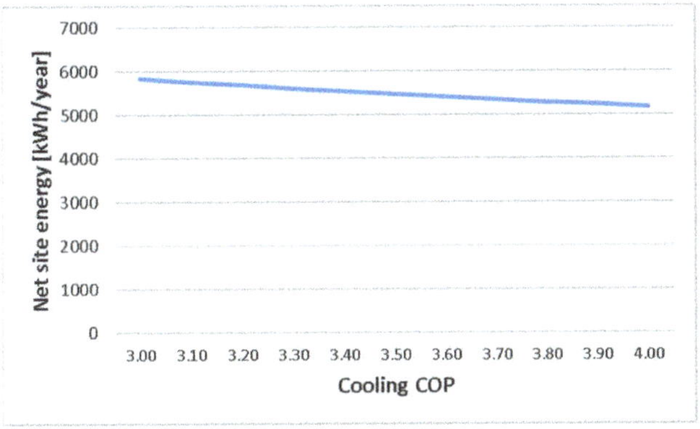

Figure 6. Parametric analysis of the net site energy at variations in the cooling COP.

The operation schedule of the cooling system has a significant impact on the energy demand. The lowest consumption is obtained with the C1 schedule (always "Off"). In this case, the energy demand is attributed only to the use of electrical appliances. The maximum value is instead reached by considering the cooling always "On" (C6 option) in the rooms where it is present. The other schedules that were considered produce an energy demand between these two extremes, as illustrated in Figure 7.

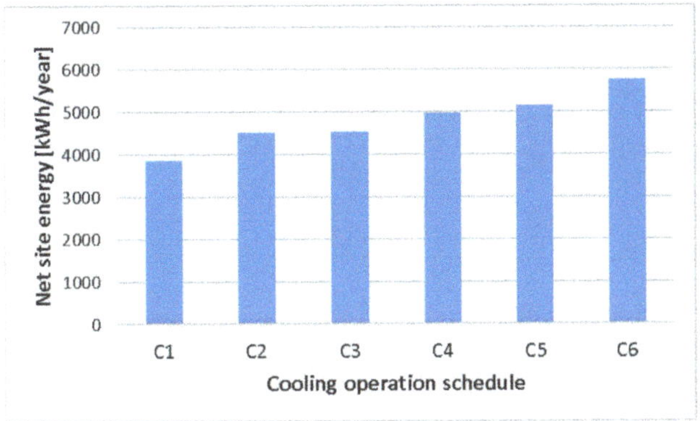

Figure 7. Parametric analysis of the net site energy upon variations in the cooling operation schedule.

The cooling set-point temperature also has a significant impact. The net site energy is halved, passing from a set-point temperature of 16 °C to 28 °C (Figure 8).

The measure that most affects the energy balance is the integration of the photovoltaic system. This, at the smallest size, allows for an energy balance close to zero. By using the medium and large sizes, the balance becomes that of a positive energy building, with generation being greater than consumption.

A parametric analysis was also conducted to explore the combined effect of the set-point temperature and the cooling schedule (Figure 9). As expected, the highest consumption is associated with the lowest cooling temperature (16 °C) and the most prolonged use of the cooling system (Always "On" operating schedule). Turning the cooling on all night (option C5—"On"—for eleven hours during the night: from 19:00 to 6:00) also involves high consumption. Consumption progressively decreases for all operating modes by increasing the set-point temperature. Maintaining a temperature of 20 °C, the most convenient operating schedule is C4 ("On" for five hours during the night: from 19:00 to 24:00). If the set-point temperature is higher than 20 °C, the most convenient operating schedule becomes C3 ("On" for nine hours during the day: from 8:00 to 17:00). Options C2 and C3 show similar consumption levels. At a temperature of 28 °C, cooling for five hours during the night (from 19:00 to 24:00—C4) involves the same net site energy demand as cooling for 24 h (C6: always "On"). By selecting a temperature of 20 °C, the C3, C2, and C4 operating modes are equivalent, approximately 40% lower than what would be achieved if cooling always in "On" mode.

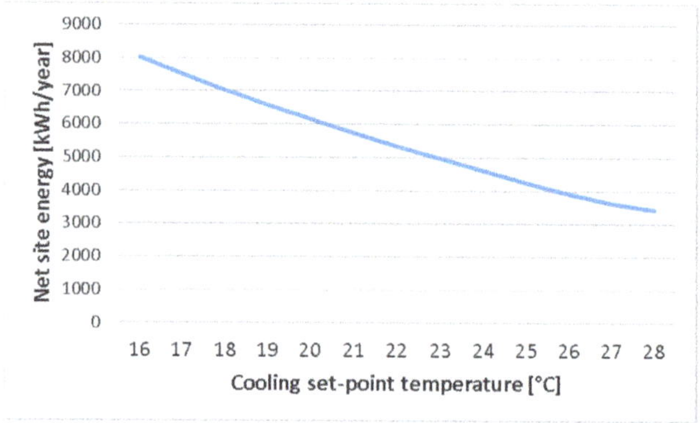

Figure 8. Parametric analysis of the net site energy upon variations in the cooling set-point temperature.

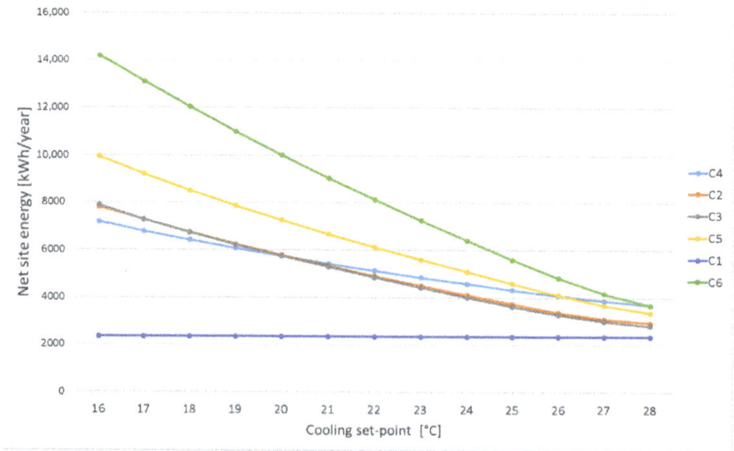

Figure 9. Parametric analysis of the net site energy upon variations in the cooling set-point temperature and cooling schedule.

3.3. *Sensitivity Analysis*

The sensitivity analysis identified the variables that had the highest impact on the net site energy requirement, which should be considered when renovating buildings in a hot–humid climate.

The LHS method was used to sample the variables, setting a number of iterations equal to 300. All the options defined for the different variables were uniformly sampled. This means that, during the various iterations, combinations of variables were explored using the different options defined for each variable.

The multiple linear regression method was used for the sensitivity analysis. This statistical method estimates the relationships between input variables. A regression analysis helps to understand how the typical value of the output changes when any one of the input variables is varied (assuming that the input variables are independent of each other). The Standardized Regression Coefficient (SRC) indicates the sensitivity of each input variable, thus identifying the most and least important variables. Other regression outputs, like "Adjusted R-squared" value and "*p*-value", help to determine the level of confidence and the reliability of the results.

Figure 10 reports the Standardized Regression Coefficient for each variable. The colors indicate the importance of the variables (High Importance: red; Medium Importance: blue; Low Importance: green).

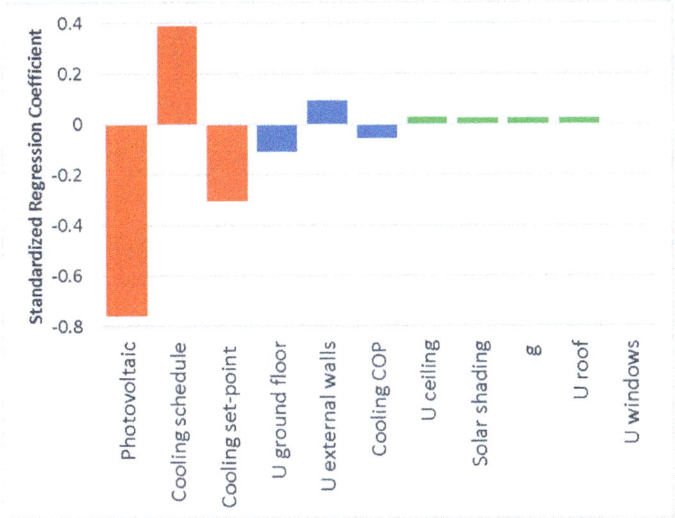

Figure 10. Standardized Regression Coefficient for each variable.

The results highlight that the net site energy demand is most strongly influenced by the photovoltaic size; however, there is an inverse relationship. This implies that increasing the PV size leads to a decrease in net site energy consumption. The net size energy is also strongly influenced by the cooling operation schedule and cooling set-point temperature. A moderate influence is obtained for the thermal transmittance of the ground floor slab, external walls, and the cooling COP. The U-value of the semi-exposed ceiling, the roof and windows, the solar shadings, and the window g-value do not have a notable influence on net site energy consumption.

The adjusted R-squared value represents the goodness of fit of the complete model. This indicates how much variation in the output is explained by the input variables.

For the analyzed output (net site energy), the adjusted R-squared value of 0.9151 is obtained, suggesting that most of the key sensitive input variables were identified.

Table 5 reports the Standardized Regression Coefficient, standard error and the *p*-value of the variables included in the analysis. The color of the variables listed in Table 5 corresponds to their importance (High Importance: red; Medium Importance: blue; Low Importance: green).

The *p*-value expresses whether the input variable has a statistically significant effect on the output. Some input variables have a *p*-value greater than 0.05, indicating that there is a low level of confidence in their respective regression result values. These are as follows: thermal transmittance of semi-exposed ceiling, shadings, windows' g-value, thermal transmittance of the roof, and thermal transmittance of the windows.

Table 5. Standardized Regression Coefficient, standard error, and the *p*-value of the analyzed variables.

Design Variables	Standardized Regression Coefficient	Standard Error	*p*-Value
Photovoltaic	−0.7587	97.5714	0.0000
Cooling schedule	0.3867	70.7853	0.0000
Cooling set-point	−0.3046	21.1847	0.0000
U ground floor	−0.1084	13.9192	0.0000
U external walls	0.0949	14.0195	0.0000
Cooling COP	−0.0543	25.4718	0.0019
U ceiling	0.0283	11.9863	0.0984
Solar shading	0.025	71.5164	0.1476
G	0.0237	25.1695	0.1672
U roof	0.0235	7.8552	0.1700
U windows	−0.0037	13.1753	0.8283

3.4. Optimization Analysis

3.4.1. Optimization of Net Site Energy

The optimization analysis was conducted in two stages. In the first phase, the optimization was carried out only as a function of the net site energy, searching for the lowest achievable value. The minimum value of −10,144.89 kWh/year was obtained, corresponding to point 1, highlighted in Figure 10, with the combination of variables illustrated in Table 6. Since the net site energy is negative, this indicates that, regarding the balance between energy consumed and energy produced on-site, the energy produced with renewable sources is greater than the energy consumed. In this configuration, the cooling system is always off and the largest photovoltaic system is used.

Table 6. Optimal solution minimizing net site energy.

Optimal Point	n°	1
Net site energy	[kWh/m^2 year]	−154.18 *
U-value external walls	[W/m^2 K]	0.8
U-value roof	[W/m^2 K]	7.0
U-value ground floor slab	[W/m^2 K]	3.8
U-value semi-exposed ceiling	[W/m^2 K]	0.8
U-value windows	[W/m^2 K]	3.0
g-value windows	[-]	0.35
Shadings	[%] of lit area	50
Cooling set-point temperature	[°C]	26
Cooling operation schedule	[-]	C1
Photovoltaic size	[kW_p]	Large
Cooling COP	[-]	3.5

* The electricity generation is higher than the electricity consumption. Convention of signs used: energy consumption is always positive.

Looking at the solutions that are closest to zero (indicated with the yellow band in the graph in Figure 11), it is possible to see that a balance close to zero can be achieved by turning on the cooling system with different schedules and by installing a small or medium-size PV system.

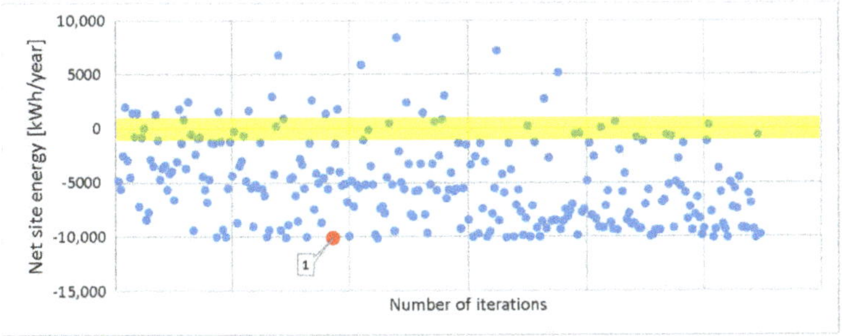

Figure 11. Single-objective optimization based on the minimization of the net site energy. The point n°1 indicates the optimal solution. The combinations closest to net "zero" site energy are highlighted with a yellow band.

3.4.2. Optimization of Net Site Energy and Discomfort ASHRAE 80% Acceptability

In the second phase, the optimization was conducted based on two objectives: minimizing the net site energy and minimizing the hours of discomfort calculated according to the ASHRAE 55, considering 80% acceptability. The point cloud obtained for the various combinations of measures is shown in Figure 12.

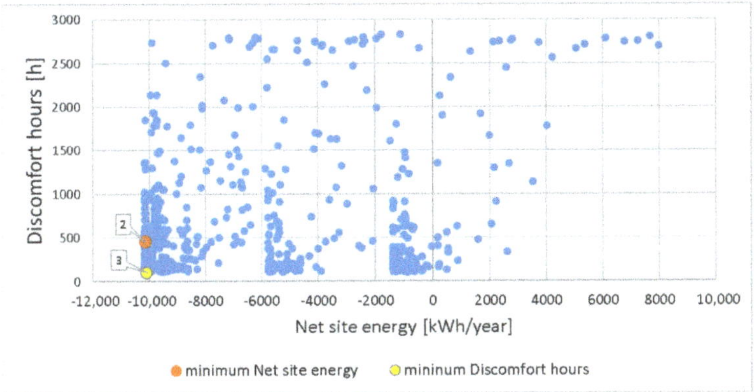

Figure 12. Optimization based on the minimization of the net site energy and minimization of discomfort hours ASHRAE 55 Adaptive 80% Acceptability. The points n°2 and n°3 correspond to the solutions with minimum net site energy and minimum discomfort hours, respectively.

The combinations of measures that generate the lowest net site energy (point 2) and the fewest hours of discomfort (point 3) are highlighted. The details of these two solutions are specified in Table 7.

Both optimal points were obtained by using the largest photovoltaic size and lead to a positive balance, with a notable energy surplus.

The graph in Figure 13 shows the grouping of points according to the size of the photovoltaic system. The figure highlights that a balance close to zero can be achieved by using a small photovoltaic system.

Table 7. Optimal solution minimizing net site energy and hours of discomfort.

Optimal Point	n°	1	2
Net site energy	[kWh/m^2 year]	**−154.50**	−153.80
Discomfort hours	[h/year]	462.6	**101.9**
U-value external walls	[W/m^2 K]	0.4	2.4
U-value roof	[W/m^2 K]	0.2	0.2
U-value ground floor slab	[W/m^2 K]	3.2	3.8
U-value semi-exposed ceiling	[W/m^2 K]	3.8	4.4
U-value windows	[W/m^2 K]	3.0	3.2
g-value windows	[-]	0.4	0.375
Shadings	[%] of lit area	30	30
Cooling set-point temperature	[°C]	27	26
Cooling operation schedule	[-]	C1	C1
Photovoltaic size	[kW$_p$]	Large	Large

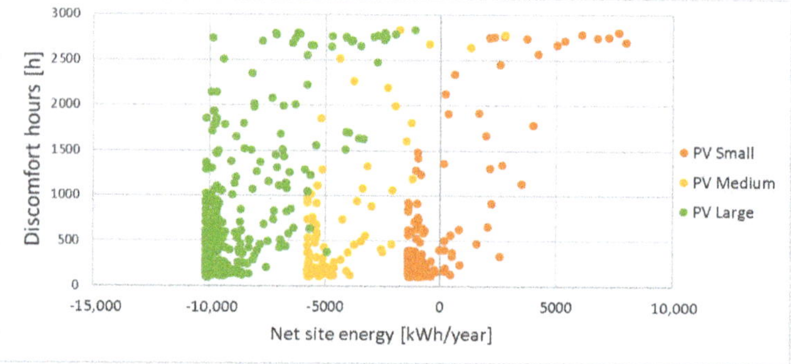

Figure 13. Optimization results grouped by PV size.

By limiting the search to a small photovoltaic size, the optimal solution can be identified at point 4, as indicated in the graph in Figure 14, which has the characteristics specified in Table 8.

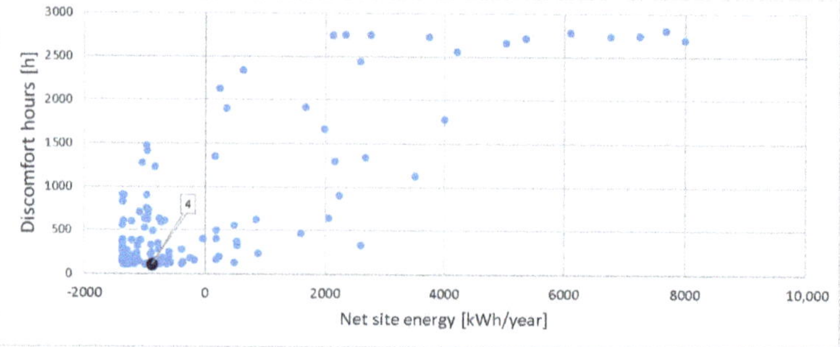

Figure 14. Optimal point considering only the combinations using small-sized photovoltaics. The point n°4 indicates the optimal solution.

Table 8. Optimal solution for minimum net site energy and fewest hours of discomfort using the small photovoltaic size.

Optimal Point	n°	4
Net site energy	[kWh/m^2 year]	−13.36
Discomfort hours	[h/year]	105.6
U-value external walls	[W/m^2 K]	2.2
U-value roof	[W/m^2 K]	0.4
U-value ground floor slab	[W/m^2 K]	3.8
U-value semi-exposed ceiling	[W/m^2 K]	4.0
U-value windows	[W/m^2 K]	3.0
g-value windows	[-]	0.375
Shadings	[%] of lit area	0
Cooling set-point temperature	[°C]	28
Cooling operation schedule	[-]	C2
Photovoltaic size	[kW_p]	small
Cooling COP	[-]	3.8

The graph in Figure 15 shows the optimization results, grouped based on the cooling schedule. The points are quite spread out and it is not possible to identify precise groupings. The points at the top, highlighted with the red band, are those that present the greatest number of discomfort hours, despite having the cooling schedule always "On". The high number of discomfort hours obtained for these points is due to the excessively low set-point temperature and the inadequate characteristics of the building envelope. On the other hand, the green band identifies points with the lowest number of discomfort hours. Among these, there are many combinations in which the cooling schedule is always "Off". In this case, despite not using the cooling system, comfort is guaranteed by the good performance of the building envelope.

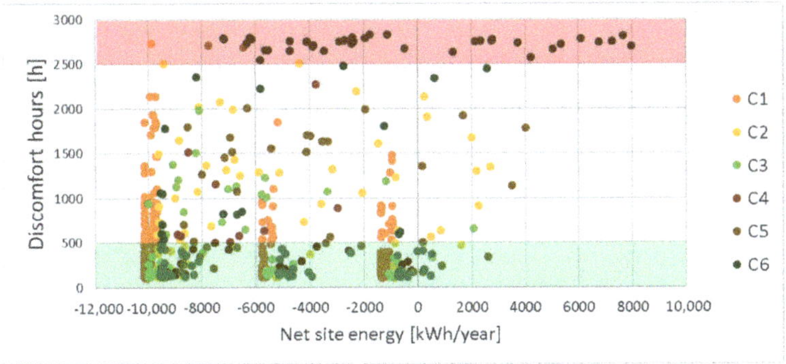

Figure 15. Optimization results grouped according to the cooling schedule. The red and green bands represent the solutions with greater and fewer discomfort hours.

4. Discussion

Building performance optimization studies generally are based on objective functions that include economic aspects (e.g., minimizing Life Cycle Costs, overall investment costs, or building operation costs, or maximizing Net Present Value), energy aspects (e.g., minimizing overall primary energy, total electrical load, etc.), and environmental aspects (minimizing life cycle environmental impact, carbon emissions, etc.). Optimization studies focused on comfort, particularly in Zero Energy buildings, are not widespread and typically consider parameters such as minimizing the Predicted Mean Vote (PMV) or Predicted Percentage of Dissatisfaction (PPD) [12]. Boutet et al. [58] conducted an optimization in a hot and humid climate but focused specifically on hygrothermal and daylighting behavior in a school in Argentina. The procedure developed by Chen and Yang [59] was aimed at finding the optimal solution as a function of cooling and lighting. Visual comfort was also the subject of optimization in the work carried out by Rabani et al. [60] for an office building in Norway. Differently from the studies available in the literature, the present research aimed at investigating the optimization of Net Zero Energy Buildings in tropical climate, with respect to building renovation, considering the combination of two objective functions, which are the minimization of energy demand and minimization of hours of discomfort, evaluated according to ASHRAE 55 Adaptive 80% Acceptability. Other studies in the literature have investigated the feasibility of Zero Energy Building renovations in tropical climates. An example is provided by Ohene et al. [41]; however, they only conducted a parametric analysis and not an optimization, concluding that retrofitting existing buildings into Net Zero Energy Buildings in a tropical climate is feasible by applying passive strategies to the envelope and installing renewable energy systems. The results obtained from the present study indicate the PV size, the operating schedule of the cooling system, and the cooling set-point temperature as the most influential variables for minimizing the net site energy and hours of discomfort in the considered building. The thermal transmittance of the floor and external walls and the cooling COP were found to be moderately influential. The thermal transmittance of the windows, roof, and ceiling, together with the windows' solar factor and the shading type, were found to be the least influential variables. Giouri et al. [61] found that cooling set-point temperature and PV output are among the factors with the highest impact on annual energy demand and adaptive thermal comfort by evaluating a high-rise office building in a Mediterranean climate. Furthermore, it is possible to observe that occupancy hours are among the most influential variables on electricity consumption, as assessed by Chacón et al. [62]. Chen et al. [63] obtained a low influence of cooling COP. In contrast to the results obtained in the present study, Chen et al. [64] reported the thermal transmittance of windows and the level of exterior window shading as variables that have a great impact on building energy demand in a hot and humid climate. However, in this case, the analyzed building is a high-rise residential building for 30–40 story apartments.

As was also found by Al-Saadi and Al-Jabri [35], in warm climates, the largest share of consumption is related to cooling (up to 70%). Based on the results reached by Al-Saadi and Al-Jabri [35], optimal envelope solutions include single glazing for windows and thermal insulation ranging from 5 to 10 cm in the walls and roof. For a building in a Mediterranean climate, high thermal insulation was used for the walls and roof (U = 0.23 W/m^2 K), and low-transmittance glass (U = 0.9 W/m^2 K triple low-e glass with Krypton filling) was used for the windows [16]. In contrast, in the present study, the optimal solution involves low insulation for opaque structures and double glazing for windows. Among the envelope elements, floor thermal transmittance has the greatest impact on energy demand. In particular, cooling energy decreases as floor thermal insulation decreases. Similar outcomes were obtained by Gou et al. [65], who highlighted that improving ground thermal insulation can diminish the heat exchange between the building and the ground, resulting in a decrease in the building's energy performance. Abdou et al. [17] achieved the optimal solution for a building in Morocco using a ground floor with high thermal transmittance (U = 2.875 W/m^2 K). The optimal solutions identified in the current study involve a cooling set-point temperature of 26 °C or 28 °C, which is higher than that obtained by Hwang et al. [66] for a school building in a hot and humid climate, for which the analysis returned cooling set-point temperature values between 23 and 25 °C. The use of a photovoltaic system can produce surplus energy, which, in this case, was estimated at about 13.4 kWh/m^2 year. Similarly, Acar and Kaska [23] obtained excess energy of up to 23 kWh/m^2, 9 kWh/m^2, and 3 kWh/m^2 for a hot, semi-arid climate, Mediterranean climate, and cold, semi-arid climate, respectively.

The application of the renovation measures makes it possible to reduce the energy consumption of the case study by approximately 29%, over the 16% indicated by the regulations in force for newly constructed buildings [55].

The present study shows some limitations. First, the analysis conducted at this stage only focused on a single-family house. Other building types (e.g., apartment buildings) were not considered, nor were other uses, which could significantly change the results of the analysis. The thermal mass of the structure was not considered in the optimization parameters, because this is difficult to change in existing buildings. The interaction of the Net Zero Energy Building with the grid was not considered, and consequently the aspects of optimal load management and possible income from the sale of surplus energy are neglected. Furthermore, the environmental aspects, in terms of CO_2 emissions, were not evaluated, nor were the cost issues, which could be useful in assessing the economic viability of the renovation interventions. Finally, the study is limited to the climatic conditions of the analyzed site, and the effects of future climate change are not examined. As noted by Chai et al. [67], climate change could alter NZEBs' performance over their life cycle. Plant size, cost, and comfort satisfaction levels could show differences that occur under climate change.

5. Conclusions

The present study explored strategies for renovating existing buildings into Net Zero Energy Buildings in a tropical climate and proposed a methodology that can be followed for renovations to NZEB in a hot and humid climate. The analysis, conducted for a single-family house, allowed for the definition of some guidelines. The most important role is played by renewable energy. The installation of a photovoltaic system makes it possible to cover the cooling energy demand and the energy demand other uses of the dwelling. In the case of single-family houses, it is possible to exploit the entire roof area to place photovoltaic panels, reaching a size that allows for more energy to be produced than is required. It would be worthwhile to extend the study with a view to sharing the excess energy that is produced with other buildings. Regarding active measures, the use of a high-efficiency system is recommended. In the examined case, an electric heat pump was used, which works efficiently in combination with the photovoltaic system. With reference to passive measures, the analysis in a warm climate showed the need to carefully consider the thermal insulation of the envelope. In particular, optimal solutions were found with low external wall insulation. In addition, floor slab insulation is critical. This element, by allowing heat exchange with the ground, should be poorly insulated or uninsulated to favor heat losses and the consequent reduction in cooling requirements. Passive measures also include window shading. Finally, with regard to occupant behavior, a set-point temperature of 28 °C and nighttime operation were identified as optimal solutions that can reduce energy consumption and ensure comfort.

Future studies will be directed to the analysis of other types of residential buildings and buildings with different uses (e.g., office, commercial, education). The interaction of the building with the grid is worthy of further investigation, depending on the interchange that can take place between multiple NZEB buildings connected in energy communities, with the integration of storage systems with the energy produced by photovoltaics. After analyzing optimization from the perspectives of energy and comfort, the next step certainly requires economic optimization in terms of Life Cycle Costs, as well as the achievement of zero-emission buildings, including at the construction and operational phases. Finally, evaluation under different climatic conditions is required to expand the range of possible solutions, and studies under different climate evolution scenarios would also increase the validity of the results.

Author Contributions: Conceptualization, C.C. and M.C.A.; methodology, C.C. and M.C.A.; software, C.C.; validation, M.C.A.; data curation, M.C.A.; writing—original draft preparation, C.C.; writing—review and editing, M.C.A.; visualization, D.M. and N.A.; supervision, D.M. and N.A.; project administration, M.C.A., D.M. and N.A.; funding acquisition, M.C.A. and D.M. All authors have read and agreed to the published version of the manuscript.

Funding: This research was funded by the Panamanian Institution Secretaría Nacional de Ciencia, Tecnología e Innovación SENACYT, under the project code IDDS22-30, as well as supported by the Sistema Nacional de Investigación (SNI).

Data Availability Statement: The data presented in this study are available from the corresponding author upon reasonable request. The data are not publicly available due to local server storage limitations.

Conflicts of Interest: The authors declare no conflicts of interest.

References

1. López-Ochoa, L.M.; Las-Heras-Casas, J.; Olasolo-Alonso, P.; López-González, L.M. Towards nearly zero-energy buildings in Mediterranean countries: Fifteen years of implementing the Energy Performance of Buildings Directive in Spain (2006–2020). *J. Build. Eng.* **2021**, *44*, 102962. [CrossRef]
2. Economidou, M.; Todeschi, V.; Bertoldi, P.; D'Agostino, D.; Zangheri, P.; Castellazzi, L. Review of 50 years of EU energy efficiency policies for buildings. *Energy Build.* **2020**, *225*, 110322. [CrossRef]
3. European Union. Directive 2010/31/EU of the European Parliament and of the Council of 19 May 2010 on the energy performance of buildings (recast). *Off. J. Eur. Union* **2010**, *53*, 13–35. [CrossRef]
4. Attia, S.; Kurnitski, J.; Kosiński, P.; Borodiņecs, A.; Belafi, Z.D.; István, K.; Krstić, H.; Moldovan, M.; Visa, I.; Mihailov, N.; et al. Overview and future challenges of nearly zero-energy building (nZEB) design in Eastern Europe. *Energy Build.* **2022**, *267*, 112165. [CrossRef]
5. Moran, P.; O'Connell, J.; Goggins, J. Sustainable energy efficiency retrofits as residenial buildings move towards nearly zero energy building (NZEB) standards. *Energy Build.* **2020**, *211*, 109816. [CrossRef]
6. Tang, F.; Chen, J.; Li, J.; Rodriguez, D. Energy saving actions toward NZEBs with multiple-criteria optimization in current residential buildings. *Energy Rep.* **2020**, *6*, 3008–3022. [CrossRef]
7. D'Agostino, D.; Mazzarella, L. What is a Nearly zero energy building? Overview, implementation and comparison of definitions. *J. Build. Eng.* **2019**, *21*, 200–212. [CrossRef]
8. Austin, M.C.; Carpino, C.; Mora, D.; Arcuri, N. A Methodology to identify appropriate refurbishment strategies towards zero energy buildings in a hot and humid climate. *J. Phys. Conf. Ser.* **2022**, *2385*, 012020. [CrossRef]
9. Sami, S.; Gholizadeh, M.; Dadpour, D.; Deymi-Dashtebayaz, M. Design and optimization of a CCHDP system integrated with NZEB from energy, exergy and exergoeconomic perspectives. *Energy Convers. Manag.* **2022**, *271*, 116347. [CrossRef]
10. Aste, N.; Adhikari, R.; Del Pero, C.; Leonforte, F. Multi-functional Integrated System for Energy Retrofit of Existing Buildings: A Solution Towards nZEB Standards. *Energy Procedia* **2017**, *105*, 2811–2817. [CrossRef]
11. García-Ballano, C.J.; Ruiz-Varona, A.; Bailo, C.M.; Matud, C.C. Monitoring of housing blocks in Zaragoza (Spain) to validate the energy savings calculation method for the renovation of nZEB dwellings. *Energy Build.* **2021**, *256*, 111737. [CrossRef]
12. Harkouss, F.; Fardoun, F.; Biwole, P.H. Optimization approaches and climates investigations in NZEB—A review. *Build. Simul.* **2018**, *11*, 923–952. [CrossRef]
13. Jaysawal, R.K.; Chakraborty, S.; Elangovan, D.; Padmanaban, S. Concept of net zero energy buildings (NZEB)—A literature review. *Clean. Eng. Technol.* **2022**, *11*, 100582. [CrossRef]
14. D'Agostino, D.; Tzeiranaki, S.T.; Zangheri, P.; Bertoldi, P. Assessing Nearly Zero Energy Buildings (NZEBs) development in Europe. *Energy Strat. Rev.* **2021**, *36*, 100680. [CrossRef]
15. Colclough, S.; Hegarty, R.O.; Murray, M.; Lennon, D.; Rieux, E.; Colclough, M.; Kinnane, O. Post occupancy evaluation of 12 retrofit nZEB dwellings: The impact of occupants and high in-use interior temperatures on the predictive accuracy of the nZEB energy standard. *Energy Build.* **2022**, *254*, 111563. [CrossRef]
16. Buonomano, A.; De Luca, G.; Montanaro, U.; Palombo, A. Innovative technologies for NZEBs: An energy and economic analysis tool and a case study of a non-residential building for the Mediterranean climate. *Energy Build.* **2016**, *121*, 318–343. [CrossRef]
17. Abdou, N.; EL Mghouchi, Y.; Hamdaoui, S.; EL Asri, N.; Mouqallid, M. Multi-objective optimization of passive energy efficiency measures for net-zero energy building in Morocco. *J. Affect. Disord.* **2021**, *204*, 108141. [CrossRef]
18. Khani, A.; Khakzand, M.; Faizi, M. Multi-objective optimization for energy consumption, visual and thermal comfort performance of educational building (case study: Qeshm Island, Iran). *Sustain. Energy Technol. Assessments* **2022**, *54*, 102872. [CrossRef]
19. Chung-Camargo, K.; Bencid, M.; Mora, D.; Chen-Austin, M.A. Low-consumption techniques in tropical climates for energy and water savings in buildings: A review on experimental studies. *Technol. RD J.* **2022**, *18*, 5–18. [CrossRef]
20. De Luca, G.; Ballarini, I.; Lorenzati, A.; Corrado, V. Renovation of a social house into a NZEB: Use of renewable energy sources and economic implications. *Renew. Energy* **2020**, *159*, 356–370. [CrossRef]

21. Izadi, A.; Shahafve, M.; Ahmadi, P. Neural network genetic algorithm optimization of a transient hybrid renewable energy system with solar/wind and hydrogen storage system for zero energy buildings at various climate conditions. *Energy Convers. Manag.* **2022**, *260*, 115593. [CrossRef]
22. Vujnović, N.; Dović, D. Cost-optimal energy performance calculations of a new nZEB hotel building using dynamic simulations and optimization algorithms. *J. Build. Eng.* **2020**, *39*, 102272. [CrossRef]
23. Acar, U.; Kaska, O. Energy and economical optimal of Nzeb design under different climate conditions of Türkiye. *J. Build. Eng.* **2022**, *60*, 105103. [CrossRef]
24. Liu, Z.; Fan, G.; Sun, D.; Wu, D.; Guo, J.; Zhang, S.; Yang, X.; Lin, X.; Ai, L. A novel distributed energy system combining hybrid energy storage and a multi-objective optimization method for nearly zero-energy communities and buildings. *Energy* **2022**, *239*, 122577. [CrossRef]
25. Huang, P.; Sun, Y. A robust control of nZEBs for performance optimization at cluster level under demand prediction uncertainty. *Renew. Energy* **2019**, *134*, 215–227. [CrossRef]
26. Liu, J.; Zhou, Y.; Yang, H.; Wu, H. Net-zero energy management and optimization of commercial building sectors with hybrid renewable energy systems integrated with energy storage of pumped hydro and hydrogen taxis. *Appl. Energy* **2022**, *321*, 119312. [CrossRef]
27. Adeyemo, A.A.; Amusan, O.T. Modelling and multi-objective optimization of hybrid energy storage solution for photovoltaic powered off-grid net zero energy building. *J. Energy Storage* **2022**, *55*, 105273. [CrossRef]
28. Madathil, D.; Pandi, R.V.; Nair, M.G.; Jamasb, T.; Thakur, T. Consumer-focused solar-grid net zero energy buildings: A multi-objective weighted sum optimization and application for India. *Sustain. Prod. Consum.* **2021**, *27*, 2101–2111. [CrossRef]
29. Zhang, W.; Yan, C.; Xu, Y.; Fang, J.; Pan, Y. A critical review of the performance evaluation and optimization of grid interactions between zero-energy buildings and power grids. *Sustain. Cities Soc.* **2022**, *86*, 104123. [CrossRef]
30. Belussi, L.; Barozzi, B.; Bellazzi, A.; Danza, L.; Devitofrancesco, A.; Fanciulli, C.; Ghellere, M.; Guazzi, G.; Meroni, I.; Salamone, F.; et al. A review of performance of zero energy buildings and energy efficiency solutions. *J. Build. Eng.* **2019**, *25*, 100772. [CrossRef]
31. Ferrara, M.; Della Santa, F.; Bilardo, M.; De Gregorio, A.; Mastropietro, A.; Fugacci, U.; Vaccarino, F.; Fabrizio, E. Design optimization of renewable energy systems for NZEBs based on deep residual learning. *Renew. Energy* **2021**, *176*, 590–605. [CrossRef]
32. Wang, Z.; Tao, H.; Cai, W.; Duan, Y.; Wu, D.; Zhang, L. Study on the multitime scale rolling optimization operation of a near-zero energy building energy supply system. *Energy Convers. Manag.* **2022**, *270*, 116255. [CrossRef]
33. D'Agostino, D.; Mazzella, S.; Minelli, F.; Minichiello, F. Obtaining the NZEB target by using photovoltaic systems on the roof for multi-storey buildings. *Energy Build.* **2022**, *267*, 112147. [CrossRef]
34. Costa, J.F.W.; Amorim, C.N.D.; Silva, J.C.R. Retrofit guidelines towards the achievement of net zero energy buildings for office buildings in Brasilia. *J. Build. Eng.* **2020**, *32*, 101680. [CrossRef]
35. Al-Saadi, S.N.; Al-Jabri, K.S. Optimization of envelope design for housing in hot climates using a genetic algorithm (GA) computational approach. *J. Build. Eng.* **2020**, *32*, 101712. [CrossRef]
36. Monzón-Chavarrías, M.; López-Mesa, B.; Resende, J.; Corvacho, H. The nZEB concept and its requirements for residential buildings renovation in Southern Europe: The case of multi-family buildings from 1961 to 1980 in Portugal and Spain. *J. Build. Eng.* **2021**, *34*, 101918. [CrossRef]
37. Cerezo-Narváez, A.; Piñero-Vilela, J.-M.; Rodríguez-Jara, E.; Otero-Mateo, M.; Pastor-Fernández, A.; Ballesteros-Pérez, P. Energy, emissions and economic impact of the new nZEB regulatory framework on residential buildings renovation: Case study in southern Spain. *J. Build. Eng.* **2021**, *42*, 103054. [CrossRef]
38. Austin, M.C.; Chung-Camargo, K.; Mora, D. Review of Zero Energy Building Concept-Definition and Developments in Latin America: A Framework Definition for Application in Panama. *Energies* **2021**, *14*, 5647. [CrossRef]
39. Kitagawa, H.; Asawa, T.; Kubota, T.; Trihamdani, A.R.; Sakurada, K.; Mori, H. Optimization of window design for ventilative cooling with radiant floor cooling systems in the hot and humid climate of Indonesia. *J. Affect. Disord.* **2021**, *188*, 107483. [CrossRef]
40. Berardi, U.; Manca, M.; Casaldaliga, P.; Pich-Aguilera, F. From high-energy demands to nZEB: The retrofit of a school in Catalonia, Spain. *Energy Procedia* **2017**, *140*, 141–150. [CrossRef]
41. Ohene, E.; Hsu, S.-C.; Chan, A.P. Feasibility and retrofit guidelines towards net-zero energy buildings in tropical climates: A case of Ghana. *Energy Build.* **2022**, *269*, 112252. [CrossRef]
42. Hamburg, A.; Kuusk, K.; Mikola, A.; Kalamees, T. Realisation of energy performance targets of an old apartment building renovated to nZEB. *Energy* **2020**, *194*, 116874. [CrossRef]
43. Wu, X.; Feng, Z.; Chen, H.; Qin, Y.; Zheng, S.; Wang, L.; Liu, Y.; Skibniewski, M.J. Intelligent optimization framework of near zero energy consumption building performance based on a hybrid machine learning algorithm. *Renew. Sustain. Energy Rev.* **2022**, *167*, 112703. [CrossRef]
44. Zhu, N.; Liu, X.; Dong, Q.; Rodriguez, D. Optimization of zero-energy building by multi-criteria optimization method: A case study. *J. Build. Eng.* **2021**, *44*, 102969. [CrossRef]
45. Delač, B.; Pavković, B.; Lenić, K.; Mađerić, D. Integrated optimization of the building envelope and the HVAC system in nZEB refurbishment. *Appl. Therm. Eng.* **2022**, *211*, 118442. [CrossRef]

46. Leman, A.M.; Sharif, H.Z.; Krizou, A.N.; A Al-Tarawne, M.T.; Subbiah, M.; Al Farsi, M.; Irfan, A.R. Single Objective Optimization of Net Zero Energy Residential Building in Tropical Climate. *IOP Conf. Ser. Mater. Sci. Eng.* **2020**, *864*, 012015. [CrossRef]
47. Huo, H.; Xu, W.; Li, A.; Lv, Y.; Liu, C. Analysis and optimization of external venetian blind shading for nearly zero-energy buildings in different climate regions of China. *Sol. Energy* **2021**, *223*, 54–71. [CrossRef]
48. Ascione, F.; De Masi, R.F.; Festa, V.; Mauro, G.M.; Vanoli, G.P. Optimizing space cooling of a nearly zero energy building via model predictive control: Energy cost vs comfort. *Energy Build.* **2023**, *278*, 112664. [CrossRef]
49. Usman, M.; Ali, M.; Rashid, T.U.; Ali, H.M.; Frey, G. Towards zero energy solar households—A model-based simulation and optimization analysis for a humid subtropical climate. *Sustain. Energy Technol. Assessments* **2021**, *48*, 101574. [CrossRef]
50. Li, H.; Wang, S. Coordinated optimal design of zero/low energy buildings and their energy systems based on multi-stage design optimization. *Energy* **2019**, *189*, 116202. [CrossRef]
51. Harkouss, F.; Fardoun, F.; Biwole, P.H. Passive design optimization of low energy buildings in different climates. *Energy* **2018**, *165*, 591–613. [CrossRef]
52. *ASHRAE Standard 55-2004*; Thermal Environmental Conditions for Human Occupancy. American Society of Heating, Refrigerating and Air Conditioning Engineers: Atlanta, GA, USA, 2004.
53. Peel, M.C.; Finlayson, B.L.; McMahon, T.A. Updated world map of the Köppen-Geiger climate classification. *Hydrol. Earth Syst. Sci.* **2007**, *11*, 1633–1644. [CrossRef]
54. González, J.; Mora, D.; Austin, M.C. Energy Consumption Difference Found between Typical and Standard Occupancy in Residential Buildings in a Tropical Developing Country. *Buildings* **2023**, *13*, 2235. [CrossRef]
55. Republic of Panama Ministry of Public Works Technical Board of Engineering and Architecture. *Resolución de La JTIA No.002 de 13 de Enero de 2023*; Republic of Panama Ministry of Public Works Technical Board of Engineering and Architecture: Panama City, Panama, 2023.
56. Austin, M.C.; Arnedo, L.; Yuil, O.; Mora, D. Energy consumption influenced by occupant behavior: A study in residential buildings in Panama. In Proceedings of the 19th LACCEI International Multi-Conference for Engineering, Education and Technology: "Prospective and Trends in Technology and Skills for Sustainable Social Development. Leveraging Emerging Technologies to Construct the Future", Virtual, 19–23 July 2021; Volume 1. [CrossRef]
57. ASHRAE—American Society of Heating, Refrigerating and Air-Conditioning Engineers. Measurement of Energy and Demand and Savings. In *ASHRAE Guideline 14 2002*; ASHRAE: Atlanta, GA, USA, 2002; ISSN 1049-894X.
58. Boutet, M.; Hernández, A.; Jacobo, G. Thermo—Lighting optimization proposal for school buildings in subtropical hot—Humid climates: Monitoring and computer simulation on autumn period. *Energy Build.* **2016**, *128*, 785–797. [CrossRef]
59. Chen, X.; Yang, H. Sensitivity analysis and optimization of a typical passively designed residential building with hybrid ventilation in hot and humid climates. *Energy Procedia* **2017**, *142*, 1781–1786. [CrossRef]
60. Rabani, M.; Madessa, H.B.; Nord, N. Achieving zero-energy building performance with thermal and visual comfort enhancement through optimization of fenestration, envelope, shading device, and energy supply system. *Sustain. Energy Technol. Assessments* **2021**, *44*, 101020. [CrossRef]
61. Giouri, E.D.; Tenpierik, M.; Turrin, M. Zero energy potential of a high-rise office building in a Mediterranean climate: Using multi-objective optimization to understand the impact of design decisions towards zero-energy high-rise buildings. *Energy Build.* **2020**, *209*, 109666. [CrossRef]
62. Chacón, L.; Austin, M.C.; Castaño, C. A Multiobjective Optimization Approach for Retrofitting Decision-Making towards Achieving Net-Zero Energy Districts: A Numerical Case Study in a Tropical Climate. *Smart Cities* **2022**, *5*, 405–432. [CrossRef]
63. Chen, R.; Tsay, Y.-S.; Ni, S. An integrated framework for multi-objective optimization of building performance: Carbon emissions, thermal comfort, and global cost. *J. Clean. Prod.* **2022**, *359*, 131978. [CrossRef]
64. Chen, X.; Yang, H.; Zhang, W. Simulation-based approach to optimize passively designed buildings: A case study on a typical architectural form in hot and humid climates. *Renew. Sustain. Energy Rev.* **2018**, *82*, 1712–1725. [CrossRef]
65. Gou, S.; Nik, V.M.; Scartezzini, J.-L.; Zhao, Q.; Li, Z. Passive design optimization of newly-built residential buildings in Shanghai for improving indoor thermal comfort while reducing building energy demand. *Energy Build.* **2018**, *169*, 484–506. [CrossRef]
66. Hwang, R.-L.; Liao, W.-J.; Chen, W.-A. Optimization of energy use and academic performance for educational environments in hot-humid climates. *J. Affect. Disord.* **2022**, *222*, 109434. [CrossRef]
67. Chai, J.; Huang, P.; Sun, Y. Differential evolution—Based system design optimization for net zero energy buildings under climate change. *Sustain. Cities Soc.* **2020**, *55*, 102037. [CrossRef]

buildings

MDPI

Article

A Reference Framework for Zero Energy Districts in Panama Based on Energy Performance Simulations and Bioclimatic Design Methodology

Lissette De León [1], Dafni Mora [1,2,3], Cristina Carpino [4], Natale Arcuri [4] and Miguel Chen Austin [1,2,3,*]

1 Research Group Energy and Comfort in Bioclimatic Buildings (ECEB), Faculty of Mechanical Engineering, Universidad Tecnológica de Panamá, Panama City 0819-07289, Panama
2 Centro de Estudios Multidisciplinarios en Ciencias, Ingeniería y Tecnología (CEMCIT-AIP), Panama City 0819-07289, Panama
3 Sistema Nacional de Investigación (SNI), Clayton 0816-02852, Panama
4 Department of Mechanical, Energy and Management Engineering, University of Calabria, V. P. Bucci 46/C, 97036 Arcavacata di Rende, CS, Italy
* Correspondence: miguel.chen@utp.ac.pa

Abstract: The need for efficient buildings to contribute to sustainable development has led to the proposal of goals and regulations in different countries linked to net zero energy objectives, following approaches specific to their regions. Such regulations are not yet developed in a developing country such as Panama. Thus, this study develops the first framework for Zero Energy Districts (ZED) in Panama based on passive and active solutions through dynamic simulation. For this, an existing urbanization area consisting of 34 residential buildings, called the existing case (EC), was studied. After undergoing a design based on the bioclimatic methodology, a redesigned case (RC) is proposed to maintain thermal comfort and reduce energy consumption. Key redesign factors are the buildings' shape, orientation, glazing, the use of shadows, material of the walls, and the air conditioning configuration. Results showed energy consumption decreases by 37.5% when considering all buildings with natural ventilation, reaching ranges of positive net energy (+356.50 kWh/m^{2}y). In the case of nearly ZED in Panama, the ranges could be considered between 14 and 180 kWh/m^{2}y. This balance indicates that there is potential on site to develop zero energy districts.

Keywords: bioclimatic; simulation; tropical climate; ZED; zero energy building; zero energy district

1. Introduction

Energy efficiency in buildings, globally, has become very important due to the contribution of the building sector to climate change, its high energy consumption, and carbon dioxide emissions. The current situation increases the pressure to adopt the measures set out in the Paris Agreement, which aims to limit greenhouse gas emissions by keeping the global average temperature increase below 1.5 °C. Improving the performance of buildings contributes to achieving the Sustainable Development Goals (SDG 3—good health and well-being; SDG 7—affordable and clean energy; SDG 11—sustainable cities and communities; SDG 12—responsible consumption and production; and SDG 13—climate action) [1].

In Panama, approximately 70% of electricity consumption is concentrated in the capital city, linking the effects with urban environments. Based on this and other energy problems, Panama's National Energy Secretariat has designed the National Energy Plan 2015–2050 (PEN), having among its axes the improvement of urban environments built in a disorderly and inefficient manner [2]. Likewise, in 2019, the Sustainable Buildings Regulation (RES) of the Technical Council of Engineering and Architecture implied the mandatory application of mechanisms that result in savings percentages between 15% and 20% in various types of buildings through compliance paths with passive and active strategies [3]. Moreover, in Chapter IV of Law 69 of 2012 (UREE), minimum energy efficiency indices have been established to distribute equipment in the country, such as air conditioning units, refrigerators, and luminaires [4].

These initiatives have given rise to research of the application of passive and active solutions to improve energy efficiency in the country; however, they can have greater ambitions to achieve high energy efficiency buildings, such as a Nearly Zero Energy Buildings (nZEBs). For example, in Europe, the Energy Performance of Buildings Directive (EPBD) makes it mandatory for the European Member States to comply with nZEB specifications from 2018 with respect to public use, and by the end of 2020, for all buildings, including those for private use [5]. The definition of nZEB is based on a building that requires a meager amount of energy to operate and uses renewable sources of generation, to the greatest extent possible, on site. The most used indicator for classifying buildings is the primary energy in kWh/m^{2}y. However, it is common to use parameters such as thermal transmittance values (U-value) of the envelope, net and final energy for heating and cooling, or CO_2 emissions. Each country in the European Union is responsible for setting its limits (considering climatic conditions, level of ambition, calculation methods, and building traditions). For residential buildings, the primary energy ranges from 20 kWh/m^{2}y to 180 kWh/m^{2}y [6].

Zero Energy Building (ZEB) philosophy has been extrapolated to the community and district level, even though zero-consumption technologies at the individual building scale promise favorable results. By considering the mutual influence between the urban context and buildings, it becomes possible to achieve better accuracy in evaluating energy performance based on the same principles and goals set at the individual level [7,8]. For example, the Green Building Council (GBC) implemented evaluations at the urban level with the LEED for Cities and Communities Certification (v4.1), focused on the following categories: natural systems and ecology, transportation and land use, water efficiency, materials and resources, and quality of life [9]. In Panama, the City of Knowledge obtained a pre-certification for communities, that being the first pre-certified pilot project in this category in Latin America.

Along these lines, the Energy in Buildings and Communities Program of the International Energy Agency (IEA-EBC) and the Solar Heating and Cooling Program (SHC) have also involved among their long-term projects the development of a common methodology for the study of net-zero energy districts. With the SHC Task 40/EBC Annex 52 project [10,11], progress was made in documenting modeling strategies and necessary technologies. IEA-EBC is also developing Annex 83, with the main objective of creating a shared definition of positive energy districts (PEDs) and providing guidance on the solutions needed for the implementation of this type of urbanization [12].

The nZEB concept has been adapted to define a Nearly Zero Energy District (nZED) and a Net Zero Energy District (NZED) as a bounded part of a very energy efficient city with a quantity of energy generated from renewable sources produced on site or nearby [7,13] that includes an iteration between buildings, residents, streets, sidewalks, and vegetation. Zero energy districts are not limited to individual buildings, but comprise several buildings and optimize energy efficiency, thermal energy, and renewable energy generation between those buildings to offset demand [14,15].

From the above and considering the momentum of the energy transition in Panama, this research seeks to have a reference framework for Zero Energy Districts (ZEDs) in the country based on energy performance simulations and bioclimatic design methodology. The study applied bioclimatic strategies and the multi-objective optimization of active and passive strategies through dynamic simulation, evaluating the reductions in energy consumption of the urbanization. In addition, given the lack of well-defined criteria regarding the concept of "zero energy" at the scale of urbanization, the bioclimatic methodology applied is systematized to serve as a basis for future research in the field of study in climatic regions similar to Panama. Moreover, with the results obtained, it is expected to contribute to future local energy plans to contemplate the scope of ZED in their goals.

2. Literature Review: Key Criteria for the Concept of Zero Energy Developments

The philosophy to achieve zero energy constructions is fundamentally based on first increasing energy efficiency and then applying generation techniques based on renewable energies. A literature review was conducted on studies focused on using passive and active strategies to reduce energy use both at the building scale and at the urban or district scale, at the residential level. In addition, research focused on describing the key criteria for the concept of zero energy buildings and developments were reviewed, including the most widely used renewable energy generation technologies for these purposes. The period between 2015 and 2022 was used as a search reference, using Google Scholar as a search engine, finding through these documents from different databases such as Scopus, Web of Science, and Springer Link. The main search terms used were: "ZED," "nZED," "zero energy building," "zero energy district," and "positive energy district."

Despite not having a specific methodology and common definition, it is known that to achieve zero or nearly zero energy environments, bioclimatic architecture, energy efficiency, and Internet of Things guidelines must be implemented [16]. "Zero energy" seeks, in a first approach, to reduce energy use. This can be reached by applying passive and active strategies, that is, implementing bioclimatic architecture and using equipment with higher energy efficiency rates. In Panama, there is a zero-energy house whose design sought to reduce the annual load by cooling through a lightweight envelope, taking advantage of natural ventilation, using high-efficiency glazing, controlled shading, natural lighting, and dehumidification since the climatic conditions of the site lead to dependence on mechanical cooling [17]. Other research in different climates has been based on dynamic simulations to study the application of strategies for energy reduction by mechanical cooling; for example, in [18], a saving of 32% was found by combining a bioclimatic scenario that involves the use of ecological materials with a high-performance HVAC system. In [19], the impact of cooling on energy consumption is evident by demonstrating how reductions are achieved by increasing the cooling temperature by one degree (26 °C to 27 °C) with savings between 23% and 35% for both a traditional home and nZEB. At the district level, projects have also been found that use dynamic simulation, such as the REMOURBAN project, in the city of Valladolid, Spain, which involved modifications to the façades and ceilings, renovation of the lighting system using LED technology, integral renovation of a thermal power plant, using two biomass boilers, rehabilitation of the distribution network, and the integration of sanitary water heating in a district network; with active and passive strategies, consumption is reduced by 45.4% [20]. Another perspective of interest was evaluated in [21], where a microclimate analysis was performed to evaluate the results of a Near-Zero Energy Project Building in the Settlement (nZES), adjusting scenarios with increased percentages of vegetation and solar reflectance of roofs and pavements, which implemented technologies such as extruded polystyrene insulation, low-e double-glazed windows with PVC frames, an LED lighting system, a high-efficiency air-to-water heat pump in the HVAC system for heating and cooling, and mechanical ventilation with heat recovery. The results indicated that in addition to mitigating outdoor air temperatures up to 1.5 °C comparing the scenarios, the difference in energy consumption between the traditional building and the low-consumption building is 70%.

Cabeza and Chàfer [16] describe the importance of bioclimatic architecture in increasing interior comfort and reducing heating, cooling, lighting, and ventilation requirements based on strategies such as using windows, ideal glazing, passive shading, and vegetation on roofs and balconies. The study also emphasizes that Building Integrated Photovoltaics (BIPV) is a key solution for producing clean electricity and reducing material and labor costs, compared to traditional panels (BAPV). On the other hand, in Colombia (tropical climate), the characteristics and definitions of Net Zero Energy Building (NZEB) in a house were studied, basing interest on high-efficiency appliances due to their importance and level of use in tropical regions, which suggests greater relevance to active systems, since the influence of thermal levels on the energy consumption of electrical appliances used in the home has a great impact on the increase or decrease in demand; it also indicates that in terms of energy production, solar energy (photovoltaic and thermal) is the most convenient option compared to wind energy or biomass [22]. In Quito, Ecuador, a roadmap was developed to highlight the possibility of achieving NZEB, taking technical and policy aspects as criteria. The roadmap targets at different stages the following: in the short term, a shared vision, indoor air quality research, vernacular architecture research, and prototyping; in the medium term, low-tech industrial infrastructure, modeling tools, standardization, and NZEB policies; finally, in the long term, it focuses on the development of databases, high-tech industrial infrastructure, and smart and decarbonized grids. Before proposing the roadmap, the study compared four typologies of houses with two different types each: traditional construction and construction with insulation system with double leaf façade, 4 cm insulation, double glazing, natural ventilation, and PV panels per roof area, using EnergyPlus and Python, obtaining that with these changes it was possible to have houses independent of mechanical heating or cooling and to achieve a zero or positive balance [23]. As for Net Zero Energy Buildings (NZEDs), no compliance routes were found, but the proposed stratification in [7] regarding the analysis of energy demand provides a starting point and divides the balance of the district into the following determinants:

- Buildings (passive design, active systems, urban climate, and urban morphology).
- Public spaces (lighting, infrastructure, landscape, and public uses).
- Energy production at the district scale (consumption patterns, energy production, and distribution), as shown in Figure 1.

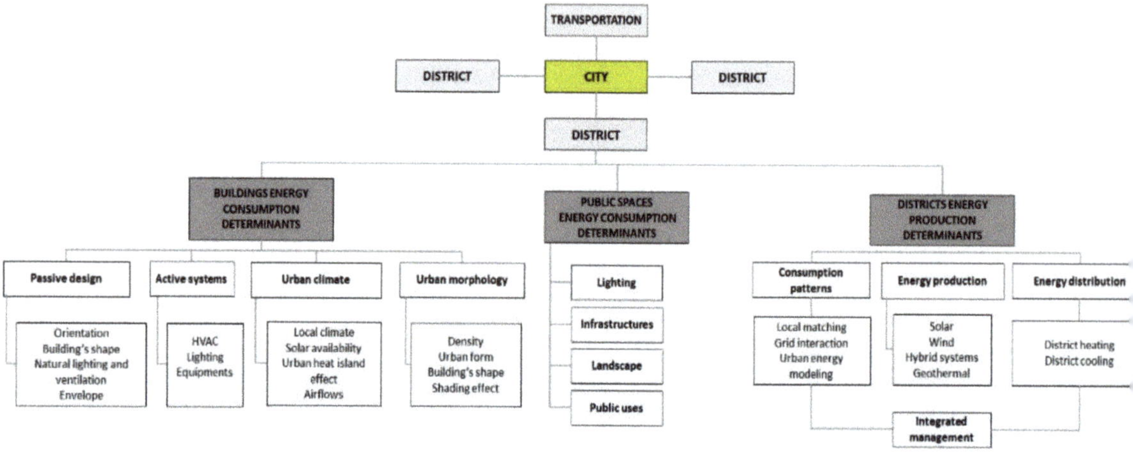

Figure 1. Energy determinants implied in each scale, focusing on the district balance towards an NZED. Reproduced with permission from [7], Elsevier, 2018.

To quantify the energy savings, in [24], by means of dynamic simulation in TRYNSYS software, a single-family house built in the 1970s was replaced with two houses of equal area on the same lot. Both active and passive strategies were used, and the houses were equipped with BIPV systems on the roof. Overall, the two houses resulted in 30% less energy consumption. On an annual basis, 44% of the consumption can be generated with the PV integration system, demonstrating that there is also an opportunity to retrofit aging buildings and convert existing developments into PED. In a case study [25] composed of 29 residential buildings with low energy performance, both envelope and systems, strategies were implemented to achieve nZED and ZED objectives, including roof-wide PV systems, reversible and efficient heat pumps, renovation of the building envelope and energy storage systems such as batteries and compressed air, sharing energy between buildings. EnergyPlus and MATLAB® found that energy sharing enables a more significant reduction in the environmental footprint of buildings compared to standard retrofit approaches.

Another important aspect is the analysis of occupant behavior at the building and district levels, taking advantage of valuable tools for capturing data from urban dynamics, such as sensors, the Internet of Things, and data intelligence. The use of physics-based energy models of urban buildings to simulate heat fluxes in buildings and thus estimate energy use is indicated. The insufficient consideration of the behavior of the occupants is one of the leading causes of the discrepancy between the actual energy consumption and that obtained through simulations because these parameters have high sensitivity concerning energy consumption, one of the most influential being the adjustment temperature of equipment for cooling and space heating [26,27].

The factors found in the literature have been shown to have an important influence both in obtaining decreases in energy consumption and achieving zero energy balance and goals. Table 1 summarizes the information described from the selected investigations. Based on this bibliography, a map was made in VOSViewer software v1.6.18.0 to illustrate the co-occurrences of terms in the title and abstract from the text data [28]. The complete count method was used, and due to the low number of references, a minimum number of two-term occurrences was selected. Figure 2 illustrates the results by the average publication year, where the most relevant terms are "data," "house," "climate," "nzeb," and "simulation," which are repeated more than seven times. It is also noted that the most recent research has a focus more closely linked to ZED, as terms such as "community" and "energy storage system" are found.

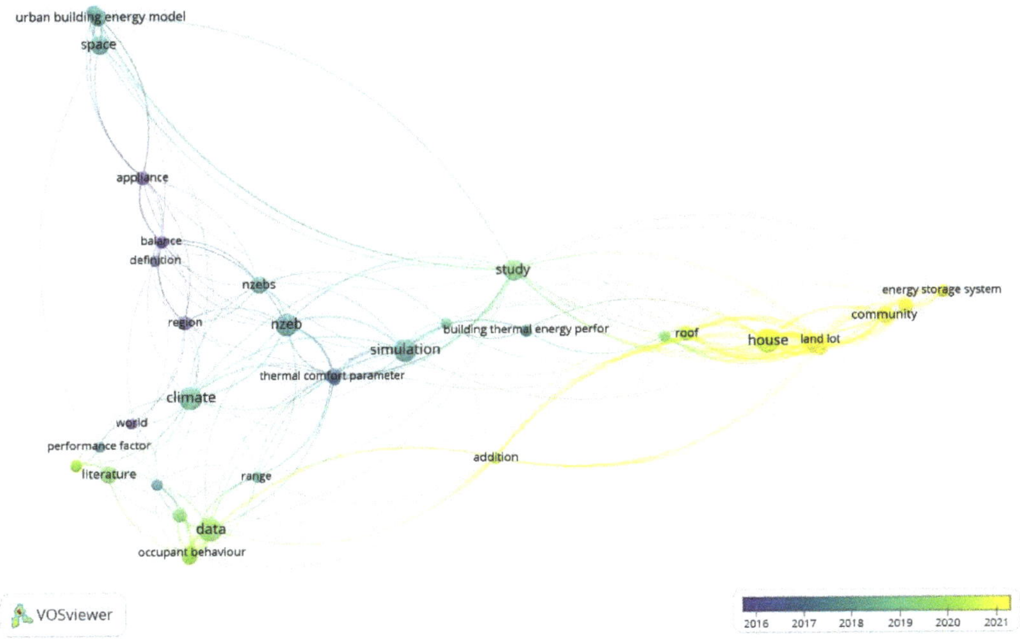

Figure 2. Overlay visualization by the average publication year of main terms in the literature review. Own elaboration in VOSViewer software version 1.6.18.0.

Table 1. Summary of key criteria for the concept of buildings and urbanizations at zero energy, based on passive and active strategies.

Place/Climate	Target	Type of Techniques	Techniques Used	Solution Detail	Type of Study	Year/Ref.
Azuero, Panama Humid Tropical	NZEB	Passive	(1) Envelope (2) Overhangs (3) Natural Ventilation (4) Dehumidification (5) Natural lighting (6) Trees and shrubs	(1) Strategic window wall ratio, low U-values, thermal mass for heat absorption (2) Protection against sunlight at specific times (3) Openings for cross ventilation (4) Cold water immersion system (5) Deep width that allows sunlight to enter (6) Windbreak belt in the East	Implementation	2015 [17]
(Spain, Italy, France), Europe Templade	nZEB	Active	(1) Heating and cooling setting temperature	(1) Temperature variation from 25 °C to 27 °C	Simulation (TRNSYS)	2017 [19]
Valladolid, Spain Mediterranean	nZED	Passive Active	(1) Envelope (2) Lighting System (3) Heating system	(1) Modification of the façade and roofs (2) Use of LED technology (3) Biomass-based boiler, district network for water heating	Simulation (DesignBuilder)	2017 [20]
Rimini, Italy Humid subtropical	nZES	Passive Active	(1) Increased vegetation (2) Increased solar reflectance (3) Envelope (4) Artificial lighting (5) HVAC System (6) Mechanical ventilation	(1) The percentage of trees is increased (2) Increased reflectance in roof and pavement (3) Extruded polystyrene insulation (4) LED technology (5) High efficiency air-water heat pump (6) Heat recovery	Simulation (ENVI-met, EnergyPlus)	2018 [21]
Not applicable	ZEB	Passive Active	(1) Bioclimatic architecture (2) Generation systems (3) Dynamic simulation	(1) Installation, passive shading, balconies and vegetation to increase thermal comfort and reduce demand for cooling, heating, lighting and ventilation (2) BIPV is an ideal solution for generation in buildings at zero energy, since in addition to producing energy it reduces material costs (3) Analyzing multiple scenarios requires dynamic simulation, multi-objective optimization and genetic algorithms	Literature Review	2020 [16]
Colombia Tropical	NZEB	Passive Active	(1) Energy efficiency (2) Envelope (3) Shading (4) Orientation (5) Natural ventilation (6) Natural lighting (7) Generation systems	(1) Use of high efficiency appliances (2) Thermal insulation in walls, floors, windows and ceilings, implementation of green roofs (3) Eaves and blinds (4) North-South or East-West orientation, depending on the climate (5) Cross ventilation (6) Use of solar tubes, skylights and low-growing trees (7) Photovoltaic and thermal panels, better options than the use of wind energy or biomass	Numerical calculation	2015 [22]

Table 1. *Cont.*

Place/Climate	Target	Type of Techniques	Techniques Used	Solution Detail	Type of Study	Year/Ref.
Quito, Ecuador Equatorial and Andean	NZEB	Passive	(1) Envelope (2) Overhangs (3) Natural ventilation	(1) Double-leaf façade, 4 cm insulation, double glazed (2) Overhangs on the north façade from 0.15 cm to 0.30 cm	Simulation (EnergyPlus and Python)	2019 [23]
Not applicable	nZED	Passive Active	(1) Energy consumption in buildings (2) Energy consumption in public spaces (3) Energy production in districts	(1) Passive design, active systems, urban climate, urban morphology (2) Public lighting, infrastructure, landscape, public use (3) Consumption patterns, energy production (solar, wind, hybrid systems and geothermal), energy distribution (district heating and cooling)	Literature Review	2018 [7]
Montreal, Canada Continental	PED	Passive Active	(1) Envelope (2) Glazing (3) HVAC System (4) Water heating (5) Shading (6) Mechanical ventilation (7) Photovolcanic generation	(1) Union of two houses on the same land, expanded polystyrene walls, wooden floors, plaster roof (2) Double with argon filling (3) Reversible heat pumps (4) CO_2 pump (5) Use of movable curtains (6) Energy recovery (7) BIPV System	Simulation (TRNSYS)	2021 [24]
Naples, Italy Mediterranean	nZED NZED	Passive Active	(1) Photovoltaic system (2) Energy storage (3) Envelope renewal	(1) PV on ceiling (2) Batteries and compressed air (3) Insulation of the outer envelope due to the high transmittance of both opaque envelopes	EnergyPlus MATLAB	2022 [25]
Not applicable	ZED	Passive Active	(1) Occupant behaviour	(1) Not considering the behaviour of occupants in energy models is one of the main causes of the discrepancy between actual consumption and that obtained through simulations	Literature Review	2017 [27], 2020 [26]

3. Materials and Methods

The research is summarized in the methodology in Figure 3. For an existing building and urbanization (hereinafter referred to as EC, "existing case"), the main characteristics of the case study are analyzed: climatic conditions, envelope composition, occupancy profiles, use of equipment, and energy consumption. Subsequently, a proposal is made to redesign the individual building and distribution on an urban scale (called RC from "redesigned case"), systematizing and applying a set of solutions found through a bioclimatic methodology and the application of parametric analysis. DesignBuilder software v6.1.6.011 [29] was used to realize dynamic simulations based on the EnergyPlus tool. The result obtained using passive and active strategies is analyzed. This allows for quantifying the energy balance and proposing a roadmap that results in the first ZED reference framework in Panama.

Figure 4 shows the location of EC, which is a residential urbanization in the district of Chitré, province of Herrera, in the Republic of Panama (7°58′51″ N 80°26′31″ W), under the climatic classification Awi (tropical savannah with low thermal oscillation).

The urbanization is bounded by a lot of 1.13 hectares of extension and consists of 34 houses of 55 m^2 of surface each and similar architectural construction. Since it was constructed between 2016 and 2019, the design does not consider the criteria of the RES [3]. The individual building comprises one floor and is divided into different architectural areas (entrance hall, living room, two bedrooms, and kitchen). The characteristics of the construction systems are summarized in Table 2.

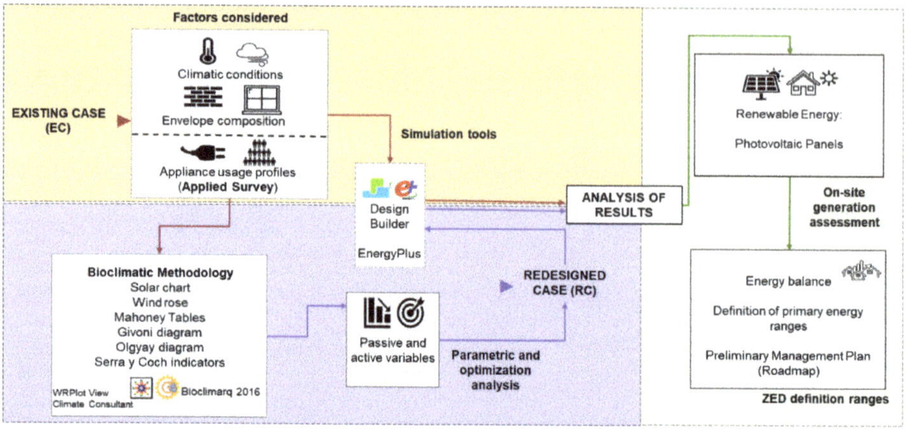

Figure 3. Methodology used for the research. Own elaboration.

Figure 4. Location of the residential urbanization under study (within the red polygon).

Table 2. Thermal properties of the enclosures that make up the building.

Envelope Elements	Layers	Thickness (m)	U-Value W/m^{2}K	Thermal Properties		
				Conductivity (W/mK)	Specific Heat (J/kgK)	Density (kg/m^3)
External and internal walls	Concrete block	0.1016	4.009	0.72	840.00	1860.00
	Mortar	0.0127		2.30	873.36	2321.40
Pitched roof	Zinc	0.00044	7.143	110	380.00	7200.00
Semi-exposed ceiling	Plaster	0.007	5.952	0.25	896.00	2800.00
Floor	Ceramic	0.006	2.890	1.30	840.00	2300.00
	Concrete	0.1		1.40	840.00	2100.00

The windows are 1.20 meters wide by 1.00 meter high and are simple glazing with a clear glass of 3 mm. The thermal properties of this glass are a U-value of 3.835 W/m^{2}K, a conductivity of 0.90 W/mK, a solar gain coefficient (SGHC) of 0.768, and a direct solar transmission (g-value) of 0.741.

Occupancy and energy use profiles were determined through a resident survey, a method widely used for its adaptability. The survey was divided into five parts considering [30]: (1) general characteristics of occupation and housing, (2) cooling mechanisms, (3) ventilation, (4) use of equipment, lighting, and appliances, and finally (5) the use of liquefied petroleum gas (LPG) for cooking. Table A1 in Appendix A shows in more detail the questions asked.

Although the survey was intended for all 34 residences, the response rate was 44%, equivalent to 15 houses. It is noteworthy that residents tend to use split air conditioning with temperatures between 16 °C and 25 °C. The ventilation schedules obtained per residence were applied to all areas except the bathroom and the portal, where 24/7 ventilation is considered. All respondents evidenced the use of fabric curtains for sun protection. In the gas section, it was obtained that cylinders of a capacity of 25 pounds and an average duration of one and a half months were used.

For residences that did not respond to the survey, a typical single-family occupancy of four inhabitants per dwelling was used [31]; for equipment usage profiles, standard schedules provided by DesignBuilder software for residential spaces were used. However, the use or not of air conditioning equipment was verified by on-site inspection, obtaining that 18 residences have this equipment. In addition, the use of energy by public lighting was evaluated, which consists of five high-pressure luminaires of 100 W each.

3.1. Thermo-Energy Evaluation by Parametric Analysis

The 3D model of EC was subjected to a parametric analysis in the DesignBuilder software, considering the data for March (the month with the highest temperatures according to climate data) to select adjustments in the design variables that allow a greater performance of the urbanization (temperature and schedule) of air conditioning use, orientation, and glass materials, walls, and roofs) and combine it with the results obtained when applying the bioclimatic methodology. The analysis evaluates multiple design options and their influence on the variables of interest, allowing for observation of the behavior of the different design combinations; in this way, the best options are detected and selected for subsequent application in RC.

The variables of interest selected for this study were electricity consumption by cooling (air conditioning) and hours of non-comfort with 80% acceptability (ASHRAE 55) [32]. Both temperature options and air conditioning usage schedules were selected based on survey results, while building materials were selected based on local market layout [33]. The design variables with the selected options to be evaluated are shown in Table 3.

Table 3. Variables of interest, design variables, and options evaluated in the parametric study.

Variables of Interest	Design Evaluation	
	Design Variables	Values/Options
Power consumption by cooling	Cooling temperature point	16 °C to 28 °C with intervals of 0.5 °C.
	Air conditioning schedule	Always on 24/7, 19:00–7:00, 13:00–21:00, 19:00–4:00, 18:00–24:00, 18.00–9:00, 18:00–10:00, 8:00–20:00, 19:00–24:00
	Type of glazing	Original Glazing; Sgl Clr 6 mm; Sgl Bronze 6 mm; Sgl Grey 6 m; Sgl LoE (e2 = 0.2) Clr 6 mm; Dbl Bronze (emissivity = 0.2) Clr 3 mm/6 mm air; Dbl Bronze (emissivity = 0.1) Clr 3 mm/6 mm air.
	Orientation	0° to 325° with intervals of 25°.
Power consumption by cooling and hours of discomfort with 80% acceptability	External walls	Original wall; super-insulated brick/block; domestic wood wall; uninsulated brick/block; super-insulated with low thermal mass; non-insulated—medium thermal mass; energy code standard—medium thermal mass; state of the art—medium thermal mass; typical reference—high thermal mass.
	Roof	Original pitched roof; clay tiles (25 mm); flat roof U = 0.25 W/m^{2}K; no insulation—medium thermal mass; no insulation—high thermal mass; state of the art—high thermal mass.

3.2. Systematization of Bioclimatic Methodology

Based on EC, the first re-design proposal for the building and its distribution (RC) considers a bioclimatic methodology and the application of the results obtained through a parametric analysis. Figure 5 presents the systematization of bioclimatic design. The input data and variables are visualized for each of the steps in the analysis: start, site location, climate analysis, sunlight analysis, wind analysis, façade analysis, bioclimatic strategy analysis, shading calculation, and design verification in terms of comfort and completion of the process. For the analysis, meteorological data provided by Solargis© (temperature, relative humidity (RH), radiation, wind, and precipitation) were used for the area specified in Figure 4.

Step 1. Site Location (Figure 4)

- Type: Residential
- Geographical coordinates: 7°58′51″ N 80°26′31″ W
- Height: 20 mbsl
- Orientation: 20°

Step 2. Climate analysis

- Minimum annual temperature: 20.6 °C
- Average annual temperature: 27.2 °C
- Maximum annual temperature: 35.6°C
- Minimum annual RH: 33%
- Average annual RH: 78%
- Maximum annual RH: 100%
- Average annual wind speed: 2.5 m/s
- Annual rainfall: 580 kg/m^2
- Average annual global radiation: 4.96 kWh/m^2day

Step 3. Sunlight analysis

- For Panama, the Sustainable Construction Guide for Energy Saving in Buildings suggests thermal comfort temperatures be obtained through the Szokolay equation [34]. The thermal comfort range is set from 23.5 °C to 28.5 °C. This range is represented in yellow on the sun chart. Temperatures below 23.5 °C are shown in light blue (cold), while temperatures above 28.5 °C are represented in orange (heat).

Step 4. Wind analysis

- WRPLOT View™ software [35] was used to produce the wind rose and a frequency distribution graph. The prevailing wind direction is north, with speeds greater than 11.10 m/s, and the most frequent speed range (37.6%) is 0.50 m/s to 2.10 m/s.

Step 5. Façade analysis

- Figure 6 represents the analysis of sunlight for the first months of the year (January to June). The solid lines represent the façades of the lot (A: 110°, B: 20°, C: 60°, D: 150° and E: 20°), and the dashed lines the azimuth for each of the façades (A′: 20°, B′: 110°, C′: 150°, D′: 240° and E′: 290°) and correspond to the angles formed by the horizontal projection of the solar ray in relation to the north. The season from July to December is also considered for the observations and recommendations.

From the evaluation of the solar chart and the wind rose, the following recommendations for the lot's façade are concluded: On façade A′, it is suggested to place windows to take advantage of ventilation and natural light in the morning hours. Use sun protection such as curtains and eaves. On façade B′, use windows to take advantage of natural light. On façade C′, it is convenient to use windows with eaves. It is not suitable to use windows on façade D′ and use thermal insulation on the walls. Finally, it is recommended to place windows on façade E′ if sun protection and thermal insulation are used on walls and glass.

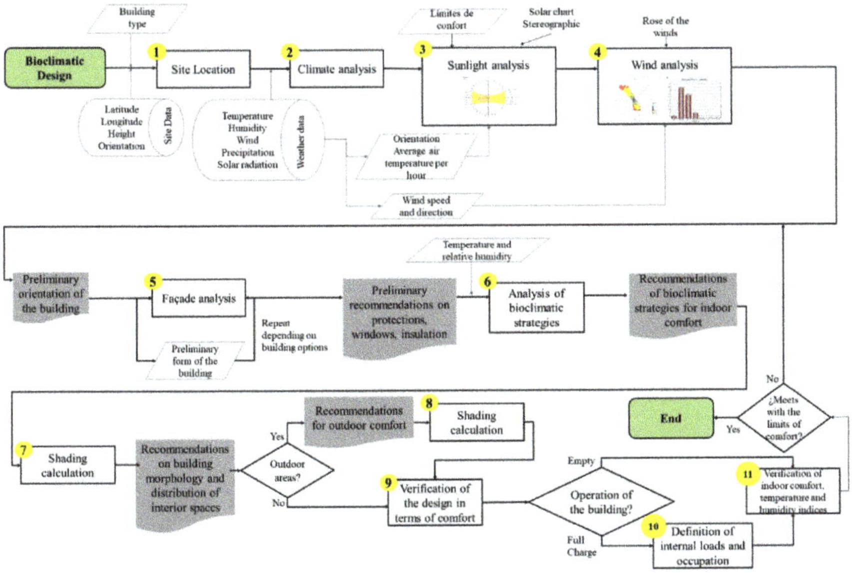

Figure 5. Flowchart of bioclimatic methodology. Own elaboration.

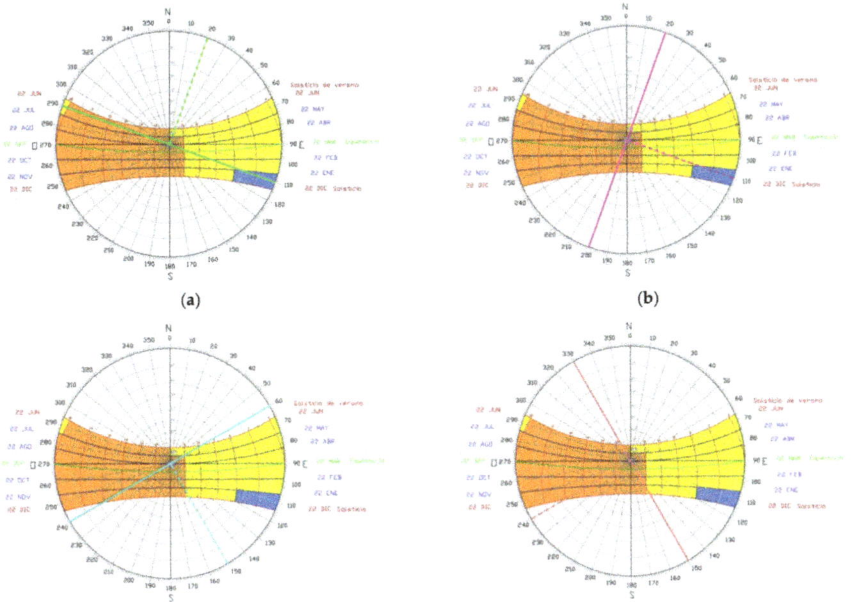

(a) (b) (c) (d)

Figure 6. *Cont.*

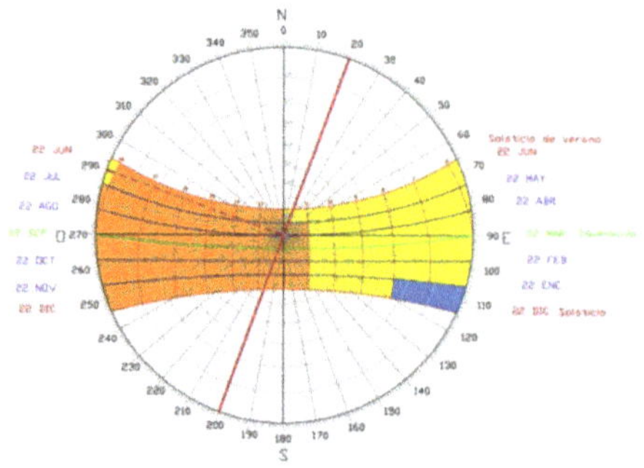

(e)

Figure 6. Façades (solid lines) and azimuth (dashed lines) of the lot represented in the solar chart for January to June: (**a**) Façade A, (**b**) Façade B, (**c**) Façade C, (**d**) Façade D, and (**e**) Façade E.

Step 6. Analysis of bioclimatic strategies

- Climate Consultant 6.0 software [36] was used to help users design more efficient buildings. The Givoni Psychrometric chart was extracted from this software, which calculates the best strategies based on the annual comfort hours that can be added to the design. The Bioclimarq 2016 spreadsheet developed by Gómez-Azpeitia was also used [37]. The input data of the spreadsheet were the location of the site, climate data (maximum, minimum, and average monthly and hourly temperatures and humidities, precipitation, radiation, and wind speed), the clothing habits of the inhabitants (very light, light, moderate, warm, and very warm), and the use of natural ventilation and artificial air conditioning. This tool extracted Olgyay's bioclimatic chart, Mahoney's recommendations, and Serra and Coch's skin form and treatment indicators.
- Givoni's chart proposes building design measures as a function of temperatures and humidity. The results indicate that the buildings under the climatic conditions studied only achieve 14 hours of comfort per year. To increase comfort hours, the software suggests using window sun protection (an additional 2379 h), dehumidification (an additional 3178 h), and cooling and dehumidification (an additional 5566 h). The set of strategies extracted is shown in Figure 7.

Step 7. Shading calculation

- Due to the results of the previous analysis in the solar chart and façades, the use of eaves on the south and west façades is considered for the redesign proposal RC. The calculation of the dimensions of the eaves was carried out considering an angle of solar height alpha (α) of 45° for both façades. Therefore, for the south façade, solar protection is achieved from 09:30 to 15:30, and for the west façade from 12:00 to 15:00 at the summer solstice and until 14:30 in the winter solstice. The dimensions of the eaves were calculated with the tangent trigonometric function, having as input data the alpha angle (α) and the dimensions of the windows. The windows have lengths of 0.75 m and 1.00 m, so when calculating the size of the eaves with an angle of 45°, results of 0.75 m and 1.00 m are obtained.

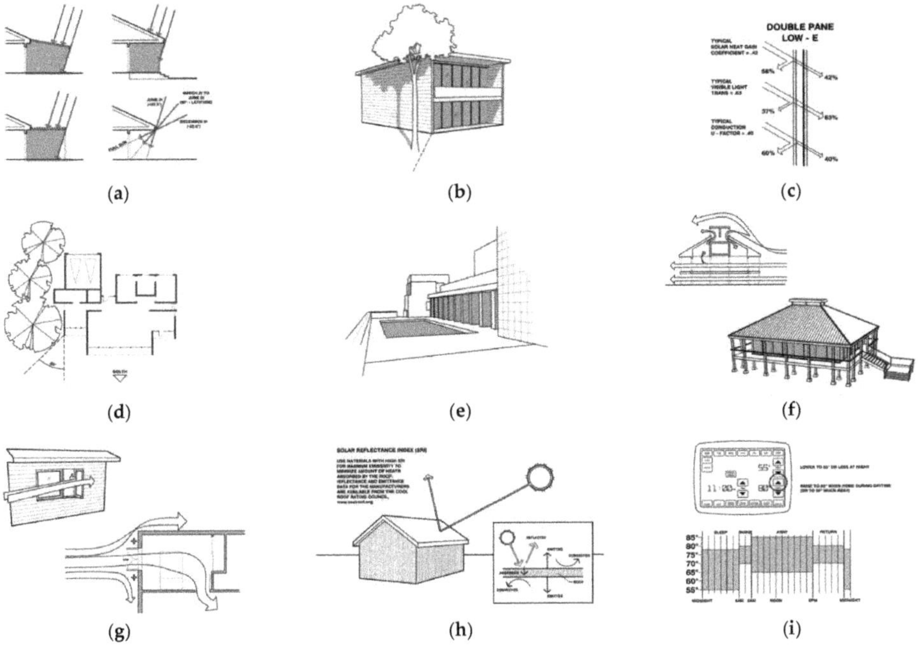

Figure 7. Recommended bioclimatic strategies: (**a**) Windows with operable overhangs or sunshades, (**b**) Minimization or elimination of glazing on west-facing façades, (**c**) High-performance glazing (insulated frames and low emissivity), (**d**) Use of plants and trees in the west orientation to minimize heat gain, (**e**) Orient most glazing to the north (using vertical fins) (**f**) Elevating the building above the ground to minimize humidity and maximize natural ventilation under the floor and roof with openings for ventilation (**g**) Shaded windows facing the prevailing breeze to promote natural ventilation, (**h**) Use of materials with light colors and ceilings with high emissivity (**i**) Increase in thermostat setting temperature.

3.2.1. Recommendations on Building Morphology

In order to select the shape of the building, the dimensions and characteristics of EC were first taken as the initial value. The indicators and coefficients of Serra and Coch were calculated from this; the adjustment of the shape and dimensions was iterated until an acceptable compactness value was reached to be considered in the proposed building for RC. Due to the high compactness of the EC buildings and the ventilation needs indicated, it was decided to iterate the compactness coefficient for the climatic incidence of this factor. Higher compactness leads to lower possibilities for radiation capture and energy loss but also complicates ventilation [38]. The results are shown in Tables 4 and 5 below:

Table 4. Main differences between the indicators of buildings EC and RC.

Building Indicators	Value	EP Value
Area of envelope surfaces (m^2)	182.67	1273.51
Volume (m^3)	135.71	1357.10
Area of surfaces that delimit patios (m^2)	150.88	1829.97
Height (m)	2.50	5.00
Sloping area (m^2)	54.28	488.52
Area of seated surfaces (m^2)	54.28	488.52

Table 4. *Cont.*

Building Indicators	Value	EP Value
Area of attached surfaces (m^2)	0.00	0.00
Mass of envelopes (kg)	25,933	1,766,616
Total area of the spans (m^2)	15.48	38.04
Glazed window area (m^2)	11.28	33.84
Average U factor (W/m^2K)	5.00	5.00
Façade fold area (m^2)	0.00	0
Average absorptivity	0.60	0.60
Variable elements	No	No

Table 5. Results obtained from the Serra and Coch coefficients for the EC and RC building.

Building Indicators	EC Value	Recommendation	RC Value	Action Required
Compactness	0.70	↓	0.47	None
Porosity	0.10	↑	0.13	Increase
Slenderness	0.52	=	0.37	None
Settlement	0.30	=	0.38	None
Annexation	0	=	0	None
Heaviness	142	=	139	None
Perforation	0.08	↑	0.03	Increase
Transparency	0.06	=	0.03	None
Insulation	0.03	↑	0	Increase
Smoothness	0	=	0	None
Color	0.6	↓	0.6	Clarify
Variability	No	=	No	None

3.2.2. Recommendations of Bioclimatic Strategies for Outdoor Comfort

The data distribution in the Olgyay diagram was analyzed using the Bioclimarq spreadsheet, in which different zones are distinguished: comfort, need for solar radiation, need for ventilation, need for humidification, and need for shading. The behavior of the temperatures and humidity of the study site indicates that it is necessary to favor shading and natural ventilation with winds of 0.5 m/s to 2.5 m/s in outdoor areas, considering the presence of free areas in the redesign.

3.3. *Inclusion of Energy Generation Solutions*

The reduced energy needs of buildings must be met with on-site renewable energy to achieve nearly zero energy targets. Once the proposed urbanization is obtained, the generation of electricity is evaluated. The use of PV technology was chosen for the following reasons: (1) The literature showed that using PV panels is the most widely employed energy solution to achieve zero or nearly zero energy objectives in residential buildings. (2) In Panama, there is a higher solar potential than wind in most countries [39]; the average radiation of the site is 4.96 kWh/m^2day. (3) Simplicity and precision in energy modeling in available tools and in future applications when studies are carried out at a constructive scale. In addition, this technology's availability in the local market facilitates its application.

DesignBuilder software allows the modeling of PV panels, considering the effect of shadows, reflections of solar radiation, and the type of performance. The analysis considered using BAPV in roofs and BIPV in the most exposed glazing.

3.4. *Nearly Zero Energy District and Energy Balance*

Case studies with zero energy targets require the definition of a boundary, determined by building systems, energy needs, the energy produced, system losses, and energy delivered and exported, as shown in the diagram in Figure 8. The dashed lines represent energy transfer within the boundary, and solid lines represent energy transfer in and out [40].

Another approach to establishing site boundaries is the definition of "district." From the description given in [41], it is established that a "district" refers to an urban block. For the present study, a district is an interrelated system that includes buildings, residents, streets, sidewalks, and vegetation in its parameters.

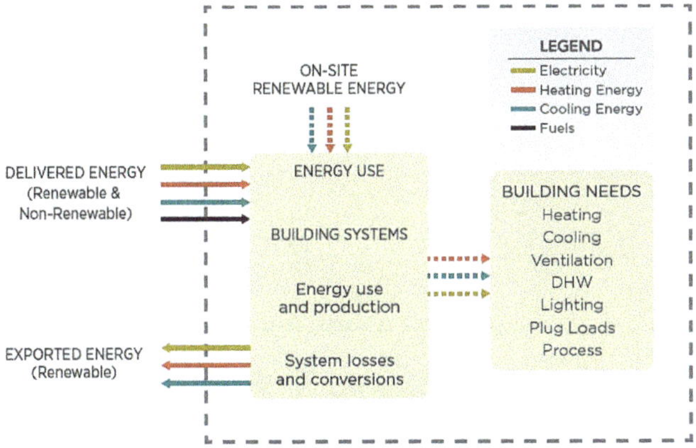

Figure 8. On-site energy transfer limit for zero energy balance [40]. Note: "The dashed lines represent energy transfer within the boundary, and the solid lines represent energy transfer entering/leaving the boundary used for zero energy accounting."

The indicator to be used for site classification is primary energy, which refers to the energy contained in natural materials or processes that have not undergone any transformation [40,42]. The conversion factors to obtain the values in terms of primary energy are obtained from ASHRAE 105 presented in Table 6, for each type of energy.

The energy balance (EB) is calculated from the energy delivered to the building (consumed) and exported (generated) for each type of energy source [40,44]. This indicator summarizes all power supplied and exported in a single value using Equation (1):

$$EB = \frac{\sum_i (E_{del,i} f_{del,i}) - \sum_i (E_{exp,i} f_{exp,i})}{A_{net}} \quad (1)$$

where $E_{del,i}$ is the energy delivered to the site by energy type "i" in kWh/m^{2}y; $E_{exp,i}$ is the energy exported from the site by energy type "i" in kWh/m^{2}y; $f_{del,i}$ is the factor to convert to primary energy the energy delivered for the type of energy "i"; $f_{exp,i}$ is the factor to convert to primary energy the energy exported for the type of energy "I"; and A_{net} is the net area used in m^2.

Table 6. Factors for conversion to primary energy. Data obtained from ASHRAE 105 [43].

Types of Energy	Conversion Factor for Primary Energy
Imported electricity	3.15
Exported renewable electricity	3.15
Natural Gas	1.09
Fuel oil (1,2,4,5,6, diesel, kerosene)	1.19
Propane and liquid propane	1.15
Steam	1.45
Hot water	1.35
Cold water	1.04
Coal or other	1.05

4. Results

4.1. Results of Parametric Analysis and Bioclimatic Methodology

Among the most relevant parametric analysis results, the optimal glazing option was Dbl Bronze (emissivity = 0.1) Clr 3 mm/6 mm air. The performance of air conditioning consumption with respect to walls and roofs was better when using super-insulated blocks/bricks and roofs with high thermal mass. However, contrary results were observed regarding the hours of non-comfort since it indicates that the best option is to use the original wall and super-insulated ceiling. The best-performing schedule turned out to be 18:00 to 9:00 only on weekdays at 23 °C. From this parametric study, a sensitivity analysis was performed. For electricity consumption for cooling, it was obtained that the buildings are more sensitive to the schedules of use compared to the selected temperature point and that, in the same way, it responds less to the types of glass and more to the orientation in which they are placed. Likewise, the results indicate that buildings have greater sensitivity to the types of walls than to the roofs and that even for the original roof, clay roofs (25 mm), and uninsulated roofs with intermediate and high thermal mass, there is no change in electricity consumption for all types of walls evaluated; the same behavior was obtained for the hours of non-comfort. In addition, for hours of non-comfort, the type of glazing and orientation were evaluated, but no relevance was found for either of these two variables.

Regarding the bioclimatic methodology, following Mahoney's recommendations and calculations made on the indicators and coefficients of Serra and Coch, the shape of the proposed building in RC was defined. It has an extended spatial and volumetric configuration, low compactness (acceptable), and low slenderness. Figure 9a shows the design of the proposed building through a rendered image. The proposed residential building consists of two floors: the first occupies three times the dimensions of the original building, while the second floor is made up of a closed surface identical to the original building and an open surface (balcony). In this way, an acceptable compactness coefficient is met, thus obtaining more natural ventilation possibilities and ease of heat dissipation. In addition, this configuration provides self-shading, which is usable in hot climates. The new building covers an area of 495 m^2, indicating that implementing these bioclimatic techniques demands greater use of land and construction materials.

The distribution of the windows is similar to that of the individual building, but most of the glazing was oriented with the prevailing wind direction (north) and minimized on the façades on the west side, thus maximizing the use of morning daylight and natural ventilation. Eaves were placed on the south- and west-facing windows. The new distribution of the 34 houses implied a greater need for land. A similar form of the land was considered, but with an extension of 10.2 hectares. All houses have a north–south orientation, an elongated east–west axis divided into three rows, and a separation of the houses of between 10 and 16 meters. A greater distance between the houses facilitates some passive design measures, such as natural ventilation and usable solar gain. The presence of trees on the west side was considered to minimize heat gains.

Because the bioclimatic methodology does not contemplate the specification of the materials to be used, these were selected from the parametric analysis, improving interior comfort conditions and electricity consumption. In addition, Dbl Bronze (emissivity = 0.1) Clr 3 mm/6 mm air with a U-value = 2.47 W/m^2K and super-insulated brick/block walls with U-value = 0.156 W/m^2K were chosen. Three-dimensional models and dynamic simulation were performed in DesignBuilder software for EC and RC. Figure 9b,c illustrate the axonometric view of both scenarios.

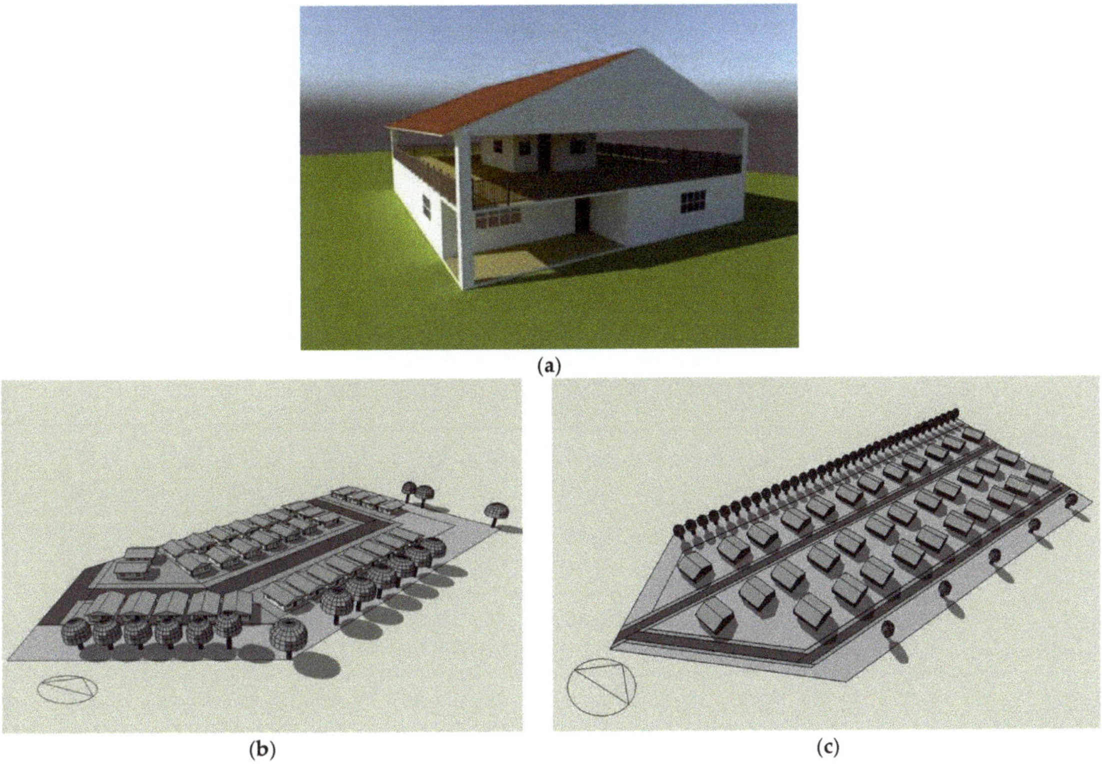

Figure 9. (**a**) Shape of the proposed building in SketchUp, (**b**) 3D model of the EC urbanization, and (**c**) 3D model of the RC urbanization in DesignBuilder software.

4.2. Verification of Thermal Comfort and Energy Consumption

As a final step of the bioclimatic methodology, the design's thermal performance and the energy consumption results at the urbanization level were verified.

4.2.1. Thermal Comfort

Figure 10 shows the average air temperature (AT), radiant temperature (RT), and operating temperature (OT) results inside the building for EC and RC. In general, temperatures are highest in March and April and lowest in November. For EC, OT ranges between 26.2 °C and 27.2 °C, presenting an annual average of 26.7 °C, while, for RC at full load, average operating temperatures of 23.2 °C were obtained. In this case, the internal load and occupancy of the individual building are the same applied to the existing urbanization, without considering the air conditioning equipment (natural ventilation).

Compared to EC, the RC obtained lower temperatures in all months, with an average reduction of 3.2 °C OT and reaching within the acceptable ranges for the site and even below (23.5 °C to 28.5 °C).

Other comfort indicators were also verified (Figure 11): percentage of relative humidity (%RH), predicted percentage of dissatisfaction (%PPD), and the value of predicted mean vote (PMV). %RH increased compared to EC (average from 86% to 94.6%). The %PPD also increased, reaching 66.4%, well above the value recommended by ASHRAE 55 (10%). On the other hand, PMV decreases, remaining in ranges between −1.10 and 0.14; that is, between slightly cool and neutral, according to the ASHRAE 55 wind chill scale between +3 and −3, although it is not recommended to use the PMV for non-mechanically conditioned areas.

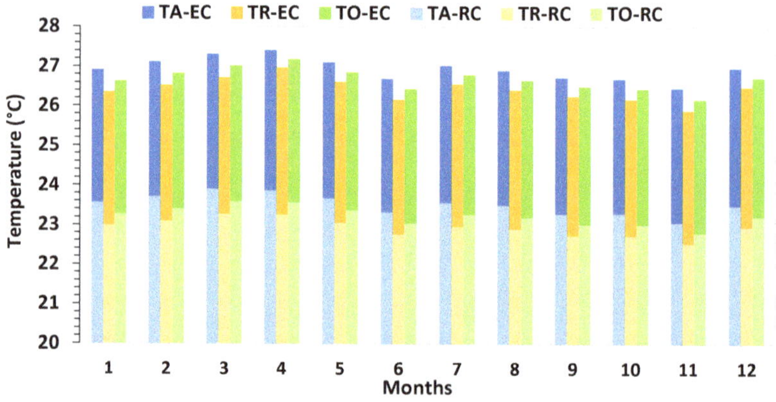

Figure 10. Monthly air temperature (AT), radiant temperature (RT), and operating temperature (OT) results for EC and RC.

In summary, with the application of the selected strategies, a significant decrease in indoor temperatures and the reach of more favorable ranges of the PMV is achieved without mechanical air conditioning. Still, a high percentage of dissatisfaction and extremely high humidity is reflected, which must be improved.

Consequently, the buildings (per floor) were reviewed individually to verify the thermal comfort of each house, given that the results presented above are monthly averages in the urbanization. With this, it was found that 10 residences (which are located on the east façade of the lot) exceeded the maximum temperature limit within the comfort zone between 0.8 °C and 1.30 °C above, considering the adaptive model of Auliciems and Szokolay, in natural ventilation. Given this, we sought to compare the thermal performance (average temperature and monthly hours of non-comfort) of the areas found using three types of walls: super-insulation, project wall, and brick/block wall without insulation, because the literature indicates that it is counterproductive to use super-insulation in a tropical climate. It was found that the super-insulated brick/block wall maintains higher temperatures on the upper floors of each residence, and the original wall produces better thermal performance in these areas. However, the super-insulated brick/block wall on the ground floors produces slightly lower temperatures. For all three cases, all zones remain above 29.1 °C. Additionally, the hours of non-comfort in these three types of walls were compared, concluding that regardless of the type of wall, the residences present similar hours of non-comfort. These results lead to the need for a deeper analysis, carried out by area of the building or daily, to know the critical points and hours for comfort. In addition, it provides evidence that the use of the super-insulated wall is not the main cause of non-comfort in the buildings evaluated.

4.2.2. Energy Consumption

The electrical requirements for EC and RC equipment and luminaires were defined from the surveys. For EC, the energy requirements for the use of air conditioning for cooling were also defined from the surveys; however, for RC, the 10 houses with temperatures above 28.5 °C were evaluated and configuration of the air conditioning was carried out, taking as reference the results obtained in the parametric analysis as an active solution. Figure 12 shows the consumption of both scenarios broken down by month. EC resulted in a consumption of 92.0 MWh/y at the urbanization level, considering the use of equipment, air conditioning, and luminaires. The highest demand is represented by equipment (EQ-ER) with a value of 52.3 MWh/y, followed using electricity by air conditioning with 35.3 MWh/y (AC-EC), and luminaires consume 4.41 MWh/y (LF-EC). In percentage, these data represent 56.8%, 38.3%, and 4.80% of total consumption, respectively, which is close to the data of the survey of the final use of electrical energy in households presented in PEN 2015–2050 [2]. For equipment (EQ-RC) and lighting (LF-RC), similar consumptions to those of existing urbanization were obtained, both in natural ventilation and with mechanical cooling, as expected. The consumption for cooling equipment evaluating only the 10 houses outside the comfort zone increased compared to the existing urbanization, obtaining a total expenditure of 133 MWh/y.

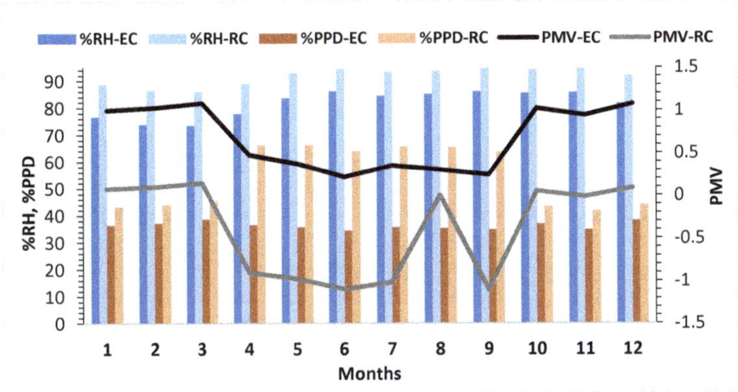

Figure 11. Relative humidity (%RH) predicted percentage of dissatisfied (%PPD) and predicted mean vote (PMV) monthly averages for EC and RC.

The increase in cooling consumption is due to the reduction in the compactness of the zones since the equipment must cool and dehumidify greater volumes of air. Considering the upper limit of comfort temperature by adding +3.5 for 80% acceptability, 29.7 °C is obtained. In this way, only the temperatures of the ground floors of C29 and C34 are above the maximum limit. Consequently, for this investigation, the operation of the urbanization was considered entirely in natural ventilation so that the total energy consumption of the urbanization would not contemplate the cost of cooling for the energy balance. In this way, the RC design has a total consumption of 58.9 MWh/y, which, compared to EC (94.2 MWh/y, including public lighting), is reduced by 37.5%. In addition, the annual expenditure for public lighting is considered, which is 2.19 MWh.

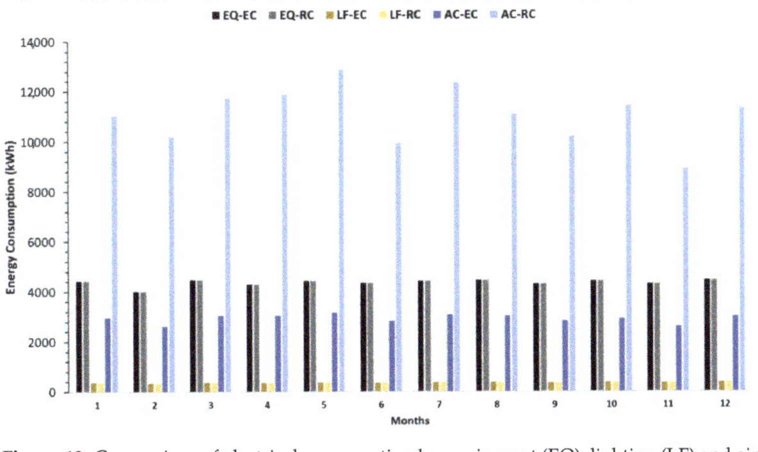

Figure 12. Comparison of electrical consumption by equipment (EQ), lighting (LF) and air conditioning (AC) in kWh between EC and RC.

4.3. Evaluation of Energy Generation

For BAPV, the criteria used to determine the application of solar panels on the roof was based on the total area available on the south side of the gable roof (200 m^2 with an inclination of 30°). Because economic factors are not considered in the analysis, the selection of the number of modules is based on the highest generation that can be obtained with an optimal orientation for Panama (southward in regions of the northern hemisphere) [45], implementing a total of 120 modules connected in series, in each residence, adding a total of 4080 modules (dimensions 1.96 m × 0.99 m) on site. The panel model was chosen based on the market availability of the region, as in [46]. The main technical specifications are listed below [46]: (1) active area: 1.68 m^2; (2) maximum rated power: 320 W; (3) number of cells: 72; (4) cell type: polycrystalline silicon; and (5) panel efficiency: 15%. These and the other electrical specifications required for the simulation were configured in the software, as in [46].

For BIPV, glazing was selected for the application of integrated panels. For its configuration, the same composition of the glazing layers applied to the bioclimatic design was chosen, chosen through parametric analysis, but adding a layer corresponding to the photovoltaic panel associated with the outer layer of the glazing. The type of performance chosen was "simple"; as the name implies, it represents a simplified model. This type of performance provides direct access to the efficiency with which photovoltaic panels convert incident solar radiation into electricity, this being 15%, and does not require specific module arrays to be defined.

The configuration of the applied and integrated panels gave way to the energy simulation in DesignBuilder. The implementation of rooftop photovoltaic modules (T-PV) turned out to have a high-generation performance. The most favorable months for production were January and December, while June had the lowest production. An average production of 159.4 MWh/month and a total production of 1912.8 MWh/y were obtained. The panels integrated into the windows (W-BIPV) have a lower generation compared to BAPV, as expected for this study, due to the low amount of surface selected for integration. These produce an average of 4.23 MWh/month and have an almost constant production throughout the year (50.8 MWh/y). In addition, the individual production of each residence (I-PV + BIPV) was analyzed, and it was found that, with the application of both systems, the generation is 4.83 MWh/month per residence. These results are presented in Figure 13 below.

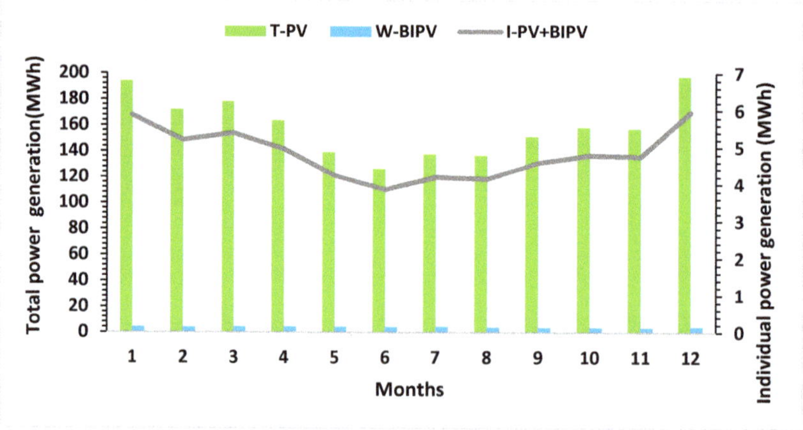

Figure 13. Average monthly photovoltaic generation for rooftop solar panels (T-PV), for panels integrated in window glazing (W-BIPV), and for both systems per individual residence (I-PV+BIPV).

5. Discussion

The results obtained from the bioclimatic methodology and the parametric analysis were the determining factors for the redesign of the urbanization. The diagram in Figure 14 shows a flow chart of the main results obtained.

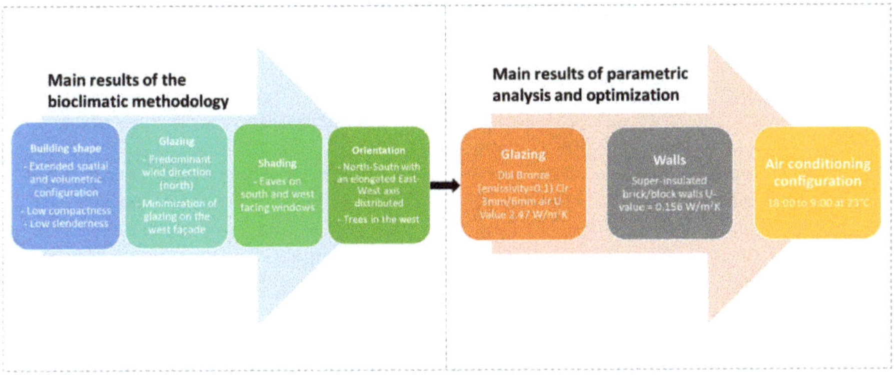

Figure 14. Main results compiled from the bioclimatic methodology and the parametric and optimization analyses.

As can be seen, bioclimatic tools are useful for recommendations on building shape, orientation, and other aspects such as shading; however, they can be complemented with parametric and optimization analysis for specific solutions in terms of passive solutions (such as envelope materials) and active solutions (in the case of the study, air conditioning configuration). Although dynamic simulation is the most used method of analysis so far in zero energy studies, as supported in the literature review section, some limitations have been found for thermal analysis. A simulation per building zone is suggested to obtain more accurate results on comfort since the results presented were performed per floor. However, this would require more time and high-capacity computers to process the data. As a reference, a computer with 8.00 Gb RAM at 2.00 GHz frequency was used for the simulations in this study.

First ZED Reference Framework in Panama

This section discusses the main results for the definition of ZED in Panama as a first reference. Power supplied to a system refers to energy coming through the power grid, heating and cooling districts, and fuels [44]. Energy flows are analyzed for RC and are limited to the electrical energy demanded using occupants' equipment and fuels (LPG). The energy needs for urbanization are summarized using electrical equipment, lighting, and cooking. The use of energy for building systems also considers the energy that is used for conversion in the systems. On the other hand, on-site renewable energy refers to the energy delivered by means of renewable technologies installed within system boundaries. The energy flows within the system are described below for the case study:

Electricity: Electricity energy needs have been calculated based on the simulation of the proposed RC model. The results indicate a consumption of 52.3 MWh/y for equipment, 6.6 MWh/y for lighting (including public lighting), and the demand for air conditioning equipment for cooling is considered null. Taking the methodology used in [47], the energy needs demanded by the refrigerator are not considered in the energy balance for renewable energy supply because it needs constant power from the network. According to the National Population and Housing Census for household electricity consumption in Panama, refrigerator represents 26%. This way, the consumption breakdown equals 13.6 MWh/y for the refrigerator and 38.7 MWh/y for other equipment.

Fuel: According to the survey of residents, dwellings use a 25-pound LPG cylinder with an average duration of 1.6 months, intended for use in kitchens. The PEN indicates that the average residential use of LPG is 1.26 cylinders/month. However, the consumption obtained through the survey will be considered in this study, equivalent to 0.625 cylinders/month. Considering the caloric value of LPG (45.67 GJ/kg), the conversion factor of energy units (3.6 GJ/MWh), and the mass of each cylinder (11.34 kg), one cylinder equals 143.86 kWh/month [2]. Therefore, 0.625 cylinders are equivalent to 89.9 kWh/month. However, the monthly energy demand will also depend on the efficiency ratio of LPG cookers and induction for domestic use, which is close to 0.47. With this, it is obtained that the consumption per residence is equivalent to 42.25 kWh/month, and for the urbanization the total consumption is 1436.5 kWh/month, equivalent to 17.2 MWh/y.

On-site renewable energy: With the production obtained by the PV panels on the roof of all the residences of the urbanization (1912.8 MWh/y) and the integration of photovoltaic glass in the windows (50.8 MWh/y) on the site, a total of 1963.6 MWh/y is produced. This amount of energy is considered in the energy balance as the input of energy generated from renewable sources on site.

Conversion to primary energy: The primary energy conversion values are documented in Table 6. A factor of 3.15 and 1.15 is considered for electricity delivered and LPG, respectively. For the renewable energy produced, a factor of 3.15 is also used. The normalized value in units of kWh/m^{2}y is obtained by dividing the energy in terms of primary energy by the occupied area, which is 1870 m^2 for the reference urbanization (34 buildings of 55 m^2) and 16,830 m^2 for the proposed urbanization (34 buildings of 495 m^2).

Table 7 summarizes the disaggregated energy demand of the different needs and presents the respective normalized values in terms of primary energy for the EC and RC. The conversion of on-site PV generation for renewable energy production is not presented in this table but results in 367.50 kWh/m^{2}y.

Table 7. Conversion of energy consumption and generation values into primary energy terms.

A1	A2	A3	A4	A5
Energy Use	**Energy Consumption (MWh/y) EC**	**Primary Energy (kWh/m^{2}y) EC**	**Energy Consumption (MWh/y) RC**	**Primary Energy (kWh/m^{2}y) RC**
Equipment	38.7	65.19	38.7	7.24
Refrigerator	13.6	22.91	13.6	2.55
Lighting	6.6	11.12	6.6	1.24
Cooling	35.3	59.46	0	0
Cooking	36.6	22.51	36.6	2.50
Total	130.8	181.19	95.9	13.52

Energy Balance: The energy balance is obtained from the subtraction of the energy delivered minus the energy generated. From Equation (1), a result of −356.50 kWh/m^{2}y is obtained. A negative balance indicates that generation is greater than energy demand. According to the results obtained, Figure 15 graphically represents the values for the models of the case study. Three regions characterize the graph: the nZED region when the energy demanded is greater than the energy supplied, the NZED region when the energy demanded equals that supplied, and the PED region when the energy demanded is less than the energy supplied. In summary, EC resulted in energy consumption of 181.19 kWh/m^{2}y (red diamond), including LPG demand. After applying passive and active strategies, RC has a reduced energy requirement of 13.52 kWh/m^{2}y by not using air conditioning equipment (yellow triangle). Finally, with the application of PV panels, the energy balance (EB) reaches a value of 356.50 kWh/m^{2}y (green square), which indicates that the energy supplied exceeds the demand, defining urbanization within the ranges of positive net energy.

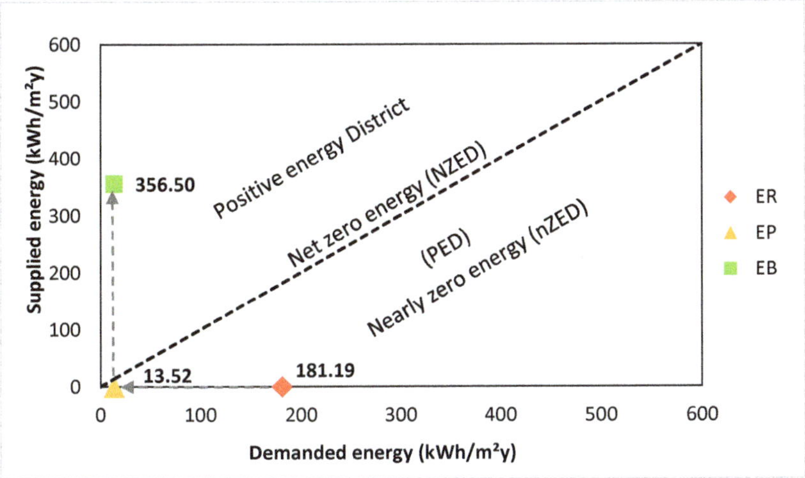

Figure 15. Graphical representation of ZED values for case study models in terms of primary energy, adapted from [48].

Figure 16 represents the system boundaries for the case study. Solid lines represent the incoming or outgoing energy of the system, and the dashed lines represent the energy transferred within the system. The energy delivered corresponds to the LPG (2.50) and the electricity through the network, in this case, the demand for the refrigerator (2.55). Of the photovoltaic generation (367.50), the demand for equipment (without considering the refrigerator) and luminaires (8.48) is used in the building. The demand for electricity in the building is 11.03 and a total of 12.20. With this numerical analysis, it is obtained that 359.02 can be exported, corresponding to the subtraction of the photovoltaic generation minus the total use of energy by equipment and luminaires.

One of the research objectives was to define the range of primary energy to consider urbanizations as almost zero energy. It was intended to obtain an nZED definition threshold, improving energy efficiency in buildings. Including in the study the use of energy for cooling, kitchens, lighting, and equipment, great potential is shown to achieve urbanizations at almost zero energy and even net positive energy. It should be noted that the reduction of the energy demanded is based solely on consumption by air conditioning and not on equipment and lighting due to the consideration of equipment that meets the efficiency required in the region. The numerical analysis performed indicates positive net energy results; that is, you can have ambitions in Panama to develop urbanizations that export energy to the grid or nearby buildings. Without photovoltaic generation, the reference urbanization has an expenditure of 181.19 kWh/m^{2}y, very close to the maximum range presented at the threshold at almost zero energy (20 kWh/m^{2}y to 180 kWh/m^{2}y) for a single building [49]. With the independence of air conditioning, the RC achieves a range of 13.52 kWh/m^{2}y, which is relatively low, even lower than the minimum range for residential buildings, which could indicate as a preliminary conclusion that an nZED range of 14 kWh/m^{2}y to 180 kWh/m^{2}y is acceptable for Panama, given the low energy requirements of homes when there is no need for mechanical cooling. However, it should be considered that this value is normalized with the total surface of the residences, which may be one of the limitations to reaching more real and achievable conclusions with fewer resources since the RC, presenting a reduced compactness value, requires a greater need for land. With the increase in surface area per building, the inclusion of more equipment per household or changes in the number and behavior of the occupants was not considered since the original conditions were maintained.

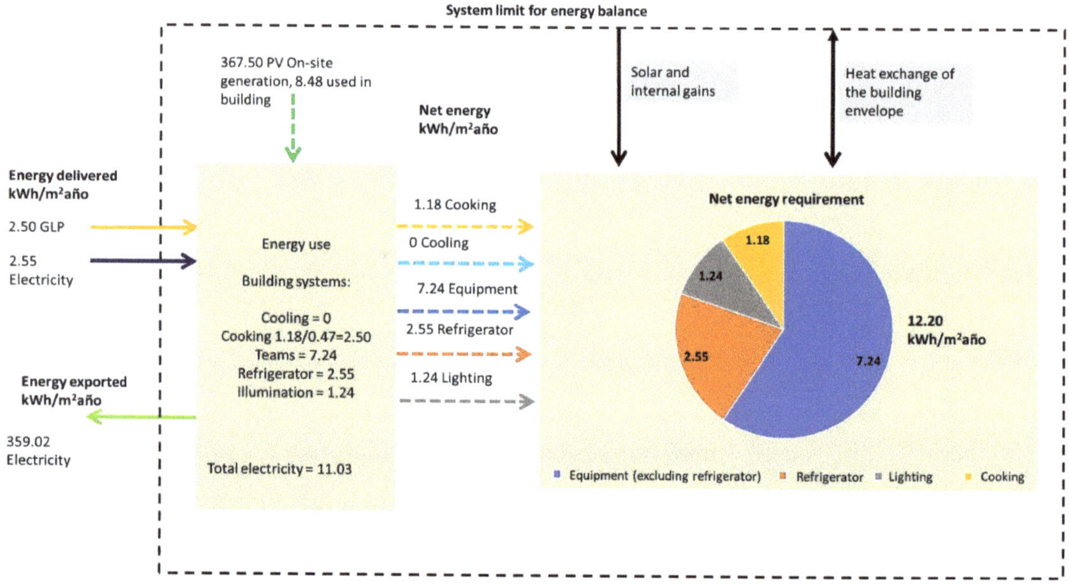

Figure 16. System limit for energy balance (in terms of primary energy), adapted from [40].

6. Conclusions

This study takes as a reference an existing residential development to apply a bioclimatic methodology and parametric analysis to evaluate the reduction in energy consumption through dynamic simulations. The scope of this work consisted of assessing the potential of the site to be defined under near-zero energy limits in the context of Panama and, at the same time, documenting and systematizing the passive and active solutions applied to obtain a reference plan for the development of future research and full-scale applications in a developing country such as Panama or with similar climatic conditions.

The application of bioclimatic tools is promising with respect to achieving the objectives and combined with dynamic simulation; guidelines on materials and configurations for active strategies can be obtained. For the case study, energy savings of 37.5% were achieved. On the other hand, the application of photovoltaic modules on the roof and glazing was sufficient to cover and exceed the on-site energy demand; however, the integration in glazing was not significant in terms of performance for the type of buildings studied.

Through the energy balance performed, it is obtained that the development can reach positive net energy ranges (356.50 kWh/m^{2}y); i.e., energy could be exported to the grid. The results show that the initiative to apply regulations and standards of this type is feasible under the climatic conditions of Panama and that they can be oriented towards a smart city market.

Among the limitations presented in this study are accurate knowledge and data set up on the software regarding occupancy and energy use profiles, as such information is subjected to the occupants' behavior and comfort demand. Likewise, the bioclimatic methodology leaves the choice of materials with a significant degree of ambiguity. This limitation was overcome by using parametric analysis.

It is recommended for future studies to consider economic factors in construction materials, land requirements, and generation technologies to have a cost–benefit relationship, since the site has the potential to achieve its objectives. On the other hand, future work could raise awareness of the effect of form indicators on design and implement other options or active low-energy solutions to improve comfort conditions. This could be used to define, for example, a percentage or fraction limit for photovoltaic generation.

Author Contributions: Original concept, formal analysis, and editing by L.D.L., D.M., C.C. and M.C.A. Introduction, figures and writing of most of the manuscript by L.D.L. and D.M. Supervision and funding by M.C.A., C.C. and N.A. All authors have read and agreed to the published version of the manuscript.

Funding: This research was funded by the Panamanian Institution Secretaría Nacional de Ciencia, Tecnología e Innovación SENACYT (https://www.senacyt.gob.pa/, accessed on 25 December 2022), under the project code FID18-056 and IDDS22-30, as well as supported by the Sistema Nacional de Investigación (SNI).

Data Availability Statement: Data supporting the findings is available upon request.

Acknowledgments: The authors would like to thank the Technological University of Panama and the Faculty of Mechanical Engineering (https://fim.utp.ac.pa/, accessed on 25 December 2022) for their collaboration, together with the Research Group ECEB (https://eceb.utp.ac.pa/, accessed on 25 December 2022). Besides, a special thanks to the Department of Mechanical, Energy and Management Engineering (DIMEG, https://www2.unical.it/portale/strutture/dipartimenti_240/dimeg/, accessed on December 2022), University of Calabria (UNICAL, https://www2.unical.it/portale/, accessed on 25 December 2022).

Conflicts of Interest: The funders had no role in the design of the study; in the collection, analyses, or interpretation of data; in the writing of the manuscript, or in the decision to publish the results.

Nomenclature

ASHRAE	American Society of Heating, Refrigerating and Air-Conditioning Engineers
AT	Air temperature
BAPV	Building applied photovoltaics
BIPV	Building integrated photovoltaics
EB	Energy balance
EC	Existing Case
EPBD	Energy Performance of Buildings Directive
GBC	Green Building Council
IEA-EBC	Energy in Buildings and Communities Program of the International Energy Agency
I-PV + BIPV	Individual power generation per house
HVAC	Heating, ventilation, and air conditioning
LED	Light-emitting diode
LEED	Leadership in Energy & Environmental Design
LPG	Liquefied petroleum gas
nZEB	Nearly zero energy building
NZEB	Net zero energy building
nZED	Nearly zero energy district
NZED	Net zero energy district
nZES	Near-zero energy Project Building in the settlement
OP	Operating temperature
PED	Positive energy district
PEN	National Energy Plan
PMV	Predicted mean vote
%PPD	Predicted percentage of dissatisfied
PV	Photovoltaic
PVC	Polyvinyl chloride
RES	Sustainable Buildings Regulation
RC	Redesign Case
%RH	Relative humidity
RT	Radiant temperature

SDG	Sustainable Development Goal
SHC	Solar Heating and Cooling Program
SGHC	Solar gain coefficient
T-PV	Rooftop photovoltaic modules
U-value	Thermal transmittance values
W-BIPV	Photvoltaic integrated to the windows
ZEB	Zero Energy Building
ZED	Zero Energy District

Appendix A

Table A1. Survey questions applied to housing residents.

Item	Open-Ended Question	Closed Question	Options to Select
Number of people living in the dwelling	✓		
Types of sun protection in the windows of the house		✓	(1) None (2) Fabric curtains (3) Blinds (4) Blackout or pore roller curtain
Factors considered for the purchase of equipment and appliances: air conditioning, refrigerator, lights.		✓	(1) Trademark (2) Price (3) Energy efficiency (4) Sales Staff Recommendations
Use of air conditioning		✓	(1) Yes (2) No
Type of air conditioning		✓	(1) Window unit (2) Split unit
Areas of the house where air conditioning is used		✓	(1) Living Room (2) Bedrooms (3) Kitchen
Cooling temperature point	✓		
Air conditioning weekday and weekend schedules	✓		
Seasonal Energy Efficiency Ratio (SEER) of air conditioning	✓		
Schedules with open windows	✓		
Type of fans owned by the house		✓	(1) Pedestal (2) Table (3) Roof (4) Wall
Number of fans owned by the house	✓		
Fan use schedule	✓		
Types of electric bulbs in the home		✓	(1) Incandescent (2) Fluorescent (3) Low consumption (4) Halogen
Number of electric light bulbs in the house	✓		
Schedule for the use of electric bulbs	✓		
Number and schedule of electrical equipment and appliances	✓		
Average monthly electricity consumption in kWh	✓		
Type of stove		✓	(1) Gas (2) Electric
Average duration of use of specified LPG tank	✓		

Table A1. *Cont.*

Item	Open-Ended Question	Closed Question	Options to Select
LPG tank capacity		✓	(1) 25 pounds (2) 60 pounds (3) 100 pounds
Use of LPG water heater		✓	(1) Yes (2) No

References

1. Parra Cortes, R. The 2030 Agenda and its Sustainable Development Goals: Background and perspectives to promote sustainable consumption and production in Chile. *Environ. Law Mag.* **2018**, *10*, 99–121. [CrossRef]
2. Secretaría Nacional de Energía. Plan Energético Nacional (PEN) 2015–2050 Panamá El Futuro Que Queremos; Secretaría Nacional de Energía: No. 28003. 2016, pp. 6–348. Available online: http://www.energia.gob.pa/energia/wp-content/uploads/sites/2/2017/06/Plan-Energetico-Nacional-2015-2050.pdf (accessed on 15 December 2022). (In Spanish).
3. Junta Técnica de Ingeniería y Arquitectura. *Reglamento de Edificaciones Sostenibles (RES)*; Gobierno de Panamá: Panama City, Panama, 2019. (In Spanish)
4. Gaceta Oficial. *Ley No. 69 del 12 de Octubre de 2012*; No. 27145; Gobierno de Panamá: Panama City, Panama, 2012; pp. 1–44. (In Spanish)
5. European Commission. Guidelines for the Promotion of Nearly Zero-Energy Buildings and Best Practices to Ensure That-Commision Recommendation (EU) 2016/1318. Official Journal of the European Union. 2016. Available online: http://data.europa.eu/eli/reco/2016/1318/oj (accessed on 23 November 2022).
6. Piderit, M.B.; Vivanco, F.; van Moeseke, G.; Attia, S. Net zero buildings. A framework for an integrated policy in Chile. *Sustainability* **2019**, *11*, 1494. [CrossRef]
7. Amaral, A.R.; Rodrigues, E.; Gaspar, A.R.; Gomes, Á. Review on performance aspects of nearly zero-energy districts. *Sustain. Cities Soc.* **2018**, *43*, 406–420. [CrossRef]
8. Marique, A.-F.; Reiter, S. A simplified framework to assess the feasibility of zero-energy at the neighbourhood/community scale. *Energy Build.* **2014**, *82*, 114–122. [CrossRef]
9. Green Building Council. LEED for Cities and Communities | U.S. Green Building Council. Available online: https://www.usgbc.org/leed/rating-systems/leed-for-cities (accessed on 6 March 2022).
10. International Energy Agency Energy in Buildings and Communities (IEA EBC) Programme. Towards Net Zero Energy Solar Buildings IEA EBC. Available online: https://www.iea-ebc.org/projects/project?AnnexID=52 (accessed on 14 March 2022).
11. Athienitis, A.K.; O'Brien, W. *Modeling, Design and Optimization of Net-Zero Energy Buildings*; Ernst & Sohn: Berlin, Germany, 2019; Volume 53, pp. 1689–1699. [CrossRef]
12. IEA Energy in Buildings and Communities. Positive Energy Districts (PED). EBC ANEXO 83. 2021. Available online: https://jpi-urbaneurope.eu/ped (accessed on 20 December 2022).
13. Baer, D.; Loewen, B.; Cheng, C.; Thomsen, J.; Wyckmans, A.; Temeljotov-Salaj, A.; Ahlers, D. Approaches to Social Innovation in Positive Energy Districts (PEDs)—A Comparison of Norwegian Projects. *Sustainability* **2021**, *13*, 7362. [CrossRef]
14. Pless, S.; Polly, B. *Communities of the Future: Accelerating Zero Energy District Master Planning Preprint*; National Renewable Energy Lab: Golden, CO, USA, 2018.
15. Polly, B.; Kutscher, C.; Macumber, D.; Schott, M. From Zero Energy Buildings to Zero Energy Districts. In Proceedings of the 2016 ACEEE Summer Study on Energy Efficiency in Buildings, Pacific Grove, CA, USA, 21–26 August 2016; pp. 1–16. Available online: https://www.researchgate.net/publication/307955877_From_Zero_Energy_Buildings_to_Zero_Energy_Districts/figures?lo=1 (accessed on 15 December 2022).
16. Cabeza, L.F.; Chàfer, M. Technological options and strategies towards zero energy buildings contributing to climate change mitigation: A systematic review. *Energy Build.* **2020**, *219*, 110009. [CrossRef]
17. Hoque, S.; Iqbal, N. Building to net zero in the developing world. *Buildings* **2015**, *5*, 56–68. [CrossRef]
18. Stasi, R.; Liuzzi, S.; Paterno, S.; Ruggiero, F.; Stefanizzi, P.; Stragapede, A. Combining bioclimatic strategies with efficient HVAC plants to reach nearly-zero energy building goals in Mediterranean climate. *Sustain. Cities Soc.* **2020**, *63*, 102479. [CrossRef]
19. Guillén-Lambea, S.; Rodríguez-Soria, B.; Marín, J.M. Comfort settings and energy demand for residential nZEB in warm climates. *Appl. Energy* **2017**, *202*, 471–486. [CrossRef]
20. García-Fuentes, M.; González, I.; Gordaliza, A.; de Torre, C. Retrofitting of a Residential District under Near Zero Energy Buildings Criteria. *Proceedings* **2017**, *1*, 686. [CrossRef]
21. Castaldo, V.L.; Pisello, A.L.; Piselli, C.; Fabiani, C.; Cotana, F.; Santamouris, M. How outdoor microclimate mitigation affects building thermal-energy performance: A new design-stage method for energy saving in residential near-zero energy settlements in Italy. *Renew. Energy* **2018**, *127*, 920–935. [CrossRef]
22. Pinto, G.A.O.; Nova, D.A.S.; Calderón, N.C.B.; Plata, G.O. Energy considerations of social dwellings in Colombia accotding to NZEB concept. *Dyna* **2015**, *82*, 120–130. [CrossRef]

23. Ordoñez, E.; Mora, D.; Gaudry, K. Roadmap Toward NZEBs in Quito. *IOP Conf. Ser. Mater. Sci. Eng.* **2019**, *609*, 072040. [CrossRef]
24. Bambara, J.; Athienitis, A.K.; Eicker, U. Residential Densification for Positive Energy Districts. *Front. Sustain. Cities* **2021**, *3*, 1–16. [CrossRef]
25. Aruta, G.; Ascione, F.; Bianco, N.; Mastellone, M.; Mauro, G.M. Optimization of Envelopes, Systems and Storage for Transition of Building Stocks to Zero Energy Districts. *Chem. Eng. Trans.* **2022**, *94*, 841–846. [CrossRef]
26. Salim, F.D.; Dong, B.; Ouf, M.; Wang, Q.; Pigliautile, I.; Kang, X.; Hong, T.; Wu, W.; Liu, Y.; Rumi, S.K.; et al. Modelling urban-scale occupant behaviour, mobility, and energy in buildings: A survey. *Build Env.* **2020**, *183*, 106964. [CrossRef]
27. Happle, G.; Fonseca, J.A.; Schlueter, A. A review on occupant behavior in urban building energy models. *Energy Build.* **2018**, *174*, 276–292. [CrossRef]
28. van Eck, N.J.; Waltman, L. *Manual for VOSviewer Version 1.6.18.0*; Univeristeit Leiden: Leiden, The Netherlands, 2022. Available online: http://www.vosviewer.com/documentation/Manual_VOSviewer_1.6.1.pdf (accessed on 23 November 2022).
29. DesignBuilder Software Ltd.-About Us. Available online: https://designbuilder.co.uk/about-us (accessed on 23 November 2022).
30. Austin, M.C.; Arnedo, L.; Yuil, O.; Mora, D. Energy consumption influenced by occupant behavior: A study in residential buildings in Panama. In Proceedings of the LACCEI International Multi-Conference for Engineering, Education and Technology, Boca Raton, FL, USA, 19–23 July 2021; pp. 1–9. [CrossRef]
31. Instituto Nacional de Estadística y Censo. *Habitantes Promedio por Vivienda Particular en la República de Panamá, Según Provincia y Comarca Indígena*; Instituto Nacional de Estadística y Censo: Panama City, Panama, 2010; (In Spanish).
32. American Society of Heating, Refrigerating and Air-Conditioning Engineers. Standard ASHRAE 55. 2021. Available online: https://www.seiscubos.com/conocimiento/estandar-ashrae-55 (accessed on 20 December 2022).
33. Araúz, J.; Mora, D.; Austin, M.C. Assessment of Different Envelope Configurations via Optimization Analysis and Thermal Performance Indicators: A Case Study in a Tropical Climate. *Sustainability* **2022**, *14*, 2013. [CrossRef]
34. Secretaría Nacional de Energía. *Guia de Construccion Sostenible Para el Ahorro de Energia en Edificaciones*; Secretaría Nacional de Energía: Panama City, Panama, 2016; pp. 1–53. (In Spanish)
35. Lakes Environmental. Lakes Environmental | WRPLOT View [Free]-Wind Rose Plots for Meteorological Data. 2018. Available online: https://www.weblakes.com/products/wrplot/index.html (accessed on 4 February 2022).
36. Climate Consultant 6. 2016. Available online: http://www2.aud.ucla.edu/energy-design-tools/ (accessed on 4 February 2022).
37. Gómez-Azpeitia, G. Bioclimarq 2016. 2016.
38. Serra Florensa, R.; Coch Roura, H. *Arquitectura y Energía Natural*; Universitat Politècnica de Catalunya: Barcelona, Spain, 1995.
39. International Renewable Energy Agency. *Evaluación del Estado de Preparación de las Energías Renovables: Panamá*; IRENA: Abu Dhabi, United Arab Emirates, 2018. (In Spanish)
40. (DOE) US Department of Energy. *A Common Definition for Zero Energy Buildings*; U.S. Departament of Energy: Washington, DC, USA, 2015; p. 22.
41. Koutra, S.; Pagnoule, C.; Galatoulas, N.-F.; Bagheri, A.; Waroux, T.; Becue, V.; Ioakimidis, C.S. The zero-energy idea in districts: Application of a methodological approach to a case study of epinlieu (Mons). *Sustainability* **2019**, *11*, 4814. [CrossRef]
42. Medved, S.; Domjan, S.; Arkar, C. *Sustainable Technologies for Nearly Zero Energy Buildings*; Springer: Cham, Switzerland, 2019; ISBN 978-3-030-02821-3. [CrossRef]
43. Emerson, K.I.; Conover, D.R.; Deru, M.P.; Heizer, M.P. *ASHRAE Stand 105-2014*; Standard methods of determining, expressing, and comparing building energy performance and greenhouse gas emissions. ASHRAE: Peachtree Corners, GA, USA.
44. Kurnitski, J.; Allard, F.; Braham, D.; Goeders, G.; Heiselberg, P.; Jagemar, L.; Kosonen, R.; Lebrun, J.; Mazarella, L.; Railio, J.; et al. How to define nearly net zero energy buildings Nzeb. *REHVA J.* **2011**, *48*, 6–12.
45. Asea Brown Boveri, S.A. *Cuaderno de Aplicaciones Técnicas no. 10: Plantas Fotovoltaicas*; ABB: Zúrich, Suiza, 2011; p. 116. Available online: https://library.e.abb.com/public/e703d99268365a43c125791f002ce826/1TXA007109G0701_CT10.pdf (accessed on 23 November 2022). (In Spanish)
46. Chacón, L.; Austin, M.C.; Castaño, C. A Multiobjective Optimization Approach for Retrofitting Decision-Making towards Achieving Net-Zero Energy Districts: A Numerical Case Study in a Tropical Climate. *Smart Cities* **2022**, *5*, 405–432. [CrossRef]
47. Austin, M.C.; Chung-Camargo, K.; Mora, D. Review of zero energy building concept-definition and developments in latin america: A framework definition for application in Panama. *Energies* **2021**, *14*, 5647. [CrossRef]
48. Alonso, M.J.; Stene, J.; Ronneseth, O.; Georges, L. State of the Art Analysis of Nearly Zero Energy Buildings-Norway. *Country Report IEA HPT Annex 49: Norway*. 2013. 978-82-536-1584-4. Available online: http://www.zeb.no/index.php/reports/item/406-state-of-the-art-analysis-of-nearly-zero-energy-buildings-country-report-iea-hpp-annex-40-task-1-norway (accessed on 20 December 2022).
49. Attia, S. *Net Zero Energy Buildings (NZEB), Frameworks and Roadmap for Project Analysis and Implementation*; Elsevier: Amsterdan, The Netherlands, 2018; ISBN 978-0-12-812461-1.

buildings

MDPI

Article

Energy Benefits of Tourist Accommodation Using Geodesic Domes

Ángel Benigno González-Avilés *, Carlos Pérez-Carramiñana, Antonio Galiano-Garrigós and María Isabel Pérez-Millán

Building Construction Department, Polytechnic School Alicante, University of Alicante, 03690 Alicante, Spain; c.perez@ua.es (C.P.-C.); antonio.galiano@ua.es (A.G.-G.); isabel.perez@ua.es (M.I.P.-M.)
* Correspondence: angelb@ua.es; Tel.: +34-65-315-48-91

Abstract: Over the last decade there has been a proliferation of glamping architecture. This study analyses the energy performance of geodesic domes for use in tourist glamping compared to more conventional prismatic architectural solutions. The energy analysis of geodesic domes applied to this type of singular construction project currently lacks detailed studies that provide conclusions about their relevance and suitability with respect to other types of architecture. The main objective of this research is to demonstrate the energy benefits of tourist accommodations that use geodesic structures compared to those with a simple geometry. A comparative study of a traditional and a geodesic geometry accommodation is carried out, considering that they share the same characteristics and they are built with the same construction solution. An energy simulation of both architectures is carried out by using DesignBuilder software. The most influential strategies, such as Direct Passive Solar Gain, Heating, Natural Ventilation Cooling, Fan-Forced Ventilation Cooling and Window Solar Shading are considered. After demonstrating the greater efficiency of geodesic domes, this study analyses the relevance of subdividing the accommodations into several geodesic dome spaces. The results quantify an energy benefit of 52% for cooling consumption using the geodesic dome solution compared to a traditional prismatic solution.

Keywords: energy efficiency; thermal comfort; glamping; geodesic domes; simulation

1. Introduction

Glamping, or glamorous camping [1], is a current word that started as a new form of accommodation in Africa [2]. It is defined as the global phenomenon that combines the experience of camping in the open air with the luxury and conditions of the best hotels.

It is also worth noting that the relationship between the concepts of pandemic and travel has been amply demonstrated throughout history [3]. The latest research shows that since COVID-19, the purchase of glamping-type travel plans has reached 45.9%, compared to 24.7% [4] for hotel/resort stays. The European trend for the use of this new camping style is booming [5], and has a monetary return in Europe of USD 965.7 M (Figure 1). For the period 2023–2030, a compound annual growth rate (CAGR) of 10.9% [6] is expected.

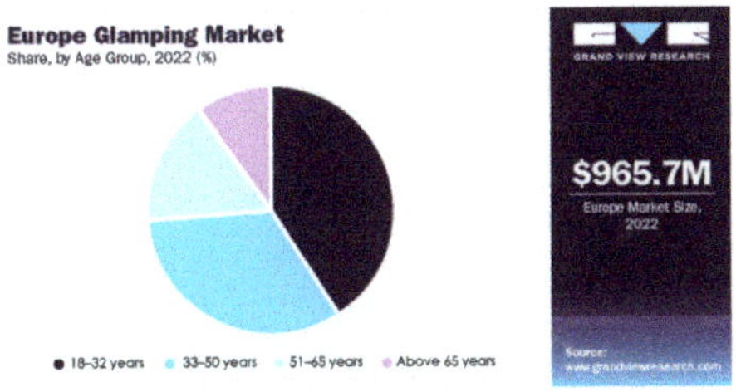

Figure 1. European glamping market in 2022 [6]. Source: Europe Glamping Market Size, Share and Growth, 2023.

The research by Claver-Cortes et al. [7] has significantly related these performance levels to environmental proactivity and their implications for policy makers. An increasing number of hotels are engaging in greater environmental stewardship [8]. This leads to economic benefits, in terms of resource savings and an image of commitment to the environment. Another reason is the new public pressure exerted by many environmentally conscious consumers [9]. In general, middle-aged tourists are more environmentally conscious and are willing to pay more for accommodations that consider renewable energy sources [10]. Travelers interested in this type of tourism want quality services, different experiences and open spaces in quiet places [11]. As opposed to mass tourism, they are looking for an individual or small-group approach in sustainable areas that protects and respects nature [12]. This evolution of the tourism market requires new permanent accommodation structures, which usually have a significant environmental impact [13]. These accommodations tend to have a higher price tag; however, considering their increasing occupancy rates, it is noted that their high price does not diminish their attractiveness [14].

This research focuses on two types of permanent housing based on a clear geometrical distinction (Table 1). On the one hand, prismatic architecture, more closely related to industrialised huts with straight geometries, is studied. On the other hand is the geodesic dome, on which this study focuses.

Table 1. Examples of recent prismatic architecture and geodesic domes. Source: Archdaily.

	Tenir Eco Hotel [15] Levelstudio		DOM(E) [16] NRJA Architects
	Labt 20 Modular home [17] Borrachia + GB Architects		Geodesic house [18] Ecoproyecta
	M+J House [19] Manuel Cerdá Architect		Sazae Sauna [20] Kengo Kuma & Associates
	Ranwu Lake campsite [21] Xiao Yin Architecture Design Firm		Two domes and a plinth: House 8 [22] B+V Architects
	Cambará Container Housing [23] Saymon Dall Alba + Mégui Dal Bó		In progress: Domo Cluster [24] Arketiposchile

The current literature has presented scattered data regarding the reduced use of materials, particularly from a structural point of view [25,26]. However, these studies have mostly focused on large-span spaces, like planetariums [27] or public areas [28]. Some energy studies may have exhibited a bias towards roof solutions [29] or the advantages or benefits of the application of a dome structure for tourist education facilities [30]. The novelty of this research lies in its quantification of the energy savings of this type of architecture when applied to domestic or tourist accommodation spaces.

1.1. The Relationship between Efficiency and Geometry as a Research Objective

The main objective of this research is to demonstrate the energy benefits of housing with geodesic structures compared to a parallelepiped geometry. This demonstration is carried out by comparing three housing units: one with a traditional parallelepiped geometry and two housing units with geodesic geometry, all executed with the same construction solution, the same surface area and in the same climate. An energy simulation of the three designs is carried out for the volume of the parallelepiped geometry, the single geodesic dome and the double geodesic dome. The software used is DesignBuilder [31] (version v.7.0.1.006, DesignBuilder Software Limited, Stroud, UK). This software allows for the simulation of energy efficiency and thermal comfort in buildings using the EnergyPlus v.23.1.0 calculation engine [32]. This software also provides computational fluid dynamics (CFD) calculations with which to simulate the average radiant temperature distribution, the operating temperature and the estimated thermal comfort in different zones within the building. For the computer simulations, the thermal transmittances of the façade and the air infiltration through the window frames, measured in situ, are considered.

The air infiltration is low, due to the small proportion of the openings in the façade and the characteristics of the wooden frames used.

It will be concluded whether this type of construction solution really provides an energy/economic benefit compared to a more traditional geometrical solution.

1.2. History of the Geodesic Dome

We can define a geodesic dome as part of a geodesic sphere. A polyhedron generated from an icosahedron or a dodecahedron can also be generated from any of the Platonic solids [33]. The faces of the regular Platonic solids are regular polygons equal to each other. Based on this description and Euler's theorem, the cube, tetrahedron, octahedron, dodecahedron and icosahedron are Platonic solids, being the only five bodies that fulfil the properties described by the theorem and that maintain the constant [34].

The Jena planetarium designed by Walther Bauersfeld in 1922 is considered to be the first geodesic dome in the world [35]. However, it is Richard Buckminster Fuller who is considered the father of geodesic structures [36]. In 1949, Fuller erected a dome capable of supporting its own weight without limit [37], and patented it in 1951 [38]. This structure was supported by the principles of tensegrity structures. It comprised a diameter of 4.2 m, was constructed of aluminium tubes (Figure 2) and covered with a vinyl coating. Years later, the US Army took advantage of the qualities of these structures. After several commissions for the army, Fuller built his own house in Carbondale (IL, USA). However, at the time, Fuller's aim was not a concern for the environment but a strategy to reduce the cost of housing [39]. Fuller also designed the US pavilion for the World Expo in Montreal in 1967, with a structure 76 m in diameter and 62 m high. The dome consisted of an interior division of seven levels, with the structure and envelope made of steel and polymers [40].

Figure 2. Demonstration of the strength of the dome; Buckminster Fuller and students hang from sections of the dome [41].

In June 1979, the American Institute of Architects (AIA) awarded Fuller the gold medal [42], considering the geodesic dome to be "the strongest, lightest and most efficient means to enclosing space yet know man". Leonhard Euler, in 1750, expressed in his polyhedra theorem the existence of only five regular polyhedra, which maintain the constant $c = 2$ that relates the faces, vertices and edges of these polyhedra [43]. The regularity of the elements that make up the Platonic solids allows for the generation of geodesic domes by allowing for the division of their faces while maintaining the relationship of the elements that make them up. Depending on how it is generated, there are three types of geodesic domes: Class I, Class II and Class III [44]. Most geodesic domes constructed are either Class I or II, due to the availability of the methods that generate tessellations with a linear increase in complexity. Class I are the easiest to use, especially when near equatorial truncation is desired. In Figure 3 we can see the different models of Class I geodesic domes, depending on their frequency from one to six.

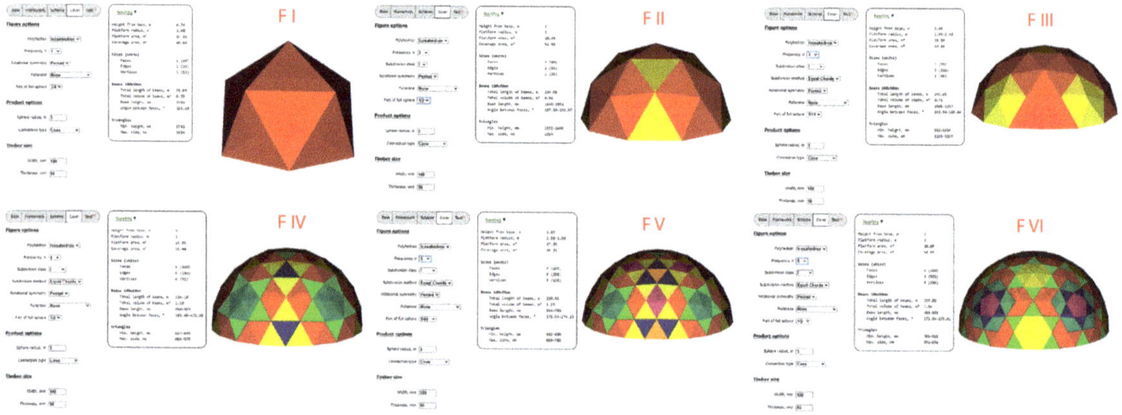

Figure 3. Models of Class I geodesic domes depending on their frequency from I to VI. Source: own elaboration from www.acidome.ru (accessed on 3 December 2023).

1.3. Geodesic Domes' Geometrical and Energetic Characteristics

The dome is the geometric shape that encloses the greatest amount of volume with the least amount of surface area. This results in important savings (greater than any other structure with a different shape) in the building materials used to enclose a usable work area.

A dome, as a spherical shape, has less surface area per unit of internal volume; thus, it reduces temperature gains or losses through its enclosure. The forces applied to the dome are distributed throughout the structure, generating a stable and solid construction [45], although it is true that hemispherical domes considerably increase the final weight compared to a 1/3 or 2/3 [46] span. It has a uniform weight distribution on the ground plane and a low centre of gravity, resulting in a structure with great seismic resistance. Its materiality is mainly based on lightweight materials, such as wood or steel, which are connected to each other by means of prefabricated or dry construction processes. This type of structure is no longer considered a high-tech construction system, as it is based on low-tech construction protocols and schemes [47], allowing for faster construction compared to that of other conventional structures.

In short, geodesic domes offer a wide range of possible uses due to the qualities described above. Geodesic domes have proven to be such a flexible architectural form, they have housed everything from radar equipment to radical back-to-landers [35]. They can solve both ephemeral needs, such as the construction of domes for one-off events, auditoriums, etc., or permanent needs for residential use; they can also be used in agriculture, as they concentrate light more efficiently than a conventional greenhouse dome [46,48]. The speed of the execution of their structure allows for the transport and construction of resistant shelters, even those with military applications. Moreover, no in-depth studies of spherical trigonometry or stereographic projections [49] are required for their design or construction. This is why, in recent years, their use as accommodation has proliferated so much. An example of their efficiency is the study of the Hotel Ecocamp in Patagonia (Chile), a pioneer project in the construction of hotel rooms with geodesic domes. Soares [42] has estimated that savings of 30% on materials and 50% on heating energy could be achieved compared to other conventional examples in the same area. Other examples in Europe are the Aurora Dome (Finland), Whitepod (Switzerland) or the geodesic domes designed by the Ecoproyecta [50] studio in the towns of Yecla and Jumilla (Murcia) in Spain.

Our research aims to demonstrate and quantify the energy savings with this type of structure compared to conventional models. This study supports the energy and economic benefits of geodesic projects designed to have a lower environmental impact on rural areas, and offers more solutions about how to improve the energy efficiency of domes and solar performance [51,52].

2. Materials and Method

Geometrical studies offer us very important variables on which to base this study. The most significant are the floor area, the envelope area and the length of the edges. The research by Haghnazar et al. [53] has already highlighted differences in edge length and other aspects which are differentiated according to frequency. Determining the optimum geometry design for geodesic domes presents difficulties, due to the fact that the height of the dome keeps on changing during the design process [54]. In order to understand the differences, depending on the frequency used, a geodesic dome with a radius of 3 m has been studied and is shown in Figure 3. Frequency I allows for a greater height from the base (4.34 m) but, on the other hand, a smaller floor area (17.12 m^2). For frequencies II, IV and VI, the height coincides with the radius of the inscribed circumference (3 m), and the surfaces are very close to 27 m^2. Frequency III has the lowest free height and would not be occupiable (2.49 m), which is very similar to frequency V (2.69 m).

Fuller's first argument for the energy efficiency of domes was the smaller surface area per unit volume guaranteed by their spherical geometry [55]. This results in a smaller surface area exposed to cold in winter and heat in summer compared to other architectures in the same space. Additionally, the continuous air flow inside the dome, with no stagnant corners, requires less energy to circulate the air and maintain uniform temperatures [56].

The dome combines the inherent stability of triangles with the advantageous volume/surface area ratio of a sphere, which results in less building materials to include more space. There is an estimated 30% reduction in materials and 50% reduction in energy compared to a conventional masonry construction [57] on the same built area, or 25–30% compared to metal [58].

2.1. Case Studies

Three different case studies located in the same climate zone BSh (cold semi-arid) of the Köppen climate classification have been analysed for this research:

- Case study 1: reference block. It is a simplified model of a traditional construction with dimensions of 10 m wide, 7.72 m deep and 3.5 m high, with an occupation surface of 77.2 m^2, 201.3 m^2 of envelope and 77.2 m^2 in contact with the ground. It consists of a single living area, an access door and glazing corresponding to 10% of the surface area of each façade. This parallelepiped shape has been very common in modern architecture in recent years. Many examples can be found on the leading architecture portals, such as Archdaily [59]. Some projects of this type of prismatic volumetry are the Tenir Eco hotel, the Labt 20 modular housing, the M + J house, the prototypes of the Lago Ranwu campsite or the Cambará Container House, among others (Table 1). The energy efficiency of this type of architecture has been widely studied [60], as well as its spatial uses [61], its structural capacity [62] and its reuse [63].

These architectures seek to minimise the resources employed to create residential units that meet energy and comfort standards, but also offer a cost-effective option in the architectural market.

- Case study 2: geodesic dome with a 5 m radius and frequency IV, with a surface occupation of 77.2 m^2 equal to the reference block (case study 1), 154.11 m^2 of envelope and 77.2 m^2 in contact with the ground.
- Case study 3: two smaller geodesic domes (3 m radius with 27.8 m^2 of surface area and another with a 4 m radius and 49.4 m^2 of surface area) with a total surface area of 77.2 m^2 (27.8 + 49.4) equal to the reference block (case study 1). In this case study, a differentiation of uses is considered; to this end, one dome is designed for night use (bedroom) and the other for day use (living room). This distinction means that both the electronic devices capable of generating heat for occupancy during the day and night hours will be different, making it possible to quantify the differentiable impact on energy consumption with respect to other case studies (with a single envelope).

For the comparison between the different case studies (Figure 4), it was decided to disregard the existence of interior partitions, in order to check for the volumetric implication of the energy analysis. The same orientation was used in all three case studies. Every façade was designed with the same proportion of openings in each orientation.

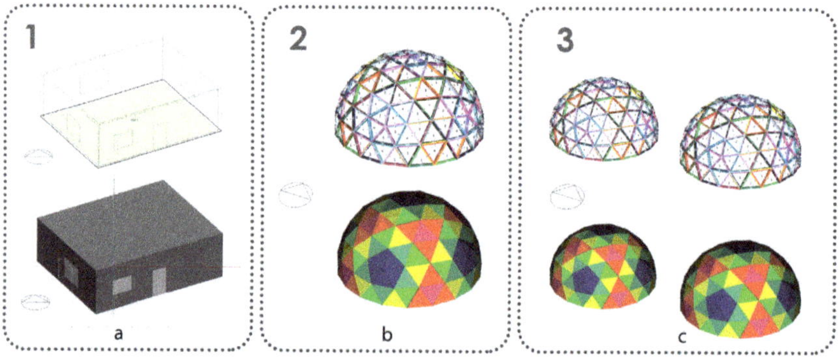

Figure 4. Case studies. (**a**) Case 1: reference block, (**b**) case 2: geodesic dome and (**c**) case 3: double geodesic dome. Source: own elaboration using DesignBuilder software.

2.2. Morphological Analysis of the Geodesic Dome

Several recent studies have calculated the optimal geometry for this type of single-layer architecture and its algorithm [54], or have studied its optimisation in the face of earthquakes [64]. In the field of structures, this type of architecture has been extensively studied [65–68]. However, in the residential field, due to their thermal characteristics, domes have a great potential that has yet to be studied.

In chronological order, the first dwellings to use this type of geometry were the Easy Domes by architect Kári Thomsen and engineer Ole Vanggaard in 1992. This was followed by the Genesis Project as a temporary home for homeless people in Los Angeles in 1993, Aso Farms in Japan in 1995, Domo House in Oregon, the hemispherical houses of Solaleya and The Inn Place in Brenham in Texas in 2006 [69] and, more recently, the superposition of several domes, such as the Domo Cluster by Arketiposchile [24] in 2012 and the geodesic dwelling by Ecoproyecta [50] in 2016.

This work aims to identify whether the use of domes produces energy savings and the thermal qualities inherent to their morphology. For this purpose, different analyses are carried out by modifying different calculation parameters for the materials used and the construction solutions employed. The demand for air conditioning, cooling and comfort conditions is included in two models (case studies 2 and 3). This will be compared to the geometry of a parallelepiped house with the same surface area (case study 1).

This study is based on a frequency IV 1/2 dome for a single-family house with the same surface area as for cases 1 and 2. The progress of this research will make it possible to determine the convenience of dividing the surface area into smaller domes, thus, sectioning the uses of the house, as in case 3. In addition, a detailed study is also carried out on the behaviour of the materials used, the construction solutions employed and how these decisions affect the thermal behaviour of the architectural complex.

All this is carried out using the energy analysis software DesignBuilder v.7.0.1.006, assessing which is the best solution and whether it really represents a sustainable improvement compared to traditional construction.

To do this, it is considered essential to determine the most appropriate length for the crosspieces that make up the structure of the dome, as this will condition the existing joint metres for this morphological solution and, therefore, the infiltrations. It is important to add, at this point, that this aspect can be reduced depending on the type of frequency used. Frequency IV 1/2 uses a greater number of edges (712) than frequencies III 5/12 (225) or III 7/12 (315). Frequency IV 1/2 is chosen as the most unfavourable case. It is also considered essential to consider the spatial suitability of the total volume generated. At frequency IV 1/2 the clear height is 5 m.

At the construction level, the geodesic dome is made up of triangular pieces pre-assembled in a workshop. These pieces are placed edge by edge on site. The construction process is carried out in situ and dry-jointed (Figure 5).

Figure 5. Wooden joints used in case studies 2 and 3.

The virtual representation of the dome under study is a fundamental element for understanding its application and behaviour. However, in addition to its morphological information, it is essential to add specific structural information in order to obtain a more exhaustive and operative control both of the manufacture and of the subsequent construction of the geodesic dome. This information is obtained by introducing different variables into the Geodesic Dome Calculator web platform [70], obtaining a structural BIM model with the complete dimensioning of each of the crosspieces, their layout, cutting angles and vertex junctions, as well as the representation of the flat faces of the dome (Figure 6a). It is also necessary to draw in detail the heights of each of the bars with respect to the 0 level (base plane) on which they must be placed (Figure 6b).

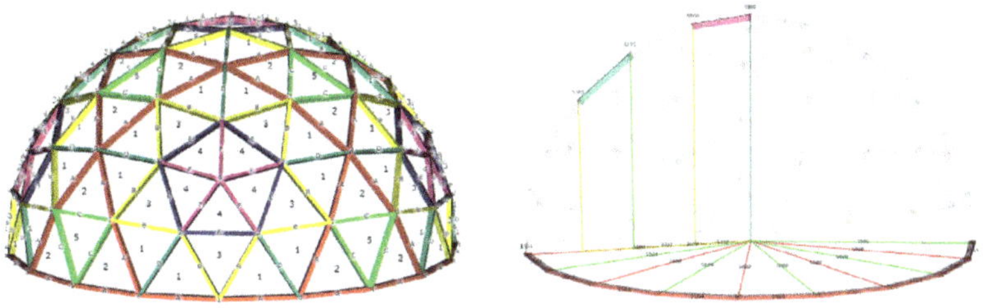

(**a**) (**b**)

Figure 6. Virtual representation of the dome: (**a**) designation of bars, faces and vertices; (**b**) 3D bar placement scheme. Source: own elaboration from www.acidome.ru (accessed on 3 December 2023).

As mentioned above, the hemisphere used is of frequency IV, so that the base is completely parallel to the horizontal plane. An icosahedron is, therefore, used as the base platonic solid. Each of the above representations provides specific information for the machining and prefabrication of the construction elements, and can be used to automate the processes or to carry them out in a more efficient and controlled manner.

In addition, the data related to the real construction are also integrated into the model, considering a basic façade module (Figure 7a) formed by the following:

- A triangular piece generated by wooden crosspieces.
- Interior thermal insulation (cellulose).
- Interior wood panel (Oriented Strand Board-OSB).
- Exterior wood panel (Oriented Strand Board-OSB).
- Interior finish (according to project requirements).
- Exterior finish (mineral-based paint).
- Joint coating (structural silicone).

The enclosure modules can also be replaced by carpentry modules to generate the façade openings (Figure 7b). Considering this, it is possible to interact with the enclosure by modifying its conditions, materiality and components, by means of the geodesic dome modulation.

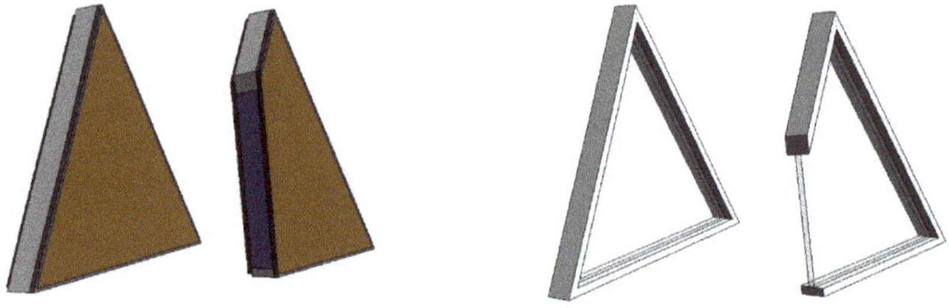

(**a**) (**b**)

Figure 7. (**a**) Representation of facade constructive module; (**b**) representation of carpentry constructive module. Source: own elaboration using Autodesk Revit software v.2020.1.

As a result, a virtual model is obtained with which to document and visualise the construction of the dome, to quantify and prefabricate the resulting modules, and also to provide morphological and spatial relationships for further comparison to traditional geometries.

When considering a similar geometrical solution, it is essential to bear in mind that the usable space in the interior of a dome will be smaller. This is due to the fact that the lower ring, 97 cm high and 35 cm deep, is a non-habitable volume to be subtracted from the total (Figure 8). Fuller, in his Carbondale dwelling, used this depth as a bookcase.

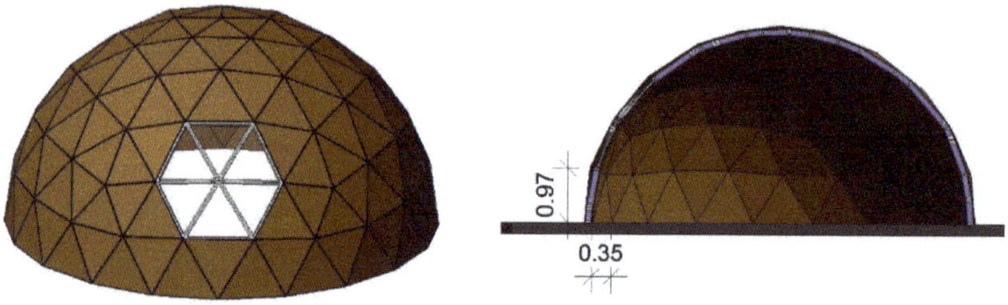

Figure 8. Global view and section of the virtual model. Source: own elaboration using Autodesk Revit software v.2020.1.

2.3. Energy Analysis of the Geodesic Dome

In recent years, computational geometry has allowed for a new formulation of the geometric bases and numerical procedures used to generate any type of spatial dome [71]. Today, digital tools are crucial for the design and manufacture of this type of non-rectilinear geometry [72]. This is why, in addition to the energy analysis of the geodesic dome, this research considers how construction costs have a direct influence on whether a house is considered to be sustainable or not.

Therefore, the Autodesk software REVIT 2020.1 and the visual programming application Revit Dynamo v.2.6.1 are used as the parametric modelling tools for the development of this research (Figure 9).

Figure 9. Code for frequency IV geodesic dome generation. Source: own elaboration using Dynamo.

Autodesk REVIT, like its counterparts from other commercial companies, allows for the creation, visualisation and documentation of parametric virtual models which are very close to the construction reality. By means of the Autodesk REVIT software, the envelope studied is modelled "AsBuild", obtaining material calculation tables that are faithful to the construction reality and allow for the costs and construction times to be determined with greater precision.

Different simulation options were considered for carrying out the proposed energy analysis (Open Studio, Therm, ECOTECT, Lider-Calener HULC, etc.); after rejecting those tools that had certain limitations when analysing complex geometries and/or compatibility issues with other modelling software, the Climate Consultant v.6.0 and DesignBuilder v.7.0.1.006 applications were finally chosen.

Climate Consultant provides detailed climate information for specific locations by reading EPW [32] (Energy Plus Weather) formats. It allows for an understanding of how lighting, ventilation and thermal comfort affect the design. DesignBuilder is a software that integrates the EnergyPlus [33] calculation module, and is one of the most advanced tools in this field of architectural design.

EnergyPlus was developed in the United States and is continuously updated; it contains climate data files even up to hourly and sub-hourly levels [73]. This energy calculation and simulation engine requires graphic interfaces, such as DesignBuilder, OpenStudio, Ecotect, etc., for its understanding and use, due to the complexity of the data.

With regard to the air conditioning/cooling demand, the energy requirements necessary to satisfy the comfort conditions (both for air conditioning or cooling) will be obtained. Energy Plus allows for the detailed incorporation of HVAC (heating, ventilation and air conditioning) installations to assess energy performance, thermal comfort and energy efficiency. The programme considers the summer and winter periods to establish calculation priorities. All indoor environmental conditions within the living area were studied, according to the climate data collected from Climate Consultant.

In relation to the comfort conditions, the calculation of the comfort/discomfort hours inside the living spaces will be carried out by means of the standardised calculation parameters of the ASHRAE (American Society of Heating, Refrigerating and Air-Conditioning Engineers) Standard 55 [74]. For the thermal balance, the thermal gains are shown considering different factors: solar, occupancy, thermal transmittance of the envelope, etc. The value of the renovations/hour required for each space is also obtained, as well as the necessary infiltration flow rate due to natural ventilation.

In short, for this study, first of all, a geometric model is created in DesignBuilder and the data obtained are analysed according to certain established parameters (thermal bridges, materiality, thermal resistance, meeting solution and carpentry solution); subsequently, a series of comparative analyses are carried out, modifying the different parameters mentioned above to adapt them to the construction and economic needs that may exist for a project with these characteristics. It also studies the different constructive casuistry that can damage the energy performance of this type of housing, making a comparison to a traditional house with similar characteristics.

3. Results

3.1. Climatic Study of the Area

The geographical location of the project is fundamental for the analysis of the EPW metadata associated with the site, as well as the actual orientation and the orientation of the project. The location of the case studies is the high plateau area of Murcia (Spain), as it is the actual location of two geodesic dome projects executed by the architectural firm Ecoproyecta. The area in which they are located corresponds to the climate zone Bsh (cold semi-arid) of the Köppen climate classification. For the climatic analysis, the data used are those collected by the nearest climatic station in the region of Murcia, very close to the work area (Figure 10).

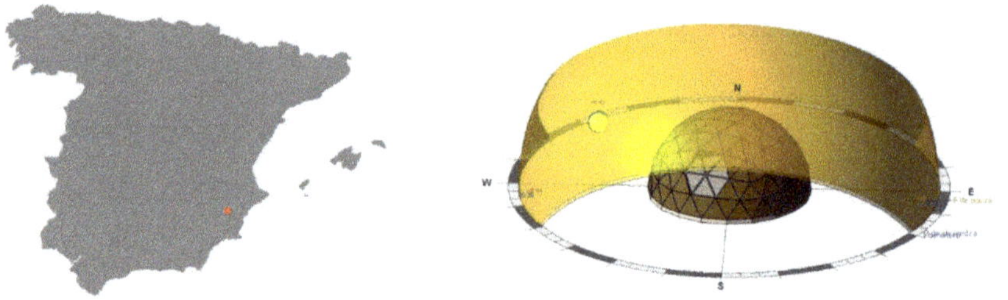

(**a**) (**b**)

Figure 10. Location (**a**) and orientation (**b**) of case studies. Source: own elaboration using Autodesk Revit software.

The calculation assumes that the user adjusts to the climate by wearing the appropriate clothing, such as a sweatshirt and long trousers in winter and lighter clothing in the summer months. Additionally, a metabolic activity level of 1.1 met is assumed, which corresponds to sedentary activities. The rest of the values are obtained from "ASHRAE Standard 55, current Handbook of Fundamentals Comfort Model" (Figure 11).

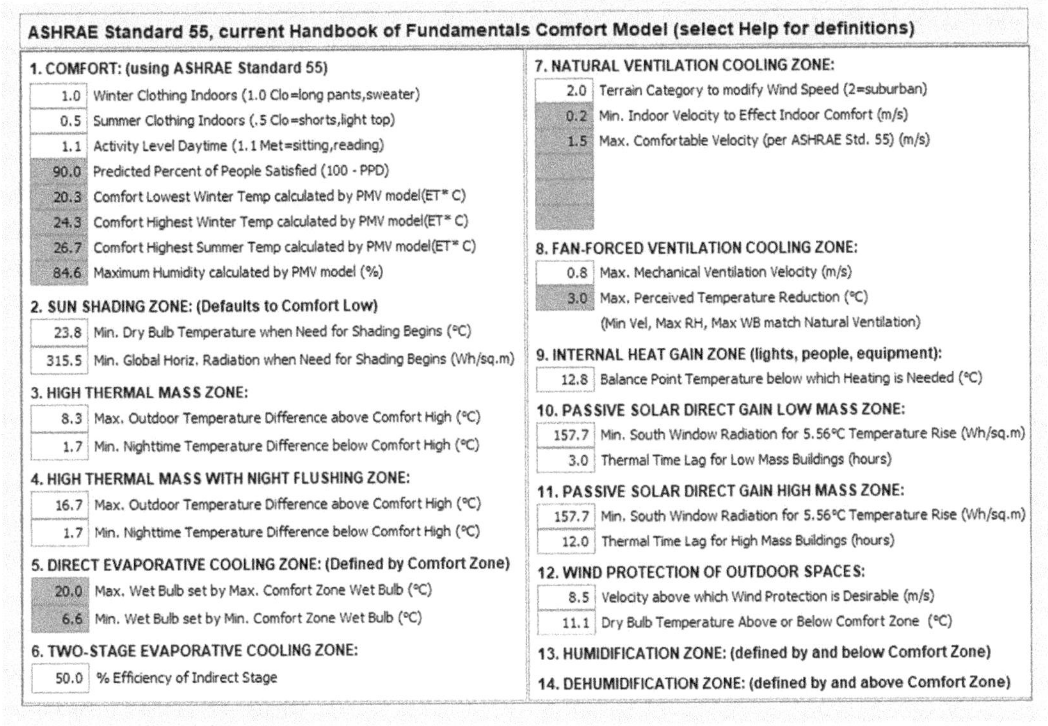

Figure 11. Editable parameters associated with the ASHRAE Standard 55 comfort model (own elaboration using Climate Consultant 6.0 software).

The range of dry bulb temperatures recorded per month in the region and the annual average range are obtained. The upper part of the grey band represents the winter comfort temperature and the lower part represents the summer comfort temperature. The green-coloured area represents the maximum and minimum temperatures. Yellow represents the average of the recorded high and low temperatures. The global average temperature can be read in the open band between the yellow shaded areas.

The software makes it possible to designate, in a simple way and for any type of project, the percentage of annual hours above or below the values recorded as comfortable. It can be seen that the average temperature in the region is around 18 °C, and that the annual temperature varies between 3 °C and 35 °C (Figure 12). It is important to add that the latest research by Espín et al. [75] warns of increasingly warmer thermal conditions in southeastern Spain, with less cold and less comfortable thresholds in winter and an increase in thermal discomfort in summer.

Figure 13 represents, in two phases, the diurnal dry and wet bulb thermal variation flux per month (blue area and red lines), and the solar radiation per square metre of surface area. The maximum dry bulb temperature reaches values of 33.7 °C in August and a minimum of 2 °C in January.

Figure 14 shows the variation in solar radiation by surface, considering normal, horizontal and diffuse radiation.

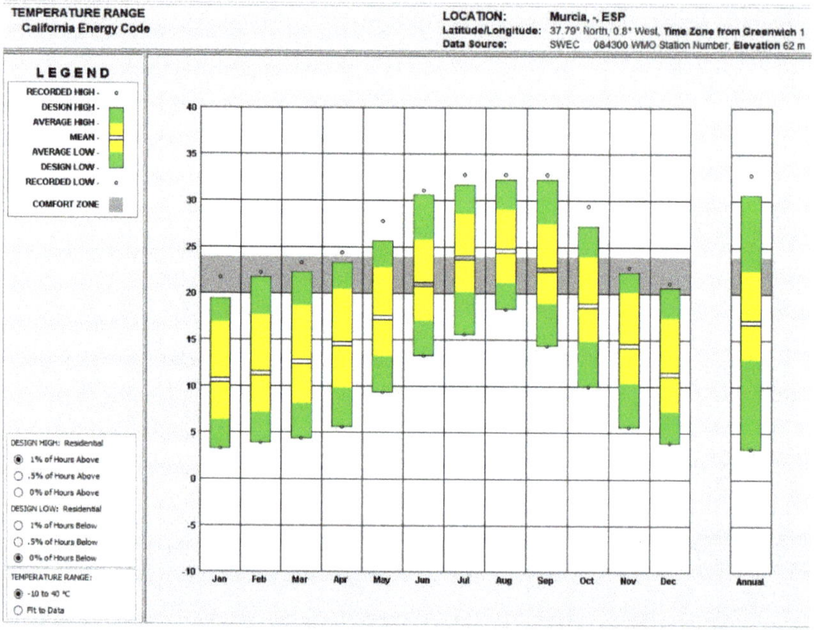

Figure 12. Simulation results: temperature range. Source: own elaboration using Climate Consultant 6.0 software.

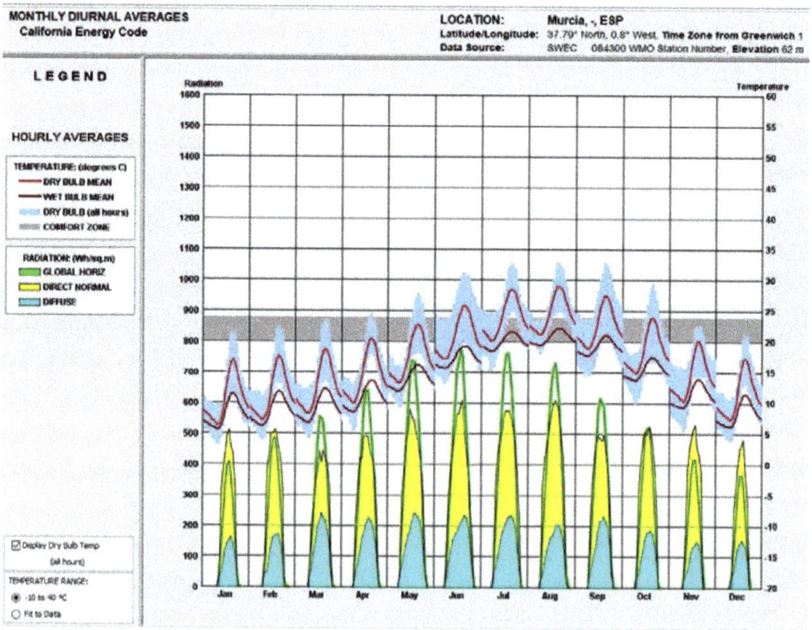

Figure 13. Simulation results: monthly average temperatures and radiation. Source: own elaboration using Climate Consultant 6.0 software.

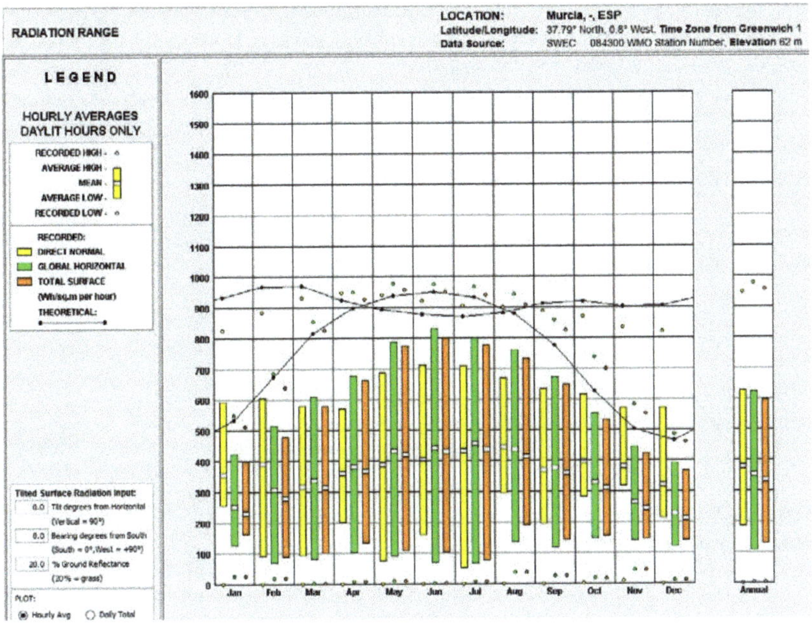

Figure 14. Simulation results: Radiation range. Source: own elaboration using Climate Consultant 6.0 software.

Figure 15a–c represent the mean sky coverage by cloud cover, the light intensity measured in lux, and the mean monthly wind speed measured in m/s, respectively.

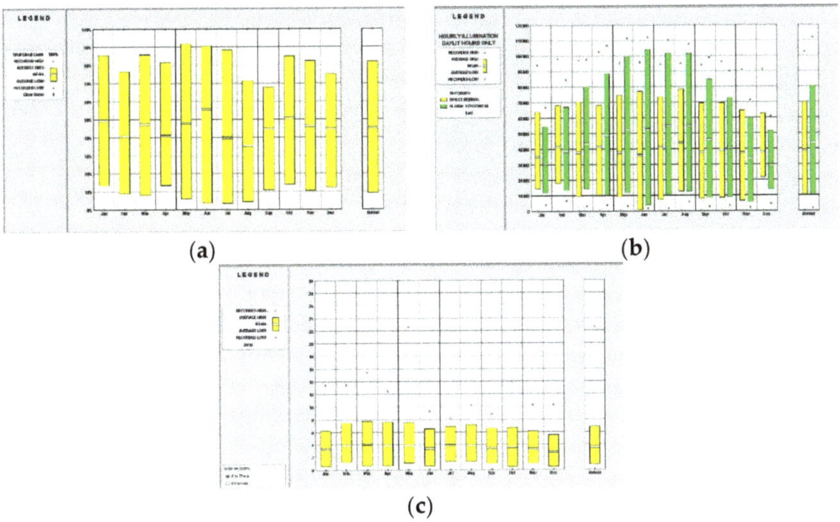

Figure 15. Simulation results: (**a**) mean sky coverage by cloudiness; (**b**) light intensity measured in lux; (**c**) mean monthly wind speed measured in m/s. Source: own elaboration using Climate Consultant 6.0 software.

Figure 16 represents the orientation and altitude of the sun for every 15 min of the year. The yellow zone indicates the comfort conditions as long as the dry bulb temperature is within the thermal comfort zone. The red zone indicates overheating when the dry bulb temperature is above the comfort zone. The blue zone represents cold conditions when the dry bulb temperature is below the comfort zone.

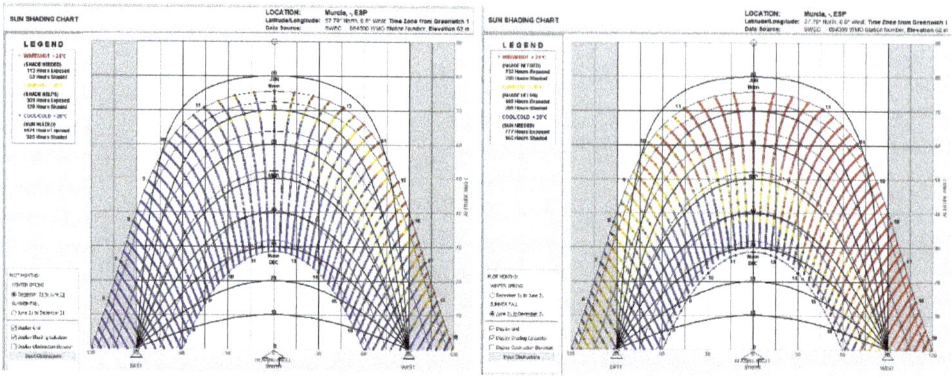

Figure 16. Simulation results: orientation and altitude of the sun for every 15 min of the year. Source: own elaboration using Climate Consultant software.

At the design level, in practice, it is important to note that openings (windows, openings, etc.) should be fully exposed to solar radiation if they are in the blue zone, while they should be fully shaded (no exposure to solar radiation) if they are in the red zone. The software itself integrates a tool that helps to make a pre-design for the windows, allowing for the configuration of the most appropriate arrangement of the necessary solar protection in each case.

Figure 17 represents the psychometric chart, as a confluence of three climatic attributes that influence the users' feelings of comfort. These three variables on the graph are the dry bulb temperature, wet bulb temperature and relative humidity. Their combination determines the different zones: above the comfort zone (red), below the comfort zone (blue) or within the comfort zone (green). On the other hand, this psychrometric chart can also be used to show how to design exterior enclosures that can modify or filter external climatic conditions to create more comfortable interior spaces.

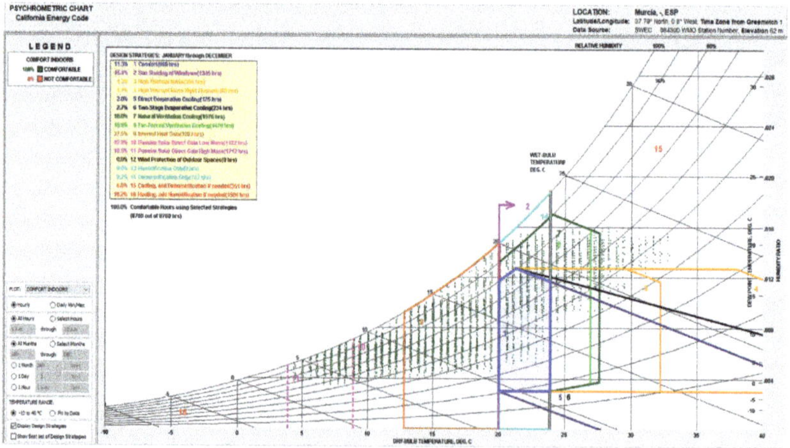

Figure 17. Simulation results: Murcia psychrometric chart. Source: own elaboration using Climate Consultant software.

Considering the site studied, the protection of the windows by shading, vegetation, eaves, etc., represents a 15.4% effectiveness in the design. This effectiveness refers to the number of hours of comfort gained within the living spaces designed in the interior. On the other hand, considering the climatic conditions, only 11.3% of the hours (988 h per year) are within the comfort range. The measures or design strategies indicated in the top left of the abacus in Figure 17 serve to improve these conditions by increasing the range of hours in comfort (Table 2).

Table 2. Design strategies from highest to lowest impact on indoor comfort.

Design Strategies				
37.50%	9	Internal heat gain	3283	hrs
19.50%	11	Passive solar direct gain high mass	1712	hrs
18.20%	16	Heating adds humidification (if needed)	1594	hrs
18%	7	Natural ventilation cooling	1576	hrs
16.80%	8	Fan-forced ventilation cooling	1470	hrs
15.40%	2	Sun shading of windows	1345	hrs
12.80%	10	Passive solar direct gain low mass	1122	hrs
11.30%	1	Comfort	988	hrs
8.20%	14	Dehumidification only	722	hrs
4.70%	4	High thermal mass night flushed	409	hrs
4.20%	3	High thermal mass	365	hrs
4.00%	15	Cooling adds dehumidification (if needed)	351	hrs
2.70%	6	Two-stage evaporative cooling	234	hrs
2.00%	5	Direct evaporative cooling	175	hrs
0.00%	12	Wind protection of outdoor spaces	0	hrs
0.00%	13	Humidification only	0	hrs

These strategies are evaluated from the highest to lowest impact on indoor comfort. Therefore, it is necessary to consider that, in relation to the gains from the domestic use of the space itself, the characteristic morphology of the geodesic domes will allow these gains to have an even greater impact, by facilitating and improving the movement of internal air in the space.

3.2. *Study of Thermal Behaviour*

3.2.1. Comparison of Parallelepiped and Geodesic Dome Geometry

The study of models 1 and 2 is carried out by applying the same environmental conditions and the same construction systems for a proper comparison of the resulting energy analysis with the Design Builder programme. Regardless of the material configuration of the envelope elements, it is essential to take into account the values of thermal transmittance (U), and to check that these values correspond to those reflected in the required thermal transmittance tables established in the CTE DB-HE 1 (Código Técnico de la Edificación, Documento Básico HE1 Limitación de la demanda energética) [76] for the climatic zone under study. The U value resulting from this envelope is 0.257 W/m^2 K. The resulting U-value of the glazed part applying the ISO 15099/NFRC standard is 1.761 W/m^2 K.

After the simulations are carried out, it can be seen that case study 2 (dome) invests 9% less in zone heating than case study 1 (reference block), due to the fact that the envelope behaves worse, generating more energy losses per square metre overall. The same is true for external ventilation, with case study 2 generating 9% less in losses than case study 1.

In relation to the cooling design, the energy and thermal envelope performance is analysed for an unfavourable summer day (July 15th) for both case study 1 (Figure 18) and case study 2 (Figure 19).

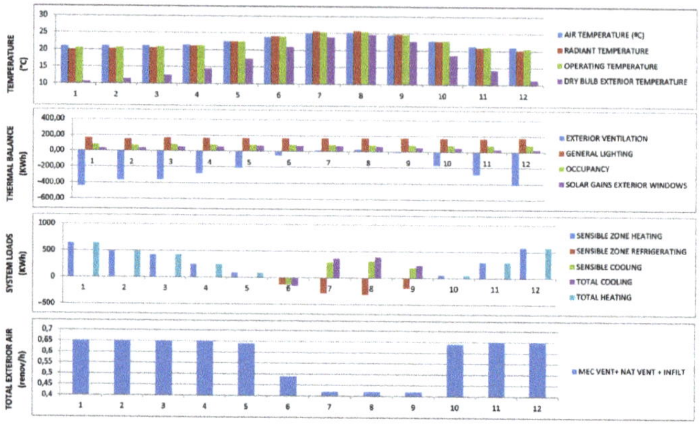

		January	February	March	April	May	June	July	August	September	October	November	December
AIR TEMPERATURE (ºC)	°C	21.06	21.11	21.21	21.48	22.47	23.89	25.06	25.17	24.59	22.84	21.32	21.07
RADIANT TEMPERATURE	°C	20.02	20.25	20.56	21.12	22.43	24.05	25.46	25.57	24.77	22.72	20.77	20.1
OPERATING TEMPERATURE	°C	20.54	20.68	20.88	21.3	22.45	23.97	25.26	25.37	24.68	22.78	21.05	20.58
DRY BULB EXTERIOR TEMPERATURE	°C	10.6	11.4	12.59	14.51	17.38	20.94	23.88	24.59	22.56	18.7	14.35	11.34
EXTERIOR VENTILATION	KWh	-442.08	-369.81	-362.26	-282.30	-210.82	-52.57	14.53	27.33	-11.75	-170.91	-282.23	-410.37
GENERAL LIGHTING	KWh	167	153.36	170.25	165.7	167	165.7	170.25	167	168.95	167	162.46	173.5
OCCUPANCY	KWh	81.41	74.75	82.99	80.77	81.41	80.77	82.99	81.41	82.36	81.41	79.19	84.57
SOLAR GAINS EXTERIOR WINDOWS	KWh	33.26	36.98	51.08	54.4	63.89	64.32	67.47	62.64	53.03	45.31	33.82	30.99
SENSIBLE ZONE HEATING	KWh	643.05	492.37	427.14	255.06	89.68	0	0	0	0	59.59	308.87	582.06
SENSIBLE ZONE REFRIGERATING	KWh	0	0	0	0	0	-134.61	-296.57	-317.08	-197.13	0	0	0
SENSIBLE COOLING	KWh	0	0	0	0	0	-134.72	296.92	317.46	197.34	0	0	0
TOTAL COOLING	KWh	0	0	0	0	0	-162.98	363.17	390.74	243.28	0	0	0
TOTAL HEATING	KWh	643.05	492.37	427.14	255.06	89.68	0	0	0	0	59.59	308.87	582.06
MEC VENT+ NAT VENT + INFILT	renov/h	0.65	0.65	0.65	0.65	0.64	0.49	0.42	0.42	0.42	0.64	0.65	0.65
RELATIVE HUMIDITY	%	40.49	41.29	41.22	44.89	53.44	63.41	67.32	69.45	69.01	58.62	48.35	40.26
DISCOMFORT HOURS SUMMER CLOTHING	hours	744	672	744	719	658.33	468.17	596.67	665.5	571	601.83	720	744
DISCOMFORT HOURS FOR WINTER CLOTHES	hours	299.67	246.5	236	146.67	50.67	513.83	744	744	647.83	98.83	171.5	285.67
DISCOMFORT HOURS ANY CLOTHING	hours	299.67	246.5	236	146.67	14.33	317.67	596.67	665.5	535.33	1.83	171.5	285.67
TOTAL LATENT LOAD	KWh	52.05	47.79	53.06	51.64	52.05	51.64	53.06	52.05	52.65	52.05	50.63	54.07

Figure 18. Simulation results for temperatures, internal and solar gains for case study 1.

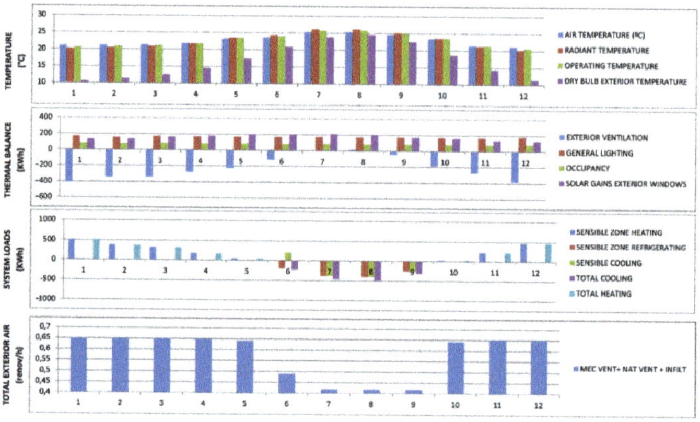

		January	February	March	April	May	June	July	August	September	October	November	December
AIR TEMPERATURE (ºC)	°C	21.07	21.18	21.36	21.83	23.24	23.68	25.35	25.43	24.65	23.52	21.52	21.1
RADIANT TEMPERATURE	°C	20.19	20.53	20.94	21.74	23.48	24.33	26.1	26.18	25.18	23.61	21.18	20.27
OPERATING TEMPERATURE	°C	20.63	20.85	21.15	21.78	23.36	24	25.73	25.81	24.92	23.56	21.35	20.68
DRY BULB EXTERIOR TEMPERATURE	°C	10.6	11.4	12.59	14.51	17.38	20.94	23.88	24.59	22.56	18.7	14.35	11.34
EXTERIOR VENTILATION	KWh	-410.37	-345.35	-341.71	-274.35	-224.71	-117.58	4.04	16.12	-42.51	-184.18	-268.81	-381.58
GENERAL LIGHTING	KWh	169.26	155.43	172.55	167.94	169.26	167.94	172.55	169.26	171.24	169.26	164.65	175.85
OCCUPANCY	KWh	82.5	75.76	84.11	81.86	82.5	81.86	84.11	82.5	83.47	82.5	80.26	85.72
SOLAR GAINS EXTERIOR WINDOWS	KWh	135.27	137.07	164.53	176.25	197.41	198.95	207.68	197.75	167.38	159.32	135.16	129.25
SENSITIVE ZONE HEATING	KWh	519.29	384.54	321.11	169.32	44.34	0	0	0	0	30.75	221.18	466.93
SENSITIVE ZONE REFRIGERATING	KWh	0	0	0	0	0	-192.2	-376.53	-395.59	-254.52	0	0	0
SENSIBLE COOLING	KWh	0	0	0	0	0	192.42	-377.14	-396.25	-254.91	0	0	0
TOTAL COOLING	KWh	0	0	0	0	0	-231.04	-458.09	-484.59	-312.04	0	0	0
TOTAL HEATING	KWh	519.29	384.54	321.11	169.32	44.34	0	0	0	0	30.75	221.18	466.93
MEC VENT+ NAT VENT + INFILT	renov/h	0.65	0.65	0.65	0.65	0.64	0.49	0.42	0.42	0.42	0.64	0.65	0.65
RELATIVE HUMIDITY	%	40.76	41.43	41.18	44.29	51.35	62.05	64.88	66.92	67.27	56.6	48.07	40.51
DISCOMFORT HOURS SUMMER CLOTHING	hours	744	670.67	727.5	672.17	491	443	553.5	602.67	534.33	473	698.83	744
DISCOMFORT HOURS FOR WINTER CLOTHES	hours	298.83	242	230.17	150.5	205.33	488.83	744	744	643.83	236.5	173	282.83
DISCOMFORT HOURS ANY CLOTHING	hours	298.83	242	227	127.17	10	240.67	553.5	602.67	475.67	20.67	161.83	282.83
TOTAL LATENT LOAD	KWh	52.75	48.44	53.78	52.34	52.75	52.34	53.78	52.75	53.36	52.75	51.31	54.8

Figure 19. Simulation results for temperatures, internal and solar gains for case study 2 (geodesic dome).

The values obtained from both analyses have a dispersion of less than 3%, which indicates that the behaviour of both architectures is similar at coping with the cooling of the area. The major difference between the two cases lies in the distinction of building elements (wall cover). Case study 2, being a dome, has a structure composed of prefabricated modules throughout the envelope, which allows for an advantage against solar radiation due to the inclination of the faces.

Table 3 presents a comparative analysis of the data associated with case study 1 (reference block) and case study 2 (dome) for the annual energy simulation. The data are represented on a monthly basis in order to identify possible seasonal differences between the two analyses (although it is also possible to choose to represent the graphs on an annual or daily basis).

Table 3. Comparison of results.

			January	February	March	April	May	June	July	August	September	October	November	December	TOTAL
GENERAL LIGHTING	Case study 1	KWh	167	153.36	170.25	165.7	167	165.7	170.25	167	168.95	167	162.46	173.5	1998.17
	Case study 2	KWh	169.26	155.43	172.55	167.94	169.26	167.94	172.55	169.26	171.24	169.26	164.65	175.85	2025.19
OCCUPANCY	Case study 1	KWh	81.41	74.75	82.99	80.77	81.41	80.77	82.99	81.41	82.36	81.41	79.19	84.57	974.03
	Case study 2	KWh	82.5	75.76	84.11	81.86	82.5	81.86	84.11	82.5	83.47	82.5	80.26	85.72	987.15
SOLAR GAINS EXTERIOR WINDOWS	Case study 1	KWh	33.26	36.98	51.08	54.4	63.89	64.32	67.47	62.64	53.03	45.31	33.82	30.99	597.19
	Case study 2	KWh	135.27	137.07	164.53	176.25	197.41	198.95	207.68	197.75	167.38	159.32	135.16	129.25	2006.02
SENSIBLE ZONE HEATING	Case study 1	KWh	643.12	492.38	427.08	255.07	89.71	0	0	0	0	59.64	308.74	581.96	2857.70
	Case study 2	KWh	519.29	384.54	321.11	169.32	44.34	0	0	0	0	30.75	221.18	466.93	2157.46
SENSIBLE ZONE REFRIGERATING	Case study 1	KWh	0	0	0	0	0	−134.55	−296.37	−316.21	−197.05	0	0	0	−944.18
	Case study 2	KWh	0	0	0	0	0	−192.2	−376.53	−395.59	−254.52	0	0	0	−1218.84
RELATIVE HUMIDITY	Case study 1	%	40.49	41.29	41.22	44.89	53.44	63.41	67.32	69.45	69.01	58.62	48.35	40.26	53.15
	Case study 2	%	40.76	41.43	41.18	44.29	51.35	62.05	64.88	66.92	67.27	56.6	48.07	40.51	52.11
DISCOMFORT HOURS SUMMER CLOTHING	Case study 1	hours	744	672	744	719	658.33	468.17	596.67	665.5	571	601.83	720	744	7904.50
	Case study 2	hours	744	670.67	727.5	672.17	491	443	553.5	602.67	534.33	473	698.83	744	7354.67
DISCOMFORT HOURS FOR WINTER CLOTHES	Case study 1	hours	299.67	246.5	236	146.67	50.67	513.83	744	744	647.83	98.83	171.5	285.67	4185.17
	Case study 2	hours	298.83	242	230.17	150.5	205.33	488.83	744	744	643.83	236.5	173	282.83	4439.82
DISCOMFORT HOURS ANY CLOTHING	Case study 1	hours	299.67	246.5	236	146.67	14.33	317.67	596.67	665.5	535.33	1.83	171.5	285.67	3517.34
	Case study 2	hours	298.83	242	227	127.17	10	240.67	553.5	602.67	475.67	20.67	161.83	282.83	3242.84
TOTAL LATENT LOAD	Case study 1	KWh	52.05	47.79	53.06	51.64	52.05	51.64	53.06	52.05	52.65	52.05	50.63	54.07	622.74
	Case study 2	KWh	52.75	48.44	53.78	52.34	52.75	52.34	53.78	52.75	53.36	52.75	51.31	54.8	631.15

The air temperature, radiant temperature and operating temperature are slightly higher in case study 2 (dome) and are also more stable, with no significant differences in the average annual temperatures.

It is important to note that, due to its layout and geometry, the dome has a higher solar gain index (2006.02 kWh) compared to the reference block (597.19 kWh). Figure 20 shows the monthly differences between both case studies. Both constructions have the same façade treatment and orientation; however, the inclination of the dome modules directly affects this analysis. For this reason, the dome has a higher efficiency in colder months, but its geometry reduces its efficiency in warmer months as it receives a larger area of solar radiation. This variable makes a significant difference between the two geometries. Depending on the climate where it is located, solar control of the geodesic dome openings will be of vital importance for the best efficiency.

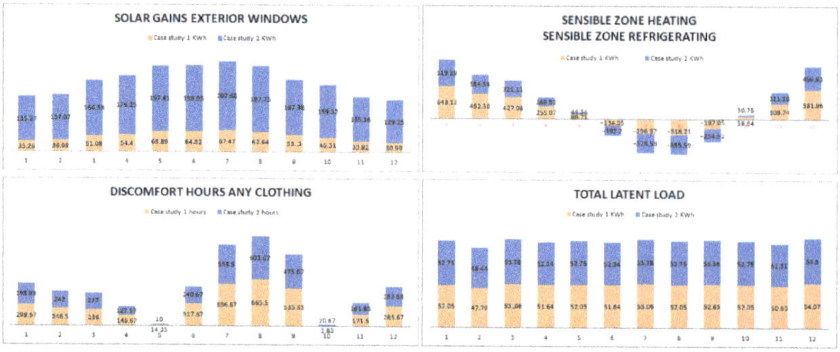

Figure 20. Simulation results for solar gains, heating and refrigerating, discomfort and latent load.

The general illumination, occupancy and latent load indices are maintained with very little dispersion between the two case studies. The relative humidity increases slightly in case 1 (reference block). The data show that the average annual relative humidity is 53.15% in case 1, while in case 2 it is 52.11% (dome).

In relation to the number of total hours of discomfort according to the type of clothing and month, it should be noted that, according to the conditions established for both cases, the result is 3517.34 total hours of discomfort for the reference block and 3242.84 total hours for case study 2 (dome). It should be noted that the hours of discomfort are more pronounced in the summer months, being lower in case study 2.

Electricity consumption between January and May and October and December is higher for the reference block than for the dome, due to heating needs. On the other hand, from June to September the cooling needs require a higher electricity consumption in the dome.

Regarding the overall consumption balance, it is possible to state that case 1 (traditional block) consumes more electricity throughout the year to maintain comfort conditions inside the house. In relation to fuel consumption, the CO_2 production graphs follow the same balance for the different seasons, i.e., the higher the fuel demand, the higher the CO_2 impact.

Thus, so far, the results of the simulations have been analysed for the two main case studies: case 1 (simplified traditional architecture block) and case study 2 (geodesic dome). Neither of them has any defined type of uses or internal partitions, in order to check the behaviour related to the global building envelope under the same conditions as the internal heat gain.

3.2.2. Comparison of the Single Geodesic Dome and Double Geodesic Dome

In this section, we will initially compare the geometries of both cases 2 and 3 to better understand their influence on energy efficiency.

Case study 2: The geodesic dome with a radius of 5 m contains an area of 77.25 m^2 (Figure 21b). When we divide this surface area into two domes with a radius of 3 m and 4 m (27.81 m^2 and 49.44 m^2, respectively), the surface area of 77.25 m^2 remains the same. It is important to remember that this surface area is the same as the reference block also studied (case study 1).

Height from base	m	3	4	5
Platform radius	m	3	4	5
Platform area,	m^2	27.81	49.44	77.25
Coverage area	m^2	55.48	98.63	154.11
Sizes (units)				
Faces		6 (160)	6 (160)	6 (160)
Edges		16 (480)	16 (480)	16 (480)
Vertices		4 (91)	4 (91)	4 (91)
Beams 150×50mm				
Total length of beams	m	429.61	872.82	716.02
Total volume of beams	m^3	2.83	3.9	4.98
Beam length	mm	760 – 975	1013– 1300	1266 – 1625
Angle between faces	°	169.9 – 172.20	169.9 – 172.20	169.9 – 172.20
Triangles				
Min. Height	mm	617 – 844	823 – 1126	1028 – 1407
Max. Side	mm	886 – 975	1181 – 1300	1476 – 1625

Figure 21. Results for the surface area, coverage area, and beams of the two geodesic domes (3 m and 4 m. diameter, respectively) in case study 3 (**a**) and the single geodesic dome (5 m. diameter) in case study 2 (**b**).

In a similar way, the envelope area remains unchanged. The 154.11 m^2 of roof in case 2 coincides with the sum of 55.48 m^2 and 98.63 m^2 of both domes in case 3 (Figure 21a).

In this way, the ratio of the occupied surface and envelope is maintained with respect to case 2 (single-dome model). However, it must be taken into account that the chimney effect in case 2 (single dome) causes more infiltration losses than in case 3 (two domes), as it has a lower height [77].

There is one aspect of the geometry that does substantially modify the behaviour of one solution or the other: the length of the joints. The dome maintains its number of edges (480), but varies in the length of its edges and, thus, the length of its joints. The 5 m radius dome has 29% less joint length than the sum of the 3 m and 4 m domes in case 3. In linear metres, we are talking about 716 m in case 2 versus 1.001 m (429 + 572) in case 3.

The division of case 2 versus case 3 allows for the distinction of uses. The simulation considers the existence of a differentiation of uses (night use and day use) and aims to check whether the division of uses into blocks of smaller volumetric dimensions affects the energy performance and, if so, analyses the implications. In addition, the internal heat gain changes when there are two spaces with different uses. In this sense, the research by Rainer Elsland et al. [78] predicted that an increased ownership of appliances is the key driver of increased internal heat gains.

In case study 3, the reduction in the occupied volume is indeed significant, with a total occupied volume of 176.6 m^3. In case study 2 (a single geodesic dome), this volume is 225 m^3, and in the reference block it is 270 m^3. With the same occupiable surface as the reference block, cases 2 and 3 reduce the interior volume by 83% and 65%.

After the geometric analysis, a detailed study of the energy performance of the two geodesic domes in case study 3 was carried out, and compared the results with those obtained for case study 2:

- Case 3 requires 24% less energy for lighting than case 2: 20,775 Wh/m^2 versus 27,060 Wh/m^2.
- Case 3 requires 52% less energy consumption for cooling than case 2: 5215 Wh/m^2 vs. 10,025 Wh/m^2.
- Case 3 requires 36% less energy loss for external ventilation than case 2: 22,086 Wh/m^2 compared to 34,293 Wh/m^2.
- The energy input for heating remains stable between case study 3 and case study 2.
- The discomfort hours remain stable between case study 3 and case study 2 for the set conditions.

As it is possible to set different zones of use in case study 3, the calculation parameters vary according to the imposed needs. In this case, considering the bedroom use and living room use separately for each block, the expenditure on lighting was recalculated, as was the expenditure for other sources of electricity consumption, obtaining an annual calculation of 60,966 Wh/m^2 for case study 3 that compared to 70,534 Wh/m^2 for case study 2, which represents an improvement of 14% in efficiency.

4. Discussion

The results presented in the previous section allow for a specific discussion of each study area, so that the geometric aspects, temperatures, solar gains and comfort hours can be analysed.

4.1. Comparative Analysis of the Geometry, Volume and Thermal Envelope

At the level of geometrical resolution, an interesting difference can be seen between case study 1 (reference block) and case study 2 (dome). Although the occupation surface is the same, 21% less envelope is required to cover the same needs. This characteristic has a direct impact on the material cost of the building per unit area. The load-bearing capacity of both structures or the need for auxiliary means is also significant, although this is not the aim of this research.

The volume delimited by reference block 1 is the largest one (270 m^3 for a clear height of 3.5 m). The use of a geodesic dome of frequency IV 1/2 with a radius of 5 m and 5 m height implies a saving of 17% in volume. Frequencies III 5/12 and III 7/12 significantly reduce this aspect compared to the frequency type IV 1/2 studied.

The comparison of the volumes between case 2 and case 3 is significant. Case study 3 has a total volume of 176.6 m^3, while maintaining the same total area for occupation. In case study 2, which consists of a single geodesic dome, the volume is 225 m^3. Case 3, with a dome of 3 m in height and another of 4 m in height, 4 m represents a 22% savings in volume compared to case 2, with a dome of 5 m in height. In other words, the use of geodesic domes with differentiated uses allows for a savings of 35% in volume for the same occupied surface area. It is true that the division into two uses, day and night, restricts the design possibilities, by reducing the free 5 m height of a single dome into separated 4 m and 3 m domes. However, at the same time, it has a very positive impact on consumption, as it means less consumption for heating and cooling.

4.2. Indoor Temperatures

Figure 17 shows that the range of dry bulb temperatures that provide comfort is between 20 °C and 24 °C, with humidity levels between 20 and 80%. Therefore, the setpoint values in DesignBuilder are set accordingly. The preset values (20 °C indoor temperature for heating with a setpoint temperature of 19 °C, and 24 °C indoor temperature for cooling with a setpoint temperature of 27 °C) strongly influence the hours of comfort and discomfort.

The maximum temperature for the extended comfort zone during summer is 29 °C, with humidity levels ranging from 20% to 80%. With mass cooling, the temperature can reach up to 31 °C, provided that the humidity remains between 5% and 80%. However, it should be noted that neither of these premises have been conclusively established.

4.3. Solar Gains

The geometry of the geodesic domes analysed maximises their capacity to capture solar radiation throughout the day. The solar gains for the month of January for reference block 1 (33.26 KWh) are almost five times lower than those for case study 2 (135.27 KWh). These solar gains are especially interesting in climates such as the BSh, due to its high annual insolation [79]. Therefore, in this type of climate, the geometric characteristics of the geodesic domes make it possible to greatly reduce the energy demand in winter. July is the month with the greatest gain in case study 1 (67.47 KWh), almost doubling the value for January. The gain for the same month in case study 2 (207.68 kWh) is more than three times higher. The annual calculation maintains a similar proportion: it is 597.19 KWh for case 1 compared to 2006.02 KWh for case study 2. This gives the geodesic dome a higher solar gain that can be used during the cold months, but also makes it necessary to protect the dome during the warm months in order to control its efficiency.

Case studies 2 and 3 generate a better distribution of sunlight and a higher uniformity compared to the reference block. This also has an impact on the lighting, as it is possible to make a better use of the natural lighting needs inside the building throughout the day over the entire surface area.

Following the data provided in the simulations, it has been found that, without entailing an increase in material cost execution, dividing the occupied surface area into uses (case study 3) allows for greater control of the resources derived from solar radiation and external ventilation compared to case 1 and case 2. This makes it possible to reduce energy consumption and obtain a more energy-efficient project.

4.4. Energy Efficiency

For warmer months, the efficiency of case 3 (60,966.47 Wh/m^2) compared to case 2 (70,534.69 Wh/m^2) is considerable in terms of cooling, as it is possible to control the solar incidence and solar radiation by separate blocks, thus optimising the available resources. It has been possible to demonstrate the estimation proposed by Soares [42] for the construction of hotel rooms with geodesic domes. The electricity consumption in case 3 is reduced in the annual calculation. The use of a double dome consumes less energy, as it has to cool a smaller volume, provided that the air infiltration and insulation are controlled. Considering this, it is not only necessary to consider the U-value of the cladding materials, but also the way in which they age. Figure 22 shows the deformation of the cellulose. It reduces in volume over time due to the effect of moisture, if there is an excessive water vapour permeability of the cladding layers.

Figure 22. Image of the deformation of the cellulose inside the enclosure.

The balance between the geometry, volume and use of the space in a dome can improve the efficiency of this type of architecture. Acting on the overall length of the joints, the optimal orientation of the windows and the adequate solar protection of its glazing in summer would allow for a reduction in heating and cooling consumption, as well as in the overall electricity consumption of case study 3 compared to case study 2.

5. Conclusions

From the studies carried out in the present research in relation to energy efficiency applied to geodesic dome envelopes, it is possible to conclude that:

- The most influential strategies for improving energy efficiency, such as direct passive solar gain, heating, natural ventilation cooling, fan-forced ventilation cooling and the solar shading of windows, allow for geodesic domes to be one of the most efficient geometries.
- In small spaces such as tourist accommodations, the examples studied allow for a greater optimisation of resources than conventional solutions. The research confirms that incorporating domes into the architectural design of tourist accommodations improves its energy performance.
- The reduction in occupied volume is indeed significant. With the same occupiable surface as the reference block, a single and a double geodesic dome reduce the interior volume by 83% and 65%, respectively. A more detailed study of the relationship between the surface area and different uses would allow for a more precise quantification of their impact on this type of architecture.
- It is confirmed that the use of differentiated day and night spaces in geodesic domes improves the energy performance by requiring 24% less energy for lighting and 52% less energy consumption for cooling, compared to other conventional examples with the same surface area.
- The most unfavourable case of domes was for the highest number of joints (712), those of frequency IV. A more detailed study of the infiltrations for dome frequencies IV, III 5/12 and III 7/12 would further improve the energy efficiency of the dome with respect to the parallelepiped geometry considered in this research. The choice of a small frequency for the design of a geodesic dome allows for less infiltrations. An intensive execution control would ensure that these infiltrations would not affect the results obtained in the software calculation.
- In order to correctly model the geometry of a geodesic dome, different software has been necessary. Autodesk REVIT needs further development to generate more complex analytical volumes. This makes it difficult to read the calculation program's results, and the associated parametric information is lost in the process. The high difficulty regarding the interoperability between different modelling and energy calculation software makes the analysis process for this type of singular geometry more difficult, due to incompatibilities in the export of files between the software.

Author Contributions: Conceptualization, C.P.-C. and M.I.P.-M.; methodology, C.P.-C. and Á.B.G.-A.; software, C.P.-C. and A.G.-G.; validation, C.P.-C. and A.G.-G.; formal analysis, M.I.P.-M.; investigation, C.P.-C., Á.B.G.-A. and A.G.-G.; resources, M.I.P.-M.; data curation, C.P.-C. and M.I.P.-M.; writing—original draft preparation, C.P.-C. and M.I.P.-M.; writing—review and editing, Á.B.G.-A. and A.G.-G.; visualization, C.P.-C. and M.I.P.-M.; supervision, C.P.-C. and A.G.-G.; project administration, C.P.-C. All authors have read and agreed to the published version of the manuscript.

Funding: This research received no external funding.

Data Availability Statement: The data presented in this study are available on request from the corresponding author.

Acknowledgments: We acknowledge the origin of this research was the final degree work of Mark Sanshi Hamaoka, directed by Angel Benigno González and Antonio Galiano.

Conflicts of Interest: The authors declare no conflicts of interest.

References

1. Beaver, A. *A Dictionary of Travel and Tourism*; Oxford University Press: Oxford, UK, 2012.
2. Horáková, H.; Boscoboinik, A. (Eds.) *From Production to Consumption: Transformation of Rural Communities*; LIT Verlag: Münster, Germany, 2012.
3. Gössling, S.; Scott, D.; Hall, C.M. Pandemics, Tourism and Global Change: A Rapid Assessment of COVID-19. *J. Sustain. Tour.* **2021**, *29*, 1–20. [CrossRef]
4. Craig, C.A.; Karabas, I. Glamping after the Coronavirus Pandemic. *Tour. Hosp. Res.* **2021**, *21*, 251–256. [CrossRef]
5. Milohnić, I.; Cvelić Bonifačić, J.; Licul, I. Transformation of Camping into Glamping—Trends and Perspectives. *ToSEE Tour. South. East. Eur.* **2019**, *5*, 457–473. [CrossRef]
6. Europe Glamping Market Size, Share & Growth. 2023. Available online: https://www.grandviewresearch.com/industry-analysis/europe-glamping-market-report (accessed on 26 May 2023).
7. Claver-Cortés, E.; Molina-Azorín, J.F.; Pereira-Moliner, J.; López-Gamero, M.D. Environmental Strategies and Their Impact on Hotel Performance. *J. Sustain. Tour.* **2007**, *15*, 663–679. [CrossRef]
8. Blanco, E.; Rey-Maquieira, J.; Lozano, J. Economic Incentives for Tourism Firms to Undertake Voluntary Environmental Management. *Tour. Manag.* **2009**, *30*, 112–122. [CrossRef]
9. Alonso-Almeida, M.-M.; Fernández Robin, C.; Celemín Pedroche, M.S.; Astorga, P.S. Revisiting Green Practices in the Hotel Industry: A Comparison between Mature and Emerging Destinations. *J. Clean. Prod.* **2017**, *140*, 1415–1428. [CrossRef]
10. Kostakis, I.; Sardianou, E. Which Factors Affect the Willingness of Tourists to Pay for Renewable Energy? *Renew. Energy* **2012**, *38*, 169–172. [CrossRef]
11. Çelik, N.; Bahar, O.; Tatar, S. The role of glamping tourism in rural development: The case of Club Amazon Bordubet. *J. Int. Soc. Res.* **2017**, *10*, 1282–1287. [CrossRef]
12. Boscoboinik, A.; Bourquard, E. Glamping and Rural Imaginary. In *From Production to Consumption: Transformation of Rural Communities*; LIT Verlag: Münster, Germany, 2012; pp. 149–164.
13. Lucivero, M. Camping and Open-Air Tourism: An Opportunity for Sustainable Tourism in Coastal Areas. In *Proceedings of the 6th Conference of the International Forum on Urbanism (IFoU): TOURBANISM, Barcelona, Spain, 25–27 January 2012*; International Forum on Urbanism: Barcelona, Spain, 2012; pp. 1–9.
14. Brooker, E.; Joppe, M. Trends in Camping and Outdoor Hospitality—An International Review. *J. Outdoor Recreat. Tour.* **2013**, *3–4*, 1–6. [CrossRef]
15. ArchDaily en Español. Hotel Tenir Eco/Levelstudio. Available online: https://www.archdaily.cl/cl/957587/hotel-tenir-eco-levelstudio (accessed on 11 January 2024).
16. ArchDaily. DOM(E)/NRJA. Available online: https://www.archdaily.com/382876/dom-e-nrja?ad_source=search&ad_medium=projects_tab (accessed on 11 January 2024).
17. ArchDaily en Español. Vivienda Modular Labt 20 (10,5) City Bell/Estudio Borrachia + GB Arquitectos. Available online: https://www.archdaily.cl/cl/1000180/vivienda-modular-labt-20-estudio-borrachia-arquitectos-plus-gb-arquitectos?ad_source=search&ad_medium=projects_tab (accessed on 11 January 2024).
18. ArchDaily en Español. Vivienda Geodésica/Ecoproyecta. Available online: https://www.archdaily.cl/cl/805836/vivienda-geodesica-ecoproyecta?ad_source=search&ad_medium=projects_tab (accessed on 11 January 2024).
19. Arquitectura. CASA M + J de Manuel Cerdá. Un Refugio Suficiente. Available online: https://arquitecturayempresa.es/noticia/casa-mj-de-manuel-cerda-un-refugio-suficiente (accessed on 11 January 2024).
20. ArchDaily en Español. Sauna Sazae/Kengo Kuma & Associates. Available online: https://www.archdaily.cl/cl/995394/sauna-sazae-kengo-kuma-and-associates?ad_source=search&ad_medium=projects_tab (accessed on 11 January 2024).
21. ArchDaily. Ranwu Lake Campsite/Archermit. Available online: https://www.archdaily.com/877966/ranwu-lake-campsite-xiao-yin-architecture-design-firm (accessed on 11 January 2024).
22. ArchDaily en Español. Dos Domos y un Zócalo: Casa 8 por B + V Arquitectos. Available online: https://www.archdaily.cl/cl/761630/dos-domos-y-un-zocalo-casa-8-por-b-plus-v-arquitectos?ad_source=search&ad_medium=projects_tab (accessed on 11 January 2024).
23. ArchDaily en Español. Casa Cambará Container/Estúdio Saymon Dall Alba + Mégui Dal Bó Arquiteta. Available online: https://www.archdaily.cl/cl/932793/casa-cambara-container-saymon-dall-alba-arquiteto-plus-megui-dal-bo-arquiteta (accessed on 11 January 2024).
24. ArchDaily en Español. Katerina Gordon. En Construcción: Domo Cluster/Arketiposchile. 2012. Available online: https://www.archdaily.cl/cl/02-165500/en-construccion-domo-cluster-arketiposchile (accessed on 20 November 2023).
25. Bysiec, D.; Jaszczyński, S.; Maleska, T. Analysis of Lightweight Structure Mesh Topology of Geodesic Domes. *Appl. Sci.* **2024**, *14*, 132. [CrossRef]
26. Kubik, M.; Augarde, C. *Structural Analysis of Geodesic Domes*; Durham University: Durham, UK, 2009.
27. Tarnai, T. Geodesic Domes: Natural and Man-Made. *Int. J. Space Struct.* **2011**, *26*, 215–227. [CrossRef]
28. Kolpakov, A.; Dolgov, O.; Korolskiy, V.; Popov, S.; Anchutin, V.; Zykov, V. Analysis of Structural Layouts of Geodesic Dome Structures with Bar Filler Considering Air Transportation. *Buildings* **2022**, *12*, 242. [CrossRef]
29. Heidari, A.; Olivieri, F. Energy Efficiency in Dome Structures: An Examination of Thermal Performance in Iranian Architecture. *Buildings* **2023**, *13*, 2171. [CrossRef]

30. Kermatigo, W.; Masruchin, F. Implementation of Domes Building in Simoketawang Village Tourism Education Facilities, Sidoarjo City. *Asian J. Soc. Humanit.* **2023**, *1*, 375–383. [CrossRef]
31. *DesignBuilder*, version v.7.0.1.006; DesignBuilder Software Ltd.: Stroud, UK, 2023. Available online: https://designbuilder.co.uk/ (accessed on 29 November 2023).
32. *EnergyPlus*; v.23.1.0; U.S. Department of Energy's (DOE) Building Technologies Office (BTO): Washington, DC, USA, 2023. Available online: https://energyplus.net/ (accessed on 29 November 2023).
33. HiSoUR Arte Cultura Historia. Cúpula Geodésica. Available online: https://www.hisour.com/es/geodesic-dome-32059/ (accessed on 29 November 2023).
34. Sutton, D. *Sólidos Platónicos y Arquimedianos*; Ediciones Oniro: Barcelona, Spain, 2005.
35. Gaspar, O. Bauersfeld's Concept for the Subdivision of the First Built Geodesic Dome Structure. In Proceedings of the IASS Annual Symposium 2020/21 and the 7th International Conference on Spatial Structures—Inspiring the Next Generation, Guildford, UK, 23–27 August 2021.
36. Chu, H.-Y. The Evolution of the Fuller Geodesic Dome: From Black Mountain to Drop City. *Des. Cult.* **2018**, *10*, 121–137. [CrossRef]
37. Díaz, E. *The Experimenters: Chance and Design at Black Mountain College*; The University of Chicago Press: Chicago, IL, USA; London, UK, 2015.
38. Buckminster Fuller, R. Building Construction. U.S. Patent US2682235A, 26 June 1954. Available online: https://patents.google.com/patent/US2682235?oq=2682235 (accessed on 13 May 2023).
39. de Lózar de la Viña, M. Casa Cúpula En Carbondale, Illinois. R. B. Fuller, 1960. *ARQ* **2013**, *84*, 14–27. [CrossRef]
40. Peñalver Oltra, M. Estudio de Alternativas de Cúpula Geodésica En Madera de Fácil Realización y Montaje. Bachelor's Thesis, Valencia Polytechnic University, Valencia, Spain, 2017.
41. South, H.; Harris, M.E. Black Mountain College Project Collection. Western Regional Archives, 3rd Floor Stacks, Row 5. p 25 Boxes. Available online: https://archives.ncdcr.gov/documents/black-mountain-college-project-inventory/open (accessed on 29 November 2023).
42. Soares, T.; Arruda, A. As estruturas geodésicas do Ecocamp na Patagônia: Um estudo sobre seus aspectos ergonômicos e sustentáveis. *Blucher Eng. Proc.* **2016**, *3*, 216–227. [CrossRef]
43. Weisstein, E.W. Polyhedral Formula. Available online: https://mathworld.wolfram.com/PolyhedralFormula.html (accessed on 11 January 2024).
44. Kitrick, C.J. A Unified Approach to Class I, II & III Geodesic Domes. *Int. J. Space Struct.* **1990**, *5*, 223–246. [CrossRef]
45. Barba Marcos, J. Cúpula Geodésica Destinada a la Implantación de Huertos Urbanos. Bachelor's Thesis, University Charles III of Madrid, Madrid, Spain, 2015.
46. CORE. Estudio Paramétrico de Cúpulas de Barras: Generación Paramétrica, Cálculo Automatizado y Comparativa del Comportamiento de Cúpulas de Haces ante Variaciones Geométricas. Available online: https://core.ac.uk/display/61917800?source=2 (accessed on 13 May 2023).
47. Stasi, G. Geometrías Geodésicas y Sistemas Constructivos: Diseño e Implementación de Sistemas Low-Tech. Ph.D. Thesis, University of Seville, Seville, Spain, 2018.
48. Shelter Domos. Domos pare Glamping. Available online: https://shelter-dome.com/case// (accessed on 13 May 2023).
49. Martín-Pastor, A.; Martín-Mariscal, A.; López-Martínez, A. Self-Built Geodesic Geometries. In Proceedings of the 3rd International Congress on Sustainable Construction and Eco-Efficient Solutions, Seville, Spain, 27–29 March 2017; pp. 168–191. [CrossRef]
50. Ecoproyecta. Cúpulas Geodésicas. Available online: https://ecoproyecta.es/cupulas-geodesicas/ (accessed on 11 January 2024).
51. Porta-Gándara, M.A.; Gómez-Muñoz, V. Solar Performance of an Electrochromic Geodesic Dome Roof. *Energy* **2005**, *30*, 2474–2486. [CrossRef]
52. 06/00879 Solar Performance of an Electrochromic Geodesic Dome Roof: Porta-Gándara, M. A. and Gömez-Muñoz, V. Energy, 2005, 30, (13), 2474–2486. *Fuel Energy Abstr.* **2006**, *47*, 129. [CrossRef]
53. Haghnazar, R.; Nooshin, H.; Golabchi, M. Improving the Regularity of Geodesic Domes Using the Concept of Stepping Projection. *Int. J. Space Struct.* **2014**, *29*, 81–95. [CrossRef]
54. Saka, M.P. Optimum Geometry Design of Geodesic Domes Using Harmony Search Algorithm. *Adv. Struct. Eng.* **2007**, *10*, 595–606. [CrossRef]
55. Fuller, R.B.; Applewhite, E.J. *Synergetics: Explorations in the Geometry of Thinking*; Macmillan: New York, NY, USA, 1975. Available online: https://philpapers.org/rec/FULSEI (accessed on 3 January 2024).
56. Laila, T.; Arruda, A.; Barbosa, J.; Moura, E. The Constructive Advantages of Buckminster Fuller's Geodesic Domes and Their Relationship to the Built Environment Ergonomics. In *Advances in Ergonomics in Design—Proceedings of the AHFE 2017 International Conference on Ergonomics in Design, Los Angeles, CA, USA, 17–21 July 2017*; Advances in Intelligent Systems and Computing; Springer: Cham, Switzerland, 2018; Volume 588, pp. 359–368. [CrossRef]
57. Soares, T.; Arruda, A. Geodesic Domes as Business Model in Hotel Management for Local Economies Development. *Cent. Estud. Diseño Comun.* **2020**, *80*, 111–122. [CrossRef]
58. Pastukh, O.; Zhivotov, D.; Vaitens, A.; Yablonskii, L. The Use of Modern Polymer Materials and Wood in the Construction of Buildings in the Form of Geodesic Domes. *E3S Web Conf.* **2021**, *274*, 1024. [CrossRef]
59. ArchDaily en Español. Available online: https://www.archdaily.cl/cl?ad_name=small-logo (accessed on 2 October 2023).

60. Pérez Aranda, A.; Arco Díaz, J.; Hidalgo García, D. Energy Study of the Envelope in Metal Containers for Building = Estudio Energético de La Envolvente En Contenedores Metálicos Para Edificación. *Build. Manag.* **2019**, *3*, 36–48. [CrossRef]
61. Freitas, G.; Rabelo, E.; Pessoa, E. Projeto Modular Com Reaproveitamento de Container Maritimo. *Braz. J. Dev.* **2023**, *9*, 24368–24404. [CrossRef]
62. Bernardo, L.; Pereira de Oliveira, L.; Nepomuceno, M.; Andrade, J. Use of Refurbished Shipping Containers for the Construction of Housing Buildings: Details for the Structural Project. *J. Civ. Eng. Manag.* **2013**, *19*, 628–646. [CrossRef]
63. Laksitoadi, B.; Hasan, M. Refurbished Shipping Containers as Architectural Module in Bandung. In *Proceedings of the 3rd International Conference on Dwelling Form (IDWELL 2020), Online, 27–28 October 2020*; Atlantis Press: Amsterdam, The Netherlands, 2020. [CrossRef]
64. Pilarska, D.; Maleska, T. Numerical Analysis of Steel Geodesic Dome under Seismic Excitations. *Materials* **2021**, *14*, 4493. [CrossRef]
65. Carvalho, J.P.G.; Lemonge, A.C.C.; Hallak, P.H.; Vargas, D.E.C. Simultaneous Sizing, Shape, and Layout Optimization and Automatic Member Grouping of Dome Structures. *Structures* **2020**, *28*, 2188–2202. [CrossRef]
66. Szmit, R. Geometry Design and Structural Analysis of Steel Single-Layer Geodesic Domes. In Proceedings of the 2017 Baltic Geodetic Congress (BGC Geomatics), Gdansk, Poland, 22–25 June 2017; pp. 205–209. [CrossRef]
67. Gholizadeh, S.; Barati, H. Topology Optimization of Nonlinear Single Layer Domes by a New Metaheuristic. *Steel Compos. Struct. Int. J.* **2014**, *16*, 681–701. [CrossRef]
68. Kaveh, A.; Rezaei, M. Topology and Geometry Optimization of Single-Layer Domes Utilizing CBO and ECBO. *Sci. Iran.* **2016**, *23*, 535–547. [CrossRef]
69. Monolithic.org. The Inn Place—New Rentals, Old Pattern. Available online: https://www.monolithic.org/rentals/the-inn-place-new-rentals-old-pattern (accessed on 1 November 2023).
70. 7/12 Kruschke GoodKarma 3V R2.2 Beams 120x40—Geodesic Dome Calculator—Acidome.ru. Available online: https://acidome.com/lab/calc/#7/12_Kruschke_GoodKarma_3V_R2.2_beams_120x40 (accessed on 2 October 2023).
71. Otero, C.; Díaz, J.; Togores Fernández, R.; Manchado, C. Chordal Space Structures, Shaped from Voronoi Diagrams. In Proceedings of the Joint International Conference on Computing and Decision Making in Civil and Building Engineering, Montreal, QC, Canada, 14–16 June 2006.
72. Vrontissi, M.; Azariadi, S. Digital Tools in the Architectural Design of a Geodesic Dome: The Case-Study of the Bearing Structure of an Artificial Sky Lighting Installation. In *Respecting Fragile Places—Proceedings of the 29th Conference on Education in Computer Aided Architectural Design in Europe (eCAADe 2011), Ljubljana, Slovenia, 21–24 September 2011*; Education in Computer Aided Architectural Design in Europe (eCAADe): Brussels, Belgium; Faculty of Architecture: Ljubljana, Slovenia, 2011. [CrossRef]
73. DesignBuilder. Aurea Consulting. Available online: https://ecoeficiente.es/designbuilder/ (accessed on 13 May 2023).
74. *ASHRAE Standard 55-2023*; Thermal Environmental Conditions for Human Occupancy. ASHRAE: Atlanta, GA, USA, 2023. Available online: https://www.ashrae.org/technical-resources/bookstore/standard-55-thermal-environmental-conditions-for-human-occupancy (accessed on 23 October 2023).
75. Espín-Sánchez, D.; Olcina-Cantos, J.; Conesa-García, C. Temporal Changes in Tourists' Climate-Based Comfort in the Southeastern Coastal Region of Spain. *Climate* **2023**, *11*, 230. [CrossRef]
76. Código Técnico de la Edificación. *Documento Básico de Ahorro de Energía*; Ministerio de Transportes, Movilidad y Agenda Urbana: Madrid, Spain, 2022.
77. Garcilópez Gracia, E.J.; Romeo Giménez, L.M. *Evaluación de Las Infiltraciones En La Edificación y Aplicación Al Edificio Del Solar Decathlon*; Universidad de Zaragoza: Zaragoza, Spain, 2010.
78. Elsland, R.; Peksen, I.; Wietschel, M. Are Internal Heat Gains Underestimated in Thermal Performance Evaluation of Buildings? *Energy Procedia* **2014**, *62*, 32–41. [CrossRef]
79. Pérez-Carramiñana, C.; Sabatell-Canales, S.; González-Avilés, Á.B.; Galiano-Garrigós, A. Influence of Spanish Energy-Saving Standard on Thermal Comfort and Energy Efficiency Owing to the War in Ukraine: Case Study of an Office Building in a Dry Mediterranean Climate. *Buildings* **2023**, *13*, 2102. [CrossRef]

buildings

MDPI

Article

Evaluating the Impact of Insulation Materials on Energy Efficiency Using BIM-Based Simulation for Existing Building Retrofits: Case Study of an Apartment Building in Kanazawa, Japan

Xiao Teng [1,2], Zhenjiang Shen [3,*] and Dara Citra Saraswati Tutuko [4]

1 Graduate School of Environment Design, Kanazawa University, Kanazawa 920-1192, Japan; riendyteng@se.kanazawa-u.ac.jp
2 Youke Communication Technology, Fuzhou 350005, China
3 International Joint Laboratory of Spatial Planning and Sustainable Development (FZUKU-LAB SPSD), Fuzhou University-Kanazawa University, Fuzhou 350025, China
4 Graduate School of Natural Science and Technology, Geoscience and Civil Engineering, Kanazawa University, Kanazawa 920-1192, Japan; daratutuko@stu.kanazawa-u.ac.jp
* Correspondence: shenzhe@se.kanazawa-u.ac.jp

Abstract: This research aims to facilitate informed decision-making to enhance building energy simulation, reduce costs, and minimize CO_2 emissions through building insulation enhancements employing BIM-based simulation. Architectural models of an apartment, a prevalent residential structure in Japan, were developed and examined under diverse insulation scenarios utilizing ArchiCAD 28. Five insulation substances were chosen based on existing guidelines to ensure conformity with local standards and were evaluated for their thermal and environmental properties: Cellulose Fiber, Glass Wool, Urethane Foam, Phenolic Board, and Rock Wool for evaluation based on thermal and environmental properties. The simulation parameters were aligned with Japan's energy efficiency standards and climate conditions. The factors addressed encompass energy performance evaluation, economic viability, and CO_2 emissions. Simulation findings highlight Urethane Foam as the most effective and environmentally friendly building insulation material. This study provides valuable perspectives for property owners, building designers, and contractors, offering a framework for insulation enhancement choices that optimizes sustainable construction, reduces environmental impact, and enhances cost-effectiveness through the implementation of BIM-based simulation.

Keywords: building energy simulation; energy performance evaluation; building insulation; carbon emission; cost-effectiveness; sustainable construction

1. Introduction

Urban areas are responsible for a significant share of global energy use and emissions. Buildings typically account for around 40% of the worldwide energy consumption [1]. The Japanese government is committed to achieving carbon neutrality by 2050 [2]. It is also supported by several actions, such as the Building Energy Efficiency Act, Zero Energy House, and Zero Energy Building initiation. Approximately 60% of existing buildings in Japan predate the 1979 Energy Saving Act. Despite advancements in energy efficiency for new constructions, the annual construction rate remains at approximately 2% [3]. The achievement of carbon neutrality is also supported by government grants, such as subsidies for

https://doi.org/10.3390/buildings15040570

buildings that meet the criteria for zero energy or effective energy buildings, emphasizing the importance of upgrading older structures to meet modern energy standards.

Renovating existing buildings to improve their energy efficiency is a complex undertaking that requires diverse strategies and factors. And considerations. Enhancing the building envelope through insulation can significantly reduce energy consumption by minimizing heat loss and gain. For instance, a study demonstrated that major retrofitting strategies could reduce energy intensity by 24.12% and CO_2 emissions by 18.56% [4]. Moreover, the Ministry of Land, Infrastructure, Transport and Tourism of Japan also initiated the Energy-Efficient Homes 2024 Campaign with financial support like subsidies for buildings that meet their requirements for energy-efficient building practices, such as wall insulation renovation. This underscores the importance of upgrading buildings to achieve energy efficiency.

Conducting energy analyses helps to examine the usage patterns within a building's designed model, supporting designers in making informed, cost-effective decisions that enhance building performance and minimize environmental impact [5]. Building Energy Modeling is a potential tool in architecture, engineering, and construction for sectors that enhance energy efficiency. BIM software allows architects to develop 3D models for multiple purposes, such as forecasting energy usage [6]. This research will leverage one BIM software, ArchiCAD 28, to perform an energy simulation of a building with several scenarios and reveal the most beneficial scenario for energy use. Determine the most efficient building insulation material for retrofitting existing buildings, with a specific focus on Kanazawa City, Japan, a region characterized by cold winters and warm summers typical of a temperate climate.

The simulation-based approach applied in this research evaluates the real-world effectiveness of five insulation materials included in Japan's guidelines for energy-efficient housing. By considering the climatic characteristics of Kanazawa City, this study aims to provide an informed decision to optimize energy savings, reduce costs, and minimize CO_2 emissions through building insulation upgrades using a BIM-based simulation. This finding is intended to support retrofitting initiatives that align with local standards and are valuable for buildings located in the same regions or climate conditions, promoting sustainable building practices.

2. Literature Review

2.1. *Utilization of Building Information Modeling to Enhance Building Energy Efficiency*

The realization of Sustainable Development Goals within the architectural sector is influenced by the advancement of artificial intelligence, smart devices, and various technological innovations [7]. Energy conservation is essential for sustainable development of the construction sector. Moreover, the energy aspect of buildings can be further implemented in broader applications, including energy monitoring systems for infrastructure monitoring systems [8]. With the increasing significance of technological advancements to aid sustainable growth for energy efficiency, one of the architectural innovation tools in this research is Building Information Modeling (BIM). BIM software enables architects to create three-dimensional models for various applications, including energy consumption prediction [6]. The shift from BIM to Building Energy Modeling (BEM) enables the refinement of building envelope designs, improving energy efficiency and economic viability [9]. Designers can utilize the approach of Building Information Modeling (BIM) and Building Energy Modeling (BEM) to evaluate design options and make design decisions efficiently during the building design process. The process of retrofitting existing structures, which involves systematic upgrading and enhancement of their energy efficiency, represents a fundamental strategy for achieving sustainability [10]. BIM is a potential tool for assessing refurbishment options through the modeling of various elements; findings indicate that the optimal refurbishment scenario can lead to a reduction of 61% in annual fuel consumption and 64% in electricity consumption for a house in Turkey [11]. A case study in China proved that evaluating the feasibility of BIM-enabled LCA by simulating scenarios for existing building refurbishment can analyze building energy factors, such as energy efficiency, environmental compatibility, and profitability tools [12]. A case study of residential buildings in Pakistan using BIM for energy simulation suggests that optimization with enhancements in window configurations and construction materials can realize savings of up to 30% over a period of 30 years [13]. Predicting accurate decision-making processes with BIM is also utilizable for a museum building in Brazil, revealing that BIM can help in the assessment of damage reconstruction planning for proper conservation methodology for project decision-making [14]. Building energy consumption is reduced by 30% when simulated with BIM in retrofitting existing public buildings in Malaysia, highlighting its potential to enhance energy efficiency [15]. BIM contributes to sustainable renovation practices across diverse building types and geographic regions.

These significant findings underscore the remarkable capacity of BIM to guide effective strategies in building energy, making BIM a valuable tool in sustainable construction practice. Within Building Information Modeling (BIM) applications, especially when appraised based on both their utilization frequency and the effectiveness of their implementation, ArchiCAD reliably stands out as prominent software [16]. Additionally, ArchiCAD complies with the ASHRAE standards, which makes it suitable as a BIM-based simulation tool for this study. Moreover, the study by Siti Birkha Mohd Ali et al. proved the leverage of ArchiCAD as a tool for effective retrofitting decision-making [17]. The energy model review range in ArchiCAD confidently integrates a wealth of information pertaining to building design, materials, environmental context, wind protection, and much more as essential components of the Energy Performance Evaluation.

Based on previous study findings, this research will further explore the energy usage of a building modeled in BIM, with a particular focus on evaluating the impact of various insulation material scenarios for residential buildings located in Kanazawa, Japan. The BIM framework exhibits significant adaptability, permitting customization for different conditions and building classifications. This versatility enhances its utility across a wide range of building types and is essential for evaluating energy consumption and refining material choices in sustainable construction methodologies. Choosing sustainable materials offers numerous benefits; thus, it will be efficient to use BIM to project their energy usage to choose the best material. The simulation will provide valuable insights, enabling informed decision-making and facilitating the selection of the most effective and sustainable options.

2.2. *Building Insulation Material for Energy Efficiency*

The building envelope is essential for assessing a building's energy efficiency by affecting the heating and cooling requirements. Insulation is a vital element of a building envelope that significantly influences energy conservation. Evaluating the building envelope is one of the significant elements for optimizing building energy [18]. Building envelope renovations and phase change material can effectively reduce the energy consumption of the building [19,20]. Since the insulation layer plays an important role in energy performance, it is important for enhancing energy efficiency and minimizing environmental impacts such as gas emissions [21]. Insulation material will play a role in maintaining thermal comfort inside the building by reducing heat transfer through the building envelope. The extent of insulation significantly influences both the internal thermal conditions and energy consumption [22]. Reducing energy loss in buildings can be achieved by applying suitable insulation to the building envelope [23]. Furthermore, the thermal barrier material has a pivotal function in regulating internal moisture, preserving sound levels, and offering flame resistance, thereby enhancing comprehensive energy, ecological, economic, and comfort optimization [24]. Various insulation materials are available in the market, and buildings employ both inorganic and organic thermal insulation materials to preserve energy and avoid thermal heat loss. Organic insulation was recently used as a choice for the fabrication of economical and lightweight polymer composites such as Polystyrene and Polyurethane Foams [25]. Another organic insulation material is cellulose-based materials, which exhibit good performance in fire safety and enhance thermal insulation by 16.66% to 17.06% [26]. Additionally, there are types of natural and mineral insulation that are widely used, such as Rockwool, Glass Wool, and Wood Wool, which have notable performance for building envelope insulation performance [27–29]. Insulation materials play a vital role in enhancing the energy efficiency of buildings and ensuring the overall building performance and occupant comfort. Applying the right insulation material to building envelopes is essential for sustainable construction practices, balancing energy savings with environmental and economic considerations.

3. Methodology

In this study, a BIM-based simulation approach was employed to evaluate the energy efficiency, cost-effectiveness, and environmental impact of various insulation materials for retrofitting residential buildings. Figure 1 provide an overview of the methodology of this study. Firstly, a building model was developed using ArchiCAD, incorporating detailed geometry, materials, and spatial configuration to serve as a versatile framework for analyzing retrofitting strategies. Secondly, the climatic conditions of Kanazawa City, located in Japan's 5th climate zone, were considered, and simulation parameters were aligned with national standards, including the Building Energy Conservation Act and the Energy-Saving Design and Construction of Housing 2023 guidelines. This ensured that the simulations adhered to the energy efficiency requirements and were applicable to buildings in similar contexts. Thirdly, five insulation materials—Cellulose Fiber, Glass Wool, Urethane Foam, Phenolic Board, and Rock Wool—were selected for simulation. Material properties, including thermal conductivity, density, heat capacity, embodied energy, and embodied carbon, were sourced from manufacturer specifications and adjusted using ArchiCAD's database in alignment with the ASHRAE Standard 140-2007 [30]. Lastly, the simulation process was conducted by defining the building's geometry, creating thermal zones and blocks based on operational profiles, and assigning utilities and material properties. Climate data, including temperature, humidity, solar radiation, and wind speed, were incorporated using Typical Meteorological Year (TMY) files. Five scenarios were generated by varying the insulation materials while maintaining consistent configurations for other building components. The simulation results were then analyzed to compare the energy consumption, cost, and CO_2 emissions for each scenario.

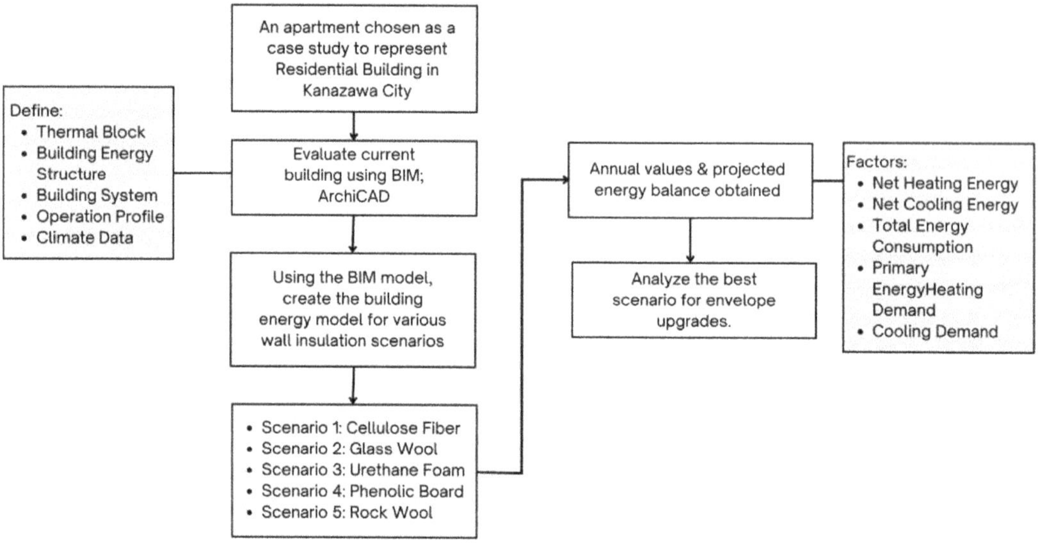

Figure 1. Methodology.

4. Case Study in Japan

4.1. Overview of the Case Study Building

This study employs ArchiCAD as a Building Information Modeling (BIM) tool to evaluate the energy efficiency of apartment buildings. The models were created in ArchiCAD and converted into energy models by defining the building geometry, zoning, utilities, occupancy, and geographic information. The model will be simulated in five scenarios of different wall insulation materials. Building energy modeling provides simulation results for energy performance, allowing for a comprehensive comparison between scenarios.

The case study in this research is an apartment building located in Kanazawa, Ishikawa Prefecture; the building model is presented in Figure 2. The building type chosen is an apartment building, which is a typical residential type in urban areas. The housing types in Kanazawa City from 2010 to 2020 reveal three main categories: detached houses, apartment buildings, and others. In 2020, detached houses constituted 56.7%, apartment buildings 42.2%, and other types 1.1% [31]. Throughout the decade, the proportion of detached houses experienced a minor decline, whereas apartment buildings exhibited a consistent increase of about 2.68%. This pattern signifies a progressive transition in housing construction preferences in Kanazawa City, with a notable increase in the prominence of apartment buildings in recent years. Japan categorizes its climate into seven zones based on yearly average temperatures [32]. The selected case study building in Kanazawa City, Ishikawa Prefecture, is located in Japan's 5th climate zone, characterized by an annual average temperature of 9–12 °C, making it suitable for assessing energy efficiency in moderately cold climates. The apartment is a three-story concrete building with a rectangular shape oriented facing south—north in a dense residential area. Built in 1999, it consists of a single entry 21 unit of 1R in a ground area of 217 m^2. Since 1980, Japan has established five energy efficiency standards for insulation in newly built residential buildings, introduced in 1980, 1992, 1999, 2013, and 2016 [32]. Furthermore, on 13 June 2022, Japan ratified the amended Building Energy Conservation Act, setting a target for all existing buildings to meet these standards on average by 2050, demonstrating the country's commitment to sustainable construction and energy conservation [33]. This case study is a representative example of older constructions that lack modern energy efficiency standards. This case study provides a valuable opportunity to explore retrofitting strategies to improve the energy efficiency of similar urban residential buildings.

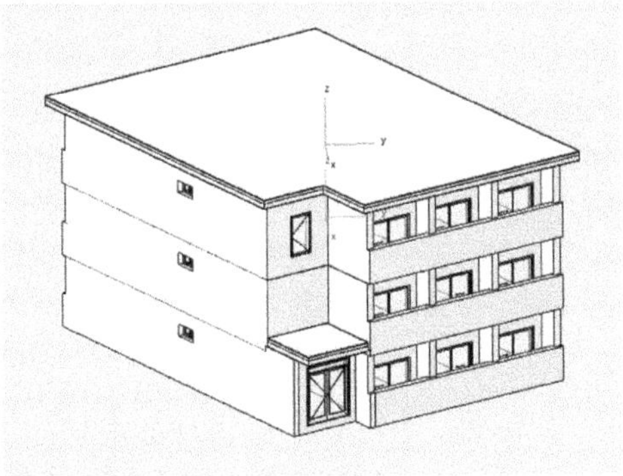

Figure 2. Case study building model.

4.2. Scenario Setting for Simulation

After the implementation of the Building Energy Efficiency Act in Japan, with several establishmenst of standards since 1980, all residential buildings comply with the latest regulations for newly constructed or retrofitting as mandatory [32]. Furthermore, in recent years, to realize decarbonization, it has become necessary to save energy not only in new houses but also in existing houses. The Ministry of Land, Infrastructure, Transport and Tourism (MLIT) created the Energy-Saving Design and Construction of Housing 2023 guideline [34]. With the newest guidelines, this research will follow a standard for 4~7 region energy-efficient housing design and construction. This standard covers some insulation materials that can be used, as shown in Table 1.

Table 1. The category of building materials listed in the Energy-Saving Design and Construction of Housing 2023 guidelines, MLIT [34].

Category	Details	Materials
By Material	- Fibrous Insulation Materials: Fine fibers that limit air movement.	Glass Wool, Rock Wool
	- Foam Plastic Insulation Materials: Contain enclosed air bubbles within their structure.	Rigid Urethane Foam, extruded polystyrene foam
By Form and Application	- Fibrous Insulation Materials: Vary in density. - High-density forms are available as boards. - Suitable for loose-fill applications.	- Felt form (low-density) - Board form (high-density) - Loose-fill form
By Permeability	- Foam Plastic Insulation Materials: Manufactured as prefabricated boards, panels, or spray-applied forms. - Moisture Permeability: Indicates the ease with which water vapor can pass through the material. Critical for external insulation designs.	- Board form, panel form, or spray-applied foam. - Low-Permeability Materials: Require a vapor barrier to prevent moisture issues.
	- Specific Note: Rigid Urethane Foam, classified as JIS A 9526 is an A-grade material for moisture resistance.	Rigid Urethane Foam under JIS A 9526

According to the criteria, five insulation materials will be simulated in this model. The materials chosen in compliance with the list of products that meet the criteria for renovation subsidy, which are listed on eligible renovation work for insulation renovation, are shown in Table 2. The value of material properties is indexed by the information on thermal conductivity and density from the product manufacture specification [35]. The heat capacity, embodied energy, and embodied carbon are listed by ArchiCAD, referring to the ASHRAE Standard 140-2007 [36], and they are adjusted to a similar product with the same thermal conductivity.

Table 2. Materials chosen for the simulation scenarios [35,36].

Scenario	Material	Properties				
		Thermal Conductivity (W/mK)	Density (kg/m^3)	Heat Capacity (J/kgK)	Embodied Energy (MJ/kg)	Embodied Carbon (kgCO_2/kg)
Scenario 1	Cellulose Fiber	0.04	50,000	2500	14.7	1.12
Scenario 2	Glass Wool	0.045	13,000	1500	40.1	1.6
Scenario 3	Urethane Foam	0.021	32,000	1450	102.1	4.84
Scenario 4	Phenolic Board	0.028	24,000	1450	102.1	4.84
Scenario 5	Rock Wool	0.04	60,000	840	16.8	0.84

5. BIM Energy Simulation

5.1. BIM-Based Building Energy Simulation Steps

Designing a building in Building Information Modeling is different from drawing a conventional 2D model; beyond that, the BIM model serves as a database system with the 3D view reflecting the data stored in the system, allowing for extraction, modification, and reinsertion of information, impacting the diverse structures of models in terms of defining elements, materials, and object descriptions [37]. The BIM Modeling process entailed the identification and simulation of different architectural and structural components, such as floor area, walls, openings (doors and windows), frames, slabs, and roofs [17]. Moreover, every geometry contains detailed information that automatically recognizes the material's essential information, such as U-Value. U-values define building energy efficiency by quantifying heat loss through the building envelope elements like walls, windows, doors, floors, and roofs, which are crucial for energy savings [38]. The steps of the BIM energy simulation in this research will be explained in the following order.

A building model is created, and prior to achieving the optimum result simulation of the energy performance evaluation, the precise building part should be defined. As presented in Figure 3, a building model is created. Adding a zone into each room with specific room usage is essential; zones in ArchiCAD's energy modeling refer to segmented areas that are analyzed for particular energy consumption and requirements. The "Thermal blocks" are collections of these zones that possess the same operational profile and mechanical systems, facilitating customized energy analysis predicated on shared attributes, as illustrated in Figure 4.

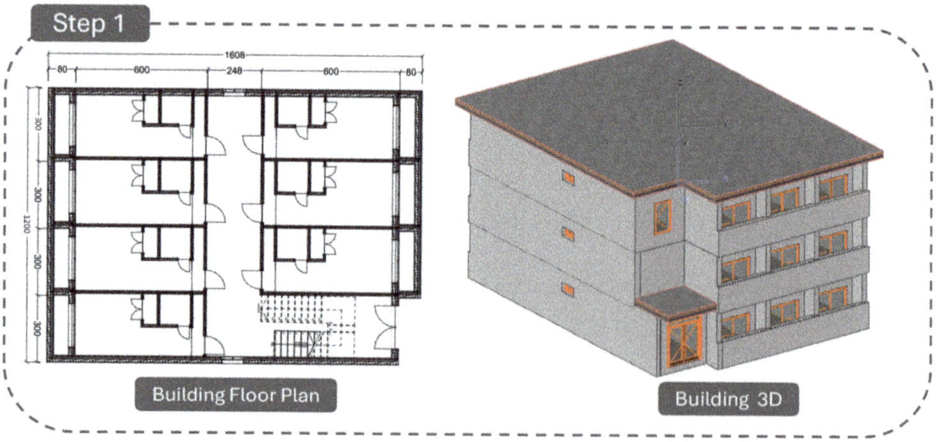

Figure 3. Step 1 involves drawing the building floor plan in BIM and creating a 3D model of the building.

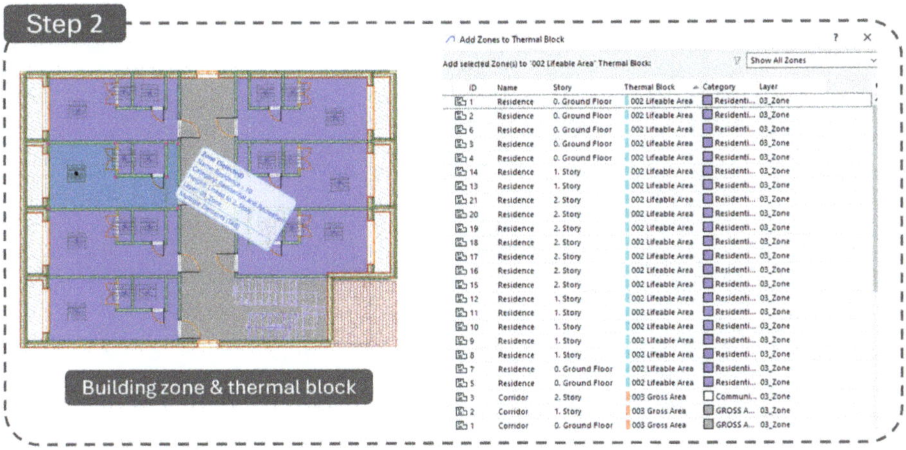

Figure 4. Step 2: Set the building zones and add them to the thermal blocks.

In the energy model review, ensure that every geometric representation contains comprehensive data that identifies the fundamental information regarding materials, such as U-Value, which will affect energy efficiency by measuring thermal dissipation through the structural envelope components, including walls, windows, doors, floors, and roofs. The focus of the simulation will be on evaluating wall insulation; thus, all building structures should be aligned with the designated wall specification, and each component of the building structure should be recognized on the building energy model, as displayed in Figure 5. The next step in building energy analysis is to assign each of the thermal blocks to its utility system, as shown in Figure 6. Utilities play a major role in the energy consumption of buildings and help shape their energy profiles [39]. Setting each system directly determines the building energy system type and source.

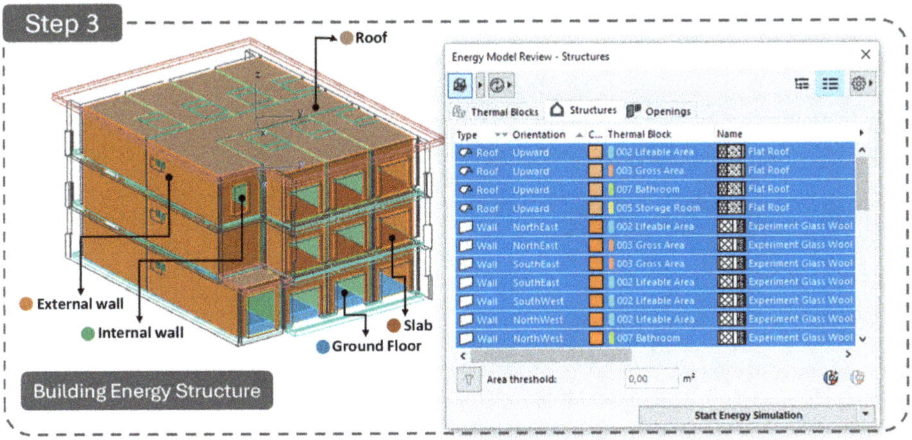

Figure 5. Step 3: Assign the model for energy model review.

After defining the building utility, define the building operation profile. Building occupancy affects energy consumption by influencing control systems. In this case, the operation profile is assigned as a residential occupancy type. Figure 7 shows the building's operational profile, which was designated as a residential building for this study. Designating an actual location for modeling building energy is essential. The location and climate of a building influence its energy use. The location and climate data that are assigned in ArchiCAD will be derived into air temperature, relative humidity, solar radiation, and wind speed served in a range of a year. The climate data is obtained from the Typical Meteorological Year (TMY) climate file, a set of hourly weather data for a specific location that represents a typical year, as shown in Figure 8.

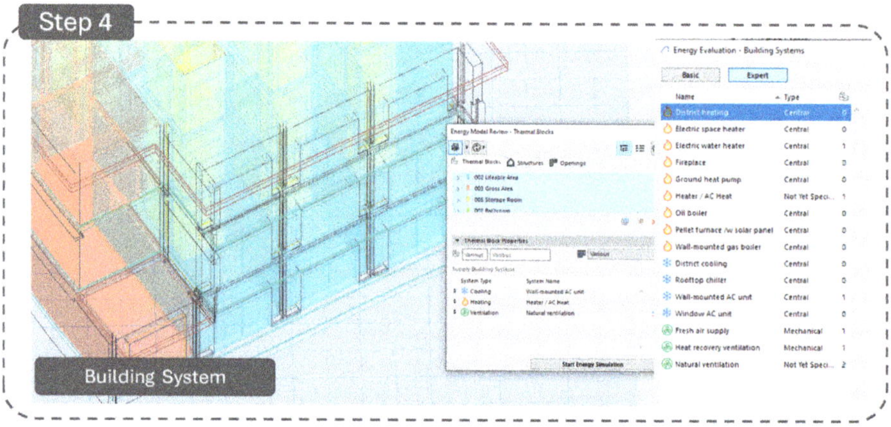

Figure 6. Step 4: Define the thermal properties of each building system.

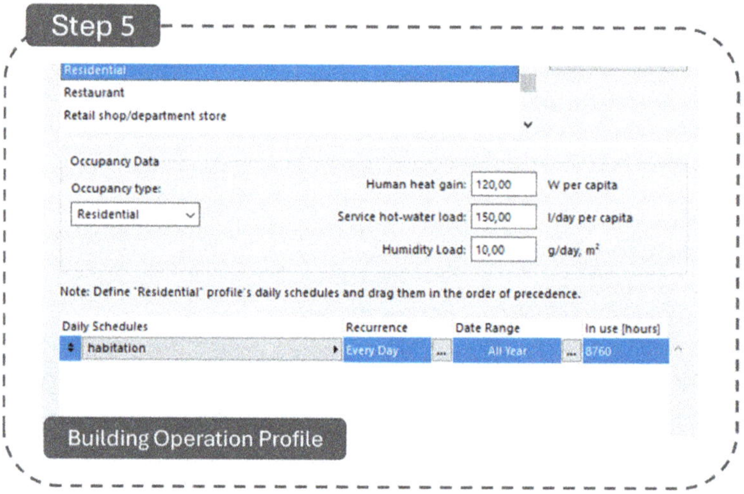

Figure 7. Step 5: Set the building operation profile.

After the building energy model is established, each model is modified into several scenarios to adjust their wall material composition. The general composition of a building consists of a structure with an exterior finish, a layer of insulation and cavity, and an interior wall with a finish. A representation of the wall composite used in this simulation is shown in Figure 9. In each scenario, the insulation layer will be replaced by the designated materials chosen. Figure 10 shows the five scenarios used in this model by replacing only the insulation material while maintaining the same wall configuration.

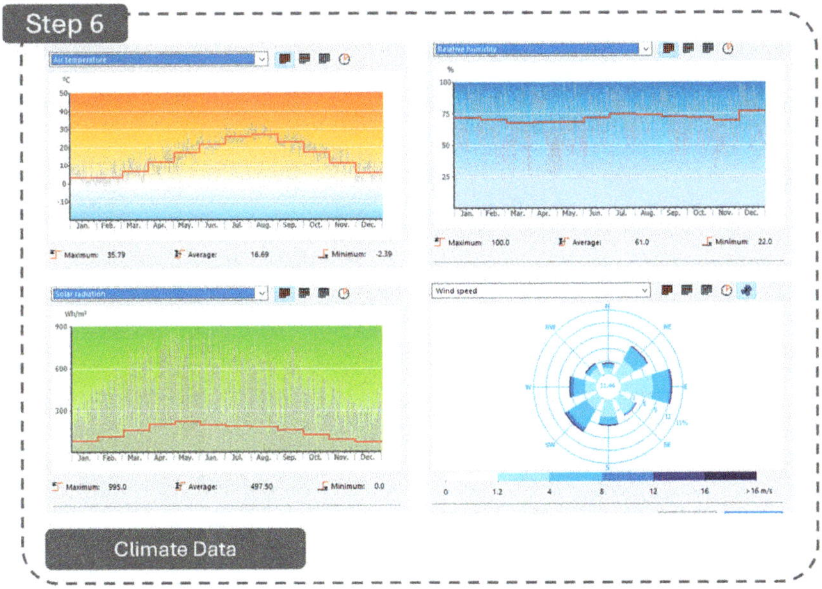

Figure 8. Step 6: Add the climate data for the building location.

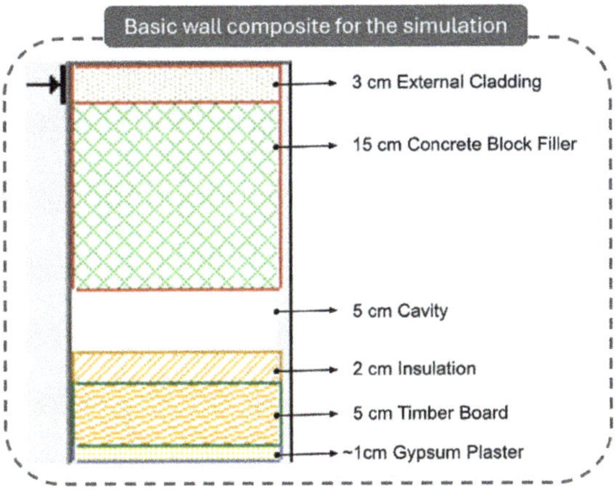

Figure 9. Basic wall composite of the simulated wall.

5.2. Energy Performance Evaluation and Simulation Results

When all the building energy modeling items, such as geometry, zones, building system, building occupancy, and geographic information, are assigned, energy performance can be generated. ArchiCAD will calculate the data contained in the model and add it to the Energy Performance Evaluation Report. The details of Simulation Scenarios 1 through 5, along with their key values and project energy balance are presented in Figure 11, contain overall information about specific annual values of energy usage. The simulation result will be focused on the value in the red box.

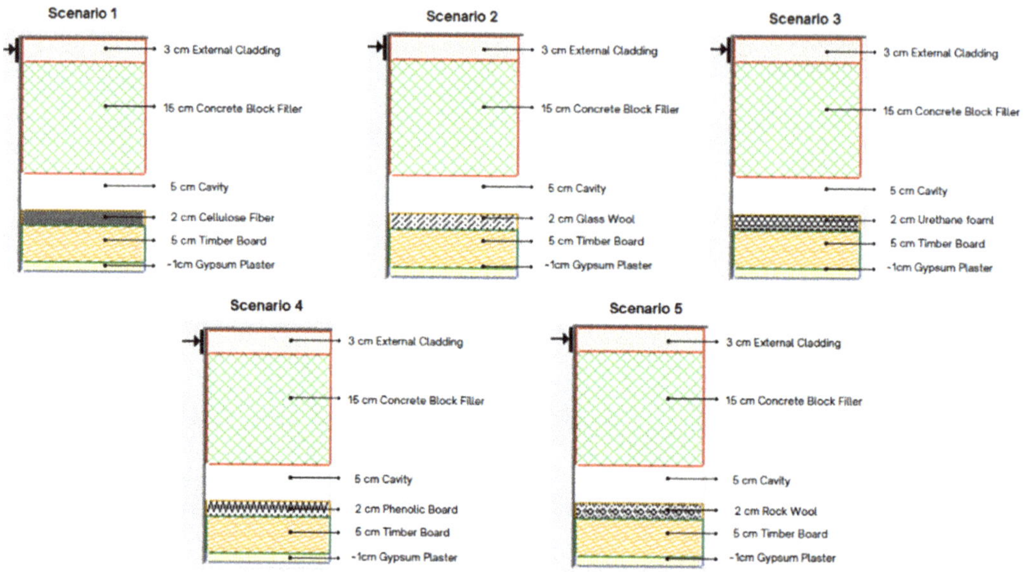

Figure 10. The wall scenarios composite.

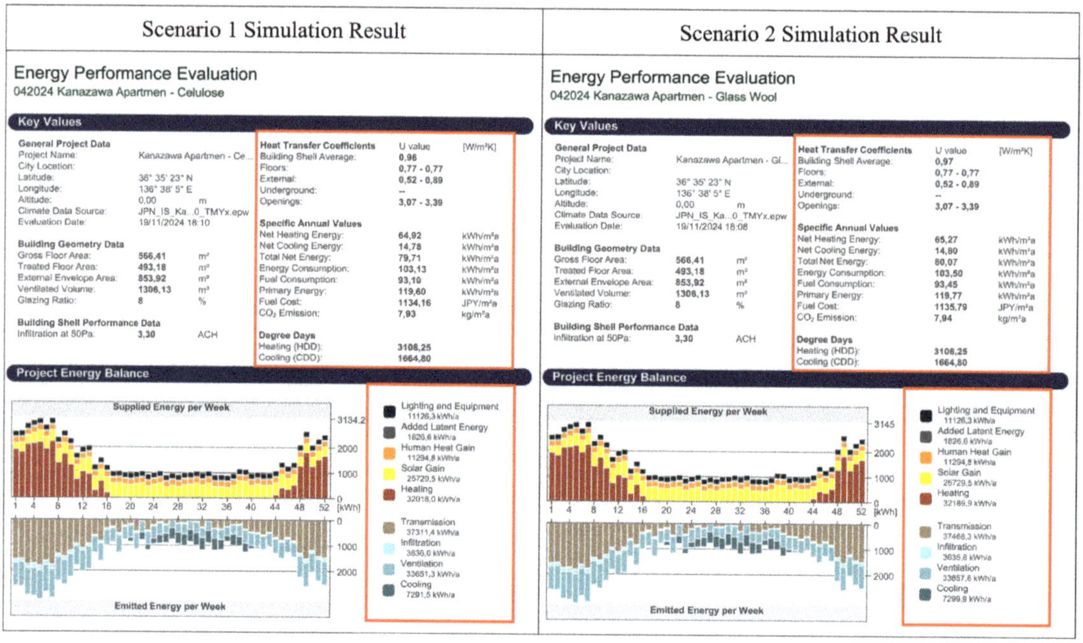

Figure 11. *Cont.*

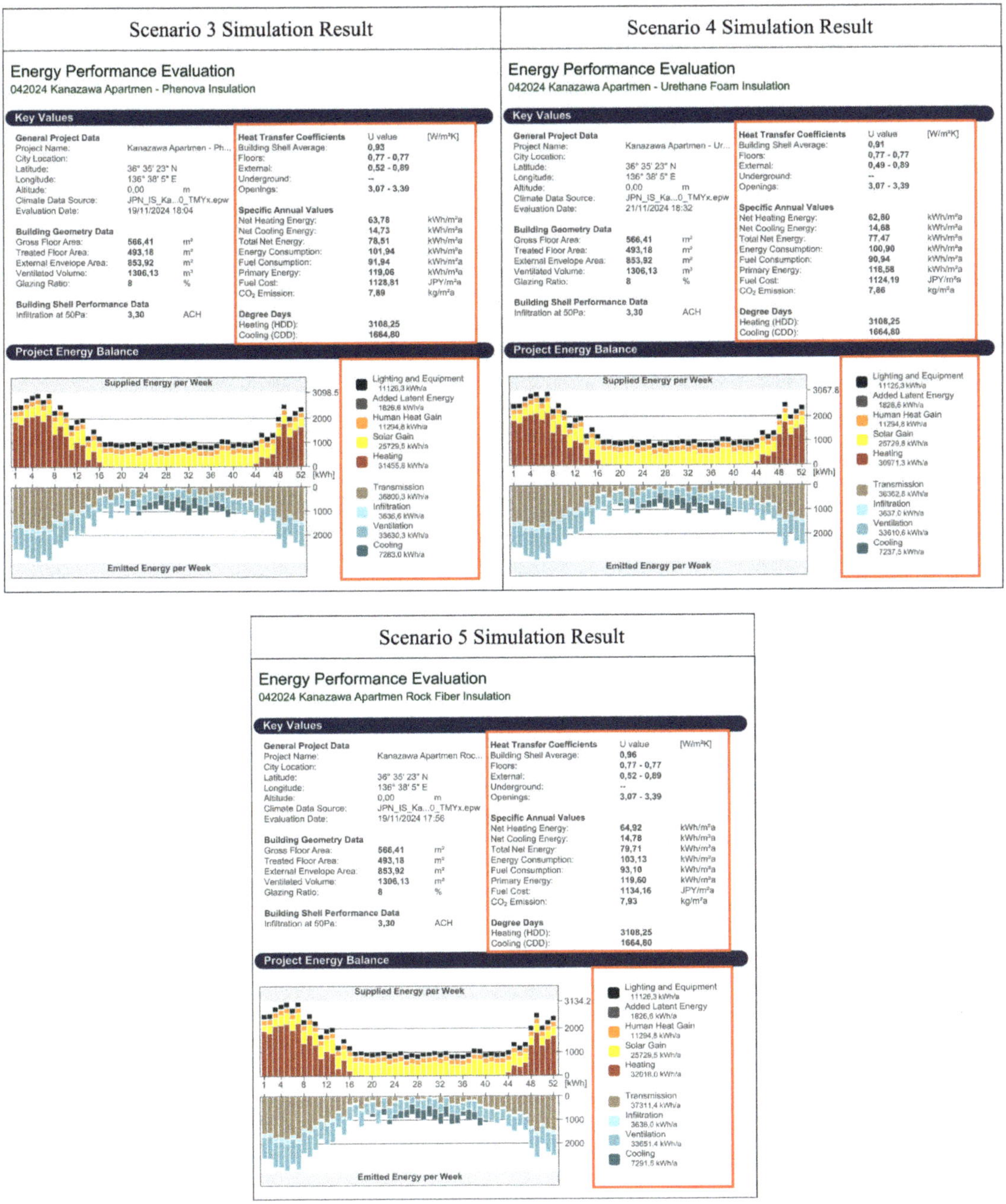

Figure 11. Simulation results for Scenario 1–Scenario 5.

6. Comparative Analysis of Insulation Materials: Energy Efficiency, Cost, and Carbon Emissions

6.1. Simulation Results and Performance Overview

Based on the simulation results, the key value and project energy balance help examine the usage patterns within a building's designed model. Some key values will be analyzed to understand which scenarios are more energy efficient, have a minimum environmental impact, and are effective in cost.

Urethane Foam is the most energy-efficient scenario with the lowest Total Energy Consumption (100.90 kWh/m^{2}a) and Primary Energy (118.58 kWh/m^{2}a). Phenolic board follows closely with slightly higher Total Energy Consumption (101.94 kWh/m^{2}a) and Primary Energy (119.06 kWh/m^{2}a). Glass Wool and Rock Wool are less efficient compared to the others, with identical Total Energy Consumption and Primary Energy levels. These differences in energy consumption highlight that foam plastic-based materials like Urethane Foam have better energy efficiency rather than fibrous materials. All scenarios have similar Net Cooling Energy requirements, with Urethane Foam being the lowest at 14.68 kWh/m^{2}a, with the lowest Fuel Cost (1124.19 JPY/m^{2}a) and CO_2 Emissions (7.86 kg/m^{2}a), indicating better cost and environmental performance. Glass Wool and Rock Wool are less efficient compared to the others, with identical Total Energy Consumption and Primary Energy levels. This suggests that while fibrous materials may offer some insulation benefits, the superior thermal performance and lower environmental impact of Urethane Foam make it a more viable option for energy-efficient building applications.

6.2. Energy Efficiency Factor

This research focuses on a case study located in Kanazawa City, Ishikawa Prefecture, which is part of Japan's 5th regional climate zone, including the Hokuriku region. The Hokuriku region stands out for having the highest intensity of electricity consumption in both residential and industrial sectors across Japan, boasting a mean usage of approximately 6620 kWh/year for residential buildings [40]. In comparison, the simulation shows that the best-performing scenario reduces energy consumption by 58.89% compared to the regional average. The building, relying on electrical energy and using Urethane Foam insulation, has an energy consumption intensity of 100.90 kWh/m^2/year, totaling approximately 57,150.77 kWh/year for the entire building. With 21 units in the building, this amounts to 2721.47 kWh/year per unit. This demonstrates that the simulated energy consumption is significantly lower than the regional average. This comparison between the simulation results and real-world data highlights that the simulated value is significantly lower than the regional average. However, it is important to note that the operational usage of each unit may vary. Nevertheless, a consumption value of 2721.47 kWh/year per unit is highly efficient, with basic utility usage settled in this scenario and well below the regional mean electricity usage.

Based on the simulation results in Table 3, the best- and least-performing materials can be ranked by order number. Urethane foam is 1st preferable option, followed by Phenolic board; 2nd, Rock Wool and Cellulose Fiber had a similar result and placed in 3rd, and Glass Wool placed in 4th as the least preferable option.

Table 3. Simulation results key values.

Scenario	Net Heating Energy (kWh/m^{2}a)	Net Cooling Energy (kWh/m^{2}a)	Total Energy Consumption (kWh/m^{2}a)	Primary Energy (kWh/m^{2}a)	Fuel Cost (JPY/m^{2}a)	CO_2 Emissions (kg/m^{2}a)
Cellulose Fiber	64.92	14.78	103.13	119.6	1134.16	7.93
Glass Wool	65.27	14.8	103.5	119.77	1135.79	7.94
Urethane Foam	62.8	14.68	100.9	118.58	1124.19	7.86
Phenolic Board	63.78	14.73	101.94	119.06	1128.81	7.89
Rock Wool	64.92	14.78	103.13	119.6	1134.16	7.93

A database from a website provided by the Sustainable Open Innovation Initiative is accessible for the Zero Energy Home (ZEH) case study inquiry search tool for the Fiscal Year 2024, providing insights into energy-saving performance, insulation, and equipment specifications from over 1500 adoption cases across Japan [41]. Table 4 shows several case studies that use similar materials to those used in this simulation, compared with the simulation result performance rank of this study.

Table 4. ZEH case study of Ishikawa Prefecture with similar insulation type in the study simulation.

Web Data	Building Type	Insulation Type	Primary Energy Reduction Rate	Simulation Result Performance Ranking
Case Study 1	2-story building	Glass Wool	21%	4
Case Study 2	2-story building	Glass Wool	26%	4
Case Study 3	2-story building	Phenolic foam	27%	2
Case Study 4	2-story building	Phenolic foam	32%	2
Case Study 5	2-story building	Phenolic foam	38%	2
Case Study 6	2-story building	Rockwool	26%	3
Case Study 7	2-story building	Sprayed Rigid Urethane foam	26%	1
Case Study 8	2-story building	Sprayed Rigid Urethane foam	37%	1
Case Study 9	2-story building	Sprayed Rigid Urethane foam	40%	1

Only four materials from the database match the material studied in this simulation. The web data case study corresponds to the priorities identified in the simulation outcomes. In general, it is evident that Glass Wool and Rockwool display comparatively low energy reduction, whereas Phenolic and Urethane Foam exhibit significant energy reduction values. Among these materials, Urethane Foam attains the highest energy reduction with a 40% reduction, while Glass Wool demonstrates the most minimal performance with a 21% reduction. Nevertheless, particular cases, such as Case Study 2, Case Study 6, and Case Study 7, reveal scenarios in which both low-performing and high-performing materials attain equivalent energy reduction values with a 26% reduction. This signifies that energy reduction is profoundly affected by other factors. Building types and floor areas may differ, prompting a comparison between real-world data and the simulated performance of these materials, including equipment and other elements affecting building energy performance. Consequently, employing Building Information Modeling (BIM) in simulations provides a proficient approach for executing comparisons customized to specific structures.

6.3. Cost Efficiency Consideration

As a consideration to choose the best insulation material, based on the simulation, Urethane Foam consistently outperforms the others, with the lowest Total Energy Consumption (100.9 kWh/m^2/year) and CO_2 emissions (7.86 kg/m^2/year). However, Urethane Foam is considered an expensive material, with an estimated cost of 5000–9000 JPY/m^2. Phenolic Board also performs well, offering a strong alternative to the material cost of approximately 3500–7000 JPY/m^2. Fibrous materials such as Glass Wool, Rock Wool, and Cellulose Fiber have low-cost material prices (approximately 2000 JPY/m^2–5000 JPY/m^2).

Renovating the building insulation may require additional labor and preparation, while new constructions allow for seamless integration. The labor and installation costs of the reconstruction project are relatively similar for each material; the difference lies in the cost per quantity of material to be installed. According to the Energy-Saving Design and Construction of Housing 2023 guideline [34], fibrous materials such as cellulose fiber, Rock Wool, and Glass Wool need to be installed with careful consideration of their moisture absorption properties. Glass Wool and Rock Wool are non-combustible; however, it requires vapor barriers to prevent moisture infiltration, as moisture can degrade their performance over time. In contrast, foam plastic materials, such as Urethane Foam and phenolic boards, are durable and can form continuous layers.

When choosing insulation material for an existing building, decision-makers must consider the compromises between efficiency, expense, and the structure's lifespan. Nevertheless, these materials require careful selection of the current building conditions and the property owner's budget. As reflected in Table 5, by evaluating the priorities of expense and efficiency, the building owner can choose the most suitable materials for their reconstruction needs.

Table 5. Decision-making selection table.

Type	Material	Feature
Low Cost	Glass Wool, Rock Wool, Cellulose Fiber	Basic performance, affordable initial cost, requires maintenance for moisture control for installment, and considered as a shorter lifespan.
Moderate Cost	Combination of Fiber-based material and Foam Based material	Moderate performance and cost: the use of mixed materials should incorporate the existing building conditions to determine which materials are best suited for specific parts of the building.
High Cost	Urethane foam and Phenolic board	High performance with high initial cost, ideal for prioritizing long-term efficiency with high durability and seamless installation.

6.4. Carbon Emission Factor

In recent years, a concept has emerged in sustainable construction practices in Japan known as LCCM (Life Cycle Carbon Minus) house certification. An LCCM (Life Cycle Carbon Minus) house is a dwelling that attains a negative CO_2 balance throughout its life cycle, encompassing the CO_2 emissions during construction, operation, and disposal by minimizing CO_2 emissions as much as feasible over its extended lifespan [42]. Achieving net-zero targets and lowering the carbon footprint of buildings throughout their life cycle is crucial, and this can be accomplished by the reduction of embodied carbon, structural optimization, and recycling of construction waste [43,44]. Thus, in a renovation project, the disposal of the building material can be processed into construction waste recycling, and the building can renew insulation with a better material and contribute to minimizing energy and carbon consumption. Accordingly, through renovation projects, building materials can be processed as recycled construction waste, and building insulation can be upgraded with better materials, contributing to reduced energy consumption and lower carbon emissions.

All scenarios exhibit slightly different energy performance levels. In contrast, to highlight the comparison between cost and environmental impact, the bar chart of Figure 12 represent a comparison between Fuel Cost and CO_2 emission. Fuel Cost (JPY/m^2a), represented as a blue bar to show the annual operational cost of fuel required per square meter. CO_2 Emissions (kg/m^2a), as the red line represents the annual carbon dioxide per square meter. The costs of materials and manufacturing profoundly influence the carbon footprint of a product, as carbon emissions are influenced by material production, transport, and disposal [45–47]. While the production of building materials significantly impacts the overall carbon footprint, the use of highly durable materials is encouraged. The BIM-based simulation tool in this research reveals the CO_2 emissions of the material by each carbon factor, making this method suitable for assessing the carbon footprint of the building.

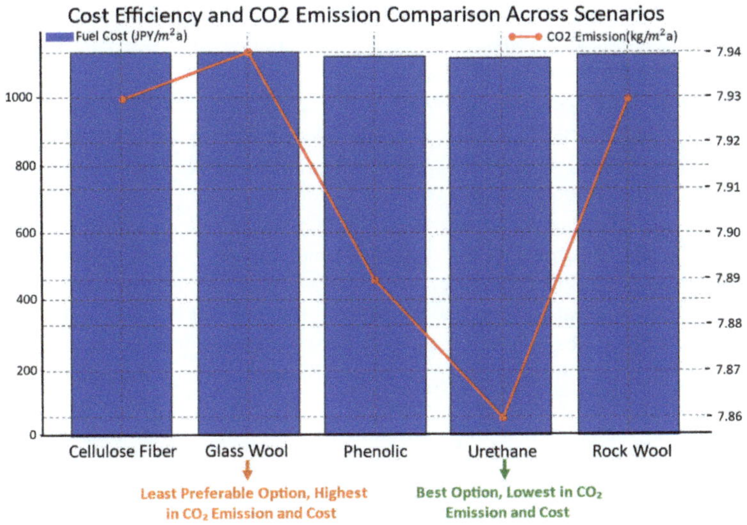

Figure 12. Cost efficiency and CO_2 emission comparison for all scenarios.

6.5. Material Selection Considering Energy Efficiency, Cost, and CO_2 Emission

Referring to the heat map in Figure 13, it provides a clearer visual representation of the difference's comparative performance of the material in terms of energy consumption, CO_2 Emission, and material cost. It is obvious that Urethane Foam emerges as the most economically advantageous and sustainable insulation alternative, consistently exhibiting the lowest fuel expenditures and CO_2 emissions across all evaluated scenarios. In contrast, Glass Wool demonstrates the least optimal performance, characterized by the highest fuel expenditure and CO_2 emissions, making it a less desirable option in terms of economic efficiency and ecological sustainability.

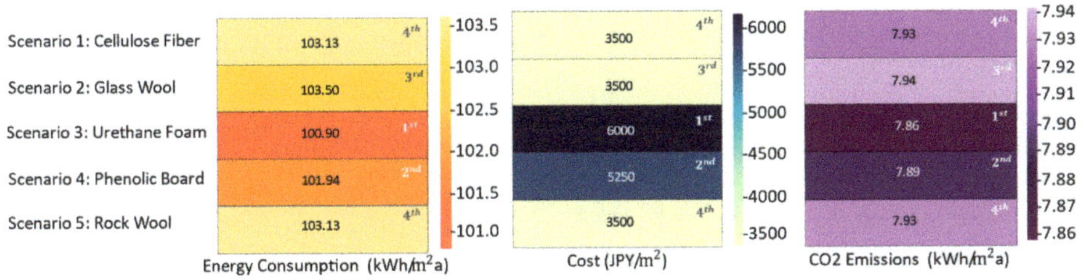

Figure 13. Heat map of each scenario for three evaluated factors: energy consumption, CO_2 emission, and cost.

This comparative analysis shows the significance of choosing materials such as Urethane Foam that have a balanced economic advantage and the least environmental impact. As the demand for sustainable construction methodologies increases, comprehending the performance metrics of insulation materials becomes important. In comparison to the investigation conducted by Siti Birkha Mohd Ali et al., which similarly employed ArchiCAD as a tool for energy simulation, it was demonstrated that the combination of window parameters markedly enhanced the infiltration rate and heat transfer coefficient, culminating in a reduction in the cooling load ranging from 3% to 6% [17]. In a parallel manner, this study underscores the significance of selecting optimal insulation strategies, with the most favorable scenario incorporating Urethane Foam resulting in a Total Energy

Consumption reduction of 58.98% in relation to the regional mean. Aligned with the study by Nasrollahzadeh, this research underscores that the insulation of the building envelope constitutes one of the paramount factors affecting energy performance [21].

The diverse insulation scenarios simulated in this research will provide guidance to property owners and contractors in selecting the most efficient insulation materials for buildings located in climates similar to those in Kanazawa City. As demonstrated in the result simulation, the best scenario will enhance 57.98% of energy consumption use compared to regional consumption intensity; it conforms to the study of [19,20] that the renovations of the building envelope and phase change material can reduce building energy consumption. The combability of BIM for energy efficiency projection and CO_2 emission underscores the effective approach for decision-making and enhances building energy and sustainability practice, which corresponds to the usage of BIM from previous studies [10–12,15]. Therefore, this research not only corroborates previous studies pertaining to the application of BIM for energy assessment but also contributes to the progressive study of building insulation materials and renovation decision-making with BIM.

7. Conclusions

In conclusion, this study provides informed decisions to optimize energy savings, reduce costs, and minimize CO_2 emissions through building insulation upgrades using BIM-based simulation. The comparative analysis of insulation materials highlights Urethane Foam as the superior choice for both economic efficiency and environmental sustainability. In contrast, Glass Wool's higher costs and environmental impact make it a less favorable option. As the demand for sustainable construction continues to grow, understanding the performance metrics of insulation materials becomes crucial. Utilizing tools like Building Information Modeling (BIM) can significantly aid designers and building owners in making informed decisions that balance cost-effectiveness, energy efficiency, and ecological responsibility. Ultimately, the choice of insulation material plays a vital role in promoting sustainable building practices and minimizing environmental impact.

Retrofitting existing buildings requires careful consideration of factors such as the building's current condition, budget constraints, and energy efficiency goals. In this research, the best retrofitting scenario is one that achieves minimal energy consumption, revealing that the insulation material will have lower energy consumption compared to the regional mean. However, this scenario may involve the use of high-cost materials compared to other options. While upgrading an existing building may require significant upfront investment, choosing the best-performing insulation material can lead to substantial cost savings over the building's lifespan by reducing energy consumption and associated operational expenses. This holistic approach ensures that retrofitting strategies are both economically viable and environmentally sustainable. For a building that is located in the same geographical condition in Japan, this research will be valuable to apply, supported by current standards and government campaign incentives for retrofitting existing buildings.

As the construction sector transitions toward sustainable methodologies, the strategic selection of insulation materials becomes a cornerstone for achieving energy-efficient and environmentally responsible buildings. Through the leveraging of tools like Building Information Modeling (BIM) and the adoption of a comprehensive framework for retrofitting, this investigation highlights the critical necessity of reconciling financial considerations, energy efficiency, and environmental impact. Ultimately, investing in high-performance insulation materials not only reduces long-term operational costs but also contributes significantly to mitigating climate change and fostering a more sustainable built environment.

Author Contributions: Conceptualization, Z.S.; Methodology, X.T. and D.C.S.T.; Software, D.C.S.T.; Validation, X.T. and D.C.S.T.; Formal analysis, X.T. and D.C.S.T.; Investigation, D.C.S.T.; Data curation, X.T. and D.C.S.T.; Writing—review & editing, X.T. and Z.S.; Visualization, X.T. and D.C.S.T.; Supervision, Z.S.; Project administration, Resource, X.T.; Funding acquisition, Z.S. All authors have read and agreed to the published version of the manuscript.

Funding: This research was supported by the Fujian Science & Technology Innovation Laboratory for Optoelectronic Information of China under the project 'Application of virtual reality technology in smart urban planning' (Grant Number: 2021ZR139).

Data Availability Statement: All data used in this research are publicly available on the official websites in Japan, https://portal-data.city.kanazawa.ishikawa.jp/, (accessed on 23 December 2024). Ministry of Land, Infrastructure, Transport and Tourism, https://kosodate-ecohome.mlit.go.jp/material/, (accessed on 23 December 2024).

Conflicts of Interest: The authors declare no conflict of interest.

References

1. Akram, M.W.; Zublie, M.F.M.; Hasanuzzaman, M.; Rahim, N.A. Global Prospects, Advance Technologies and Policies of Energy-Saving and Sustainable Building Systems: A Review. *Sustainability* **2022**, *14*, 1316. [CrossRef]
2. Kuwahara, R.; Kim, H.; Sato, H. Evaluation of Zero-Energy Building and Use of Renewable Energy in Renovated Buildings: A Case Study in Japan. *Buildings* **2022**, *12*, 561. [CrossRef]
3. Rauschen, M.; Ikaga, T.; Thomas, S.; Hoshi, H.; Karwatzki, J.; Winterseel, B. *Energy Efficiency in Buildings, Particularly for Heating and Cooling*; Wuppertal Institute for Climate, Environment and Energy: Wuppertal, Germany, 2020.
4. Sharma, S.K.; Mohapatra, S.; Sharma, R.C.; Alturjman, S.; Altrjman, C.; Mostarda, L.; Stephan, T. Retrofitting Existing Buildings to Improve Energy Performance. *Sustainability* **2022**, *14*, 666. [CrossRef]
5. Gardezi, S.S.S.; Haris Ali, S.H.; Fayaz, R.; Shah, H.H. Energy Performance Analysis of a Multi-Story Building Using Building Information Modeling (BIM). *J. Sustain. Perspect.* **2022**, *2*, 16–23. [CrossRef]
6. Del Ama Gonzalo, F.; Moreno Santamaría, B.; Montero Burgos, M.J. Assessment of Building Energy Simulation Tools to Predict Heating and Cooling Energy Consumption at Early Design Stages. *Sustainability* **2023**, *15*, 1920. [CrossRef]
7. Teng, X.; Shen, Z.; Xixi, W.; Tri, N. Pre-evaluation and visual simulation of energy consumption influencing factors of smart and healthy architecture in the design stage. In *IOP Conference Series: Earth and Environmental Science*; Institute of Physics: London, UK, 2023.
8. Teng, X.; Shen, Z. Design of a smart visiting service management system for personal information collection in order to integrate tourism management into an isolated island. *Appl. Sci.* **2020**, *10*, 6442. [CrossRef]
9. Truong, N.S.; Luong, D.L.; Nguyen, Q.T. BIM to BEM Transition for Optimizing Envelope Design Selection to Enhance Building Energy Efficiency and Cost-Effectiveness. *Energies* **2023**, *16*, 3976. [CrossRef]
10. Chen, Z.S.; Yang, L.P.; Rodríguez, R.M.; Zhu, Z.; Pedrycz, W.; Skibniewski, M.J. BIM-aided large-scale group decision support: Optimization of the retrofit strategy for existing buildings. *Appl. Soft Comput.* **2022**, *131*, 109790. [CrossRef]
11. Yildirim, M.; Polat, H. Building Information Modeling Applications in Energy-Efficient Refurbishment of Existing Building Stock: A Case Study. *Sustainability* **2023**, *15*, 13600. [CrossRef]
12. Dauletbek, A.; Zhou, P. BIM-based LCA as a comprehensive method for the refurbishment of existing dwellings considering environmental compatibility, energy efficiency, and profitability: A case study in China. *J. Build. Eng.* **2022**, *46*, 103852. [CrossRef]
13. Khan, A.M.; Tariq, M.A.; Alam, Z.; Alaloul, W.S.; Waqar, A. Optimizing energy efficiency through building orientation and building information modelling (BIM) in diverse terrains: A case study in Pakistan. *Energy* **2024**, *311*, 133307. [CrossRef]
14. González, J.; Figueiredo, K.; Hammad, A.W.A.; Tam, V.W.Y.; Haddad, A.N.; Illankoon, C. Heritage BIM (HBIM) applied in emergency scenarios: A case study of the National Museum in Brazil. *Int. J. Constr. Manag.* **2024**, 1–15. [CrossRef]
15. Hmidah, N.A.; Bin Haron, N.A.; Hizami, A.A.; Law, T.H.; Altohami, A.B.A. Energy Consumption of Retrofitting Existing Public Buildings in Malaysia under BIM Approach: Pilot Study. *Sustainability* **2023**, *15*, 10293. [CrossRef]
16. Dolzhencko, A.V.; Grebenik, A.V.; Sedashova, M.A.; Rudenskyi, D.S. Expanding the capabilities of the ARCHICAD software package for effective solution of construction design tasks. In *Journal of Physics: Conference Series*; IOP Publishing Ltd.: Bristol, UK, 2021.
17. Ali, S.B.M.; Mehdipoor, A.; Samsina Johari, N.; Hasanuzzaman, M.; Rahim, N.A. Modeling and Performance Analysis for High-Rise Building Using ArchiCAD: Initiatives towards Energy-Efficient Building. *Sustainability* **2022**, *14*, 9780. [CrossRef]
18. Fitriaty, P.; Shen, Z. Predicting energy generation from residential building attached Photovoltaic Cells in a tropical area using 3D modeling analysis. *J. Clean. Prod.* **2018**, *195*, 1422–1436. [CrossRef]

19. Rashid, F.L.; Dulaimi, A.; Hatem, W.A.; Al-Obaidi, M.A.; Ameen, A.; Eleiwi, M.A.; Jawad, S.A.; Bernardo, L.F.A.; Hu, J.W. Recent Advances and Developments in Phase Change Materials in High-Temperature Building Envelopes: A Review of Solutions and Challenges. *Buildings* **2024**, *14*, 1582. [CrossRef]
20. Riantini, L.S.; Machfudiyanto, R.A.; Rachmawati, T.S.N.; Rachman, M.D.A.; Fachrizal, R.; Shadram, F. Energy Efficiency Analysis of Building Envelope Renovation and Photovoltaic System in a High-Rise Hotel Building in Indonesia. *Buildings* **2024**, *14*, 1646. [CrossRef]
21. Nasrollahzadeh, N. Comprehensive building envelope optimization: Improving energy, daylight, and thermal comfort performance of the dwelling unit. *J. Build. Eng.* **2021**, *44*, 103418. [CrossRef]
22. Rijal, H.B.; Yoshida, K.; Humphreys, M.A.; Nicol, J.F. Development of an adaptive thermal comfort model for energy-saving building design in Japan. *Archit. Sci. Rev.* **2021**, *64*, 109–122. [CrossRef]
23. Lakatos, Á. Novel Thermal Insulation Materials for Buildings. *Energies* **2022**, *15*, 6713. [CrossRef]
24. Kumar, D.; Alam, M.; Zou, P.X.W.; Sanjayan, J.G.; Memon, R.A. Comparative analysis of building insulation material properties and performance. *Renew. Sustain. Energy Rev.* **2020**, *131*, 110038. [CrossRef]
25. Abu-Jdayil, B.; Mourad, A.H.; Hittini, W.; Hassan, M.; Hameedi, S. Traditional, state-of-the-art and renewable thermal building insulation materials: An overview. *Constr. Build. Mater.* **2019**, *214*, 709–735. [CrossRef]
26. Hwang, J.; Kim, Y.; Park, J.; Rie, D. A Study on the Evaluation of Thermal Insulation Performance of Cellulose-Based Silica Aerogel Composite Building Materials. *Polymers* **2024**, *16*, 1848. [CrossRef] [PubMed]
27. Dai, H.; Gao, H.; Yang, P.; Mo, J.; Zhang, H.; Lei, S.; Wang, L. Thermal insulation performance of rock wool reinforced kaolinite-based porous geopolymer. *Appl. Clay Sci.* **2023**, *246*, 107176. [CrossRef]
28. Hill, C.; Norton, A.; Dibdiakova, J. A comparison of the environmental impacts of different categories of insulation materials. *Energy Build.* **2018**, *162*, 12–20. [CrossRef]
29. Song, Z.; Lei, Y.; Ran, W.; Yuan, M.; Shang, S.; Cui, S. Structural properties and barrier performance of low-cost aerogel composites for building insulation. *J. Build. Eng.* **2024**, *90*, 109485. [CrossRef]
30. *ANSI/ASHRAE Standard 140-2007*; Standard Method of Test for the Evaluation of Building Energy Analysis Computer Programs (Addendum C). ASHRAE: Atlanta, GA, USA, 2011.
31. Kanazawa City Open Data Portal: Housing Construction Methods [Proportion] National Census. Available online: https://portal-data.city.kanazawa.ishikawa.jp/ (accessed on 8 January 2025).
32. Shimoda, Y.; Sugiyama, M.; Nishimoto, R.; Momonoki, T. Evaluating decarbonization scenarios and energy management requirement for the residential sector in Japan through bottom-up simulations of energy end-use demand in 2050. *Appl. Energy* **2021**, *303*, 117510. [CrossRef]
33. Huyen Nguyen, T.; Take, K.; Take, K. Reviewing of the net-zero energy buildings and housings in Japan. In *IOP Conference Series: Earth and Environmental Science*; Institute of Physics: London, UK, 2024.
34. General Incorporated Association for the Promotion of Wood Utilization in Architecture: FY 2023 (Reiwa 5) Ministry of Land, Infrastructure, Transport and Tourism (MLIT) Subsidy Project: Energy-Saving Design and Construction for Housing [Regions 4-7 Edition] [Revised] Compliance with the 2016 Energy Efficiency Standards. 2024. Available online: https://www.mlit.go.jp/jutakukentiku/house/04.html (accessed on 21 November 2024).
35. Ministry of Land, Infrastructure, Transport and Tourism. Insulation Renovation of Openings, Criteria for Applicable Construction Works. Available online: https://kosodate-ecohome.mlit.go.jp/material/ (accessed on 17 October 2024).
36. *ANSI/ASHRAE Standard 140-2017*; Standard Method of Test for the Evaluation of Building Energy Analysis Computer Programs. ANSI: Washington, DC, USA, 2017.
37. Vycital, M.; Jarský, C. N automated nD model creation on BIM models. *Organ. Technol. Manag. Constr.* **2020**, *12*, 2218–2231. [CrossRef]
38. Muhaxheri, K.; Muhaxheri, B.B. Impact of U-Values in Evaluation of Implemented Energy Efficiency Measures and Energy Savings in Public Buildings in Context of Kosovo Legislation. *WSEAS Trans. Environ. Dev.* **2023**, *19*, 11–24. [CrossRef]
39. Bass, B.; Ezell, E.; New, J. Using Measured Building Energy Data to Infer Building Characteristics for Urban Building Energy Modeling. In Proceedings of the 2022 Building Performance Analysis Conference and SimBuild, Chicago, IL, USA, 14–16 September 2022.
40. Otsuka, A. Industrial electricity consumption efficiency and energy policy in Japan. *Util. Policy* **2023**, *81*, 101519. [CrossRef]
41. Sustainable Open Innovation Initiative: ZEH Information ZEH (Detached House) Case Study Search Tool Fiscal Year 2024. Available online: https://zehweb.jp/zehinfo/example/ (accessed on 20 January 2025).
42. Institute for Built Environment and Carbon Neutral for SDGs: LCCM Housing Certification. Available online: https://www.ibecs.or.jp/rating/lccm.html (accessed on 20 January 2025).
43. Khan, S.A.; Alam, T.; Khan, M.S.; Blecich, P.; Kamal, M.A.; Gupta, N.K.; Yadav, A.S. Life Cycle Assessment of Embodied Carbon in Buildings: Background, Approaches and Advancements. *Buildings* **2022**, *12*, 1944. [CrossRef]

4. Meng, Q.; Hu, L.; Li, M.; Qi, X. Assessing the environmental impact of building life cycle: A carbon reduction strategy through innovative design, intelligent construction, and secondary utilization. *Dev. Built Environ.* **2023**, *16*, 100230. [CrossRef]
5. Seo, M.S.; Kim, T.; Hong, G.; Kim, H. On-Site measurements of CO_2 emissions during the construction phase of a building complex. *Energies* **2016**, *9*, 599. [CrossRef]
6. Zhan, Z.; Xia, P.; Xia, D. Study on Carbon Emission Measurement and Influencing Factors for Prefabricated Buildings at the Materialization Stage Based on LCA. *Sustainability* **2023**, *15*, 13648. [CrossRef]
7. Hertwich, E.G. Increased carbon footprint of materials production driven by rise in investments. *Nat. Geosci.* **2021**, *14*, 151–155. [CrossRef]

buildings

MDPI

Article

Review of the Cost-Optimal Methodology Implementation in Member States in Compliance with the Energy Performance of Buildings Directive

Paolo Zangheri [1], Delia D'Agostino [2,*], Roberto Armani [3] and Paolo Bertoldi [2]

1 ENEA, 21027 Ispra, Italy
2 European Commission, Joint Research Centre, 21027 Ispra, Italy
3 Dipartimento di Energia, Politecnico di Milano, 20156 Milano, Italy
* Correspondence: delia.dagostino@ec.europa.eu; Tel.: +39-0332-78-3512

Abstract: The building sector has a central role in achieving the European goals of a zero-emission and fully decarbonized stock by 2050. Among the provisions of the Energy Performance of Buildings Directive (EPBD) recast, the implementation of the cost-optimal methodology marked a novel approach in the establishment of minimum energy performance requirements for new and existing buildings. Member States must develop cost-optimal calculations every 5 years to verify and accordingly update the national requirements in force. This paper analyses the latest national cost-optimal reports, providing an updated assessment of the implementation of the cost-optimal methodology throughout Europe. It quantifies the progress in reaching cost-optimal levels in comparison with the previous assessment. It focuses on the conformity and plausibility of calculations in compliance with the policy framework. Furthermore, it evaluates the gap with national requirements, showing that the gap is higher than 15% only in a few Member States. The results provide a comprehensive review of the European progress towards cost-optimality in both the residential sector (average cost-optimal level 80 kWh/m^{2}y for new, 130 kWh/m^{2}y for existing buildings) and the non-residential sector (140 kWh/m^{2}y for new, 180 kWh/m^{2}y for existing buildings). An overall positive development can be inferred from the analysis of the Member States' progress in the methodology's implementation. The review also gives inputs for the cost-optimal methodology update foreseen for 2026 (e.g., cost-optimality for districts and historical buildings). The outcomes assume a crucial relevance for the ambitious energy efficiency targets established by Europe.

Keywords: cost-optimal methodology; cost-optimal analysis; nearly zero-energy buildings; energy policy; energy efficiency; building decarbonization; European Member States; Energy Performance of Buildings Directive

1. Introduction

Accounting for approximately 40% of total energy consumption and significantly contributing to global warming [1], buildings are at the core of the European strategy towards a zero-emission and fully decarbonized stock by 2050 [2]. The European Green Deal and the Renovation Wave further point to the strategic role of buildings in creating jobs, improving security, boosting green technologies, and fostering a circular economy [3,4]. The strategy foresees regulatory and financing measures that will at least double the annual energy renovation rate of buildings by 2030 and foster deep energy retrofit [5]. Within that framework, comprehensive energy and climate policy packages are progressively making the EU energy performance of buildings more efficient [6]. In particular, the Energy Performance of Buildings Directive recast (2010/31/EU—EPBD [7]) and its recent revision [8] comprise important provisions for a long-term improvement of the EU building stock [9]. Among them is the stipulation that Member States must calculate and establish cost-optimal levels of minimum energy performance requirements for new and existing buildings following the established comparative methodology framework [10] and related Guidelines [11].

The EPBD also requires Member States to develop appropriate policies and provide the necessary financing measures to foster the transition to nearly zero-energy buildings (NZEBs) [12,13], a status mandatory for all new buildings since 2021 [14,15]. The EPBD does not advocate a uniform approach for implementing NZEBs throughout Europe, requiring Member States to draw up national roadmaps that account for the variety of building types and climates [16–19].

In this context, the cost-optimal methodology appears very effective both for upgrading the energy performance requirements in force at national level and for assessing the effects of policy measures to achieve the NZEB targets [20]. A previous study [12] showed that the majority of Member States seem to adopt the cost-optimal approach in appropriate ways and use it to define NZEB requirements.

However, since the release of the cost-optimal methodology, different debates have arisen around the approach [21]. Reference building definition [22], energy performance assessment and optimization variables [23,24], uncertainty, and sensitivity analysis [25] are among the most commonly discussed topics in the literature. Recently, the importance of including the indoor environmental quality and the interactions between the building, its systems, and its occupants became evident as a way to complement energy efficiency, as conceived by the cost benefit analysis [26].

Although it has been pointed out that the methodology has mainly spread theoretically at government and scientific levels, but has not yet disseminated sufficiently among professionals [27], its introduction unquestionably signifies an important milestone towards the renovation of the existing building stock and a substantial transformation towards a zero-carbon society.

From an early assessment of the cost-optimal methodology, a heterogeneous situation was evident in European countries, as each building type and climate presented varying cost-optimal levels [21]. Regardless of comparison issues among cost-optimal levels and a non-uniform application across Europe [28], it is generally agreed that it represents an efficient and complete decision-making tool for building design that considers both energy and economic evaluations. Several literature studies focus on a specific climate or building type [28–44], but a thorough overview of the cost-optimal methodology is still missing and undoubtedly necessary.

Accordingly, this paper aims to provide a comprehensive review of the implementation of the cost-optimal methodology in Member States to assess its progress, its strengths and weaknesses, and possible future developments in the light of the recent policy developments. To accomplish this goal, the most recent cost-optimal reports (link provided in the Supplementary material section) submitted by Member States have been considered as data sources for this policy analysis [45].

After describing the methodology (Section 1.1), this paper reports the criteria adopted to assess the cost-optimal reports (Section 2), and it provides the main assessment results in Section 3. Discussion and conclusions are presented in Sections 4 and 5, respectively.

1.1. The Cost-Optimal Methodology

The calculation framework of cost-optimal levels to be used by Member States to benchmark their building standards was defined in 2012 by the European Commission [10]. It is based on the principle of the cost-benefit analysis, and it is calculated from two economic perspectives: the financial and the macroeconomic, which refer to different discount rates (lower in the macroeconomic one) and cost items [46]. While the financial perspective includes taxes, the macroeconomic perspective considers greenhouse gas emission costs [47]. The calculation approach can be summarized in the following steps [6]:

1. Establishment of reference buildings. Real or virtual buildings representing the building stock must be selected. Member States must define at least three categories, for both new and existing buildings (residential single-family, residential multi-family, of-

fices). According to [10] (Annex I), Member States must also define reference buildings for other building categories, for which specific energy performance requirements exist. For new buildings, the standard energy performance in force can be assumed as base case. For the existing stock, at least two reference buildings must be considered, which can be established on the basis of size, age, cost, structure, construction material, use pattern, or climatic zone.

2. Identification of energy efficiency and renewable measures. These must be implemented in new or existing buildings, including different packages of measures or measures of different levels (e.g., from lower to higher insulation levels), which must respect the EU and national legislation. The impact of applying packages of measures on reference buildings (starting with the current requirements and beyond, including the NZEB level) should be estimated in the cost-optimal calculation in terms of energy and financial performance.
3. Calculation of the (net) primary energy consumption. The energy performance calculation must be based on the current national or CEN standard methodologies for each selected building variant. Framework conditions for the calculations must be defined in terms of climate data, performance of energy systems, primary energy factors, and indoor air quality.
4. Calculation of the global cost. The net present value (NPV) should be used at each step, based on 30 years for residential and 20 years for non-residential buildings. The included cost categories are: initial investment costs, running costs (i.e., energy, operational, maintenance, replacement costs), disposal costs, final value, and the cost associated with CO_2 emissions (only for the macroeconomic perspective). For the assessment of the financial performance (global costs) of the chosen combinations of packages, the European Standards EN 15459 is suggested as reference [48]. This method results in a discounted value of all costs (present and future) for a defined period of calculation. The global cost formula can be written as in Equation (1)):

$$C_G(\tau) = C_I + \sum_j \left[\sum_{i=1\ldots\tau} \left(C_{a,i}(j) \times R_d(i) + C_{C,i}(j)\right) - V_{f,\tau}(j)\right] \quad (1)$$

where $C_G(\tau)$ corresponds to the global cost referred to starting year τ_0 C_I is the initial investment cost; $C_{a,i}(j)$ is the annual cost for component j at the year i (including running costs and replacement costs);$R_d(i)$ is the discount rate for year i;$V_{f,\tau}(j)$ is the final value of the component j at the end of the calculation period (referring to the starting year $\tau_{0)}$.

5. Identification of cost-optimal levels. This should be expressed in primary energy consumption (in kWh/m^2 per year) for each reference building. The cost-optimal configuration presents the lowest costs maintaining a high performance. It can be identified in the lower part of the curve that reports global costs (Eur/m^2) and energy consumption (kWh/m^{2}y) (Figure 1).
6. Evaluation of the shortfall against current minimum energy performance requirements. If the difference is higher than 15%, Member States must justify the gap or define a plan to reduce it. The national benchmark related to the final outcome of the cost-optimal calculations can be calculated for a financial or a macroeconomic perspective.

Sensitivity analysis can help in evaluating the robustness of key parameters, such as discount/interest rate, the annual increase in energy prices, and primary energy factors associated with different fuels.

Relevant additional information is given in the official Guidelines [11], as stated in Annex III [7]. They give methods for establishing reference building types and subcategories, reducing package combinations, gathering cost data, and deriving cost-optimal levels. They provide a list of CEN standards, cost categorization, indications for sensitivity analysis, equations for the calculation of shortfalls against current requirements, indoor air quality, and comfort. They also clarify the full cost approach, calculation period, replacement, energy and disposal costs, taxation, subsidies, and feed-in tariffs.

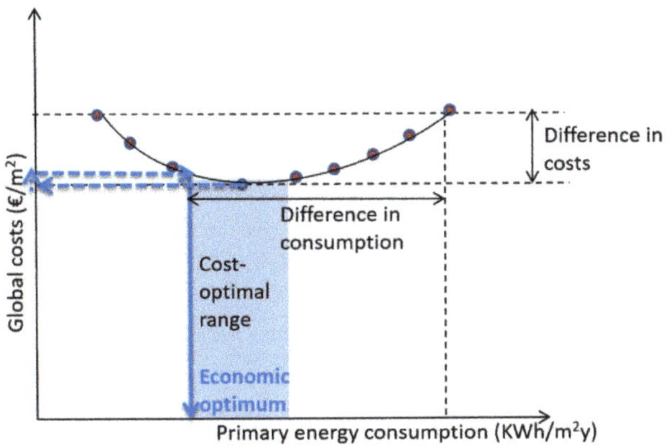

Figure 1. Different variants within the graph and position of the cost-optimal range.

Figure 2 shows the timeline of EPBD recast to implement cost-optimal calculations and NZEB requirements.

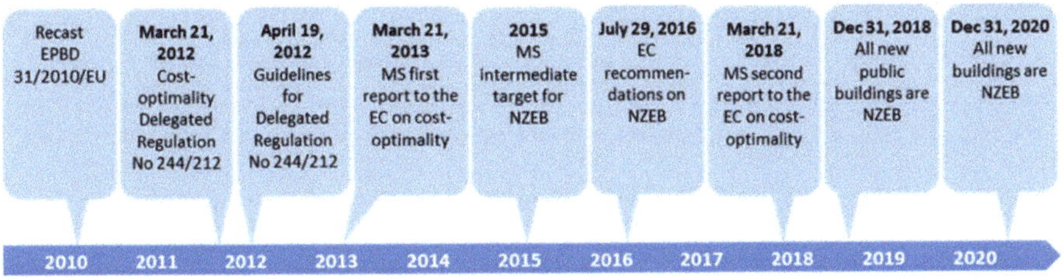

Figure 2. Timeline for cost-optimal and NZEB requirements.

In line with the EPBD timeline (Figure 2), the first cost-optimal calculations were due by 2013. According to the first assessment of 2015 [28], apart from a few cases, the overall picture seemed rather positive in terms of compliance with the EPBD requirements. Although reports were mainly assessed as plausible, energy performance levels showed space for potential improvement towards more stringent requirements to reduce primary energy use and life-cycle costs. A joint, consistent, and comparable level of ambition was desirable after the first assessment [49]. The second cost-optimal reports were submitted in 2020. Accordingly, this paper reviews the Member States' progress in reaching cost-optimal levels of minimum energy performance requirements.

2. Methodology

The methods of this cost-optimal implementation review were divided into the following main steps. In the first step, reports were collected and officially translated. Then, the template for data collection was developed and the reports analyzed (total lengths: around 6000 pages of reports and 4000 pages of annexes). Clarifications and additional information were often requested from Member States in cases of missing or non-extractable information, frequently due to unclear translations, units, tables, figures or explained methodology.

As detailed in Section 1.1, the cost-optimal methodology can be divided into different steps, to which the reference categories of Table 1 correspond. The key assessed topics developed to collect data and information for each reference category, assess the cost-optimal calculations, and guide the report evaluation are also reported in Table 1.

Table 1. Reference categories.

Number	Reference category	Assessed topics	Reference
0	Scope of report		Art. 4 [7]
1	Establishment of reference buildings	• Reference buildings and related data sources and databases • Considered climatic conditions • Established new and existing building types and properties (e.g., floor area, primary energy, reference age, climate, typical heating, and cooling systems)	Annex 1, point 1 [6]
2	Identification of energy efficiency measures, measures based on renewable energy sources, and/or packages and variants of such measures for each reference building	• Current, future/coming, and NZEB requirements for new and existing buildings • Established packages of measures and/or building variants to the different building types • Details on the measures considered for both new and existing residential and non-residential buildings	Annex 1, point 2 [6]
3a	Calculation of the primary energy demand	• Details on the performed calculations (e.g., multi-stage optimization, dynamic method, starting year, accounted energy uses) and energy demands (referring to the energy scheme of the Regulation) • Specifics on the used weather data, indoor (operative) temperature, primary energy factors	Annex 1, point 3 [6]
3b	Calculation of the global cost in terms of net present value for each reference building	• Cost categories defined in the global cost calculation • Details on the calculations performed (e.g., financial and macroeconomic perspectives, data sources for construction and renovation costs, full cost approach for cost quantification, omitted cost items, technology costs) • Specifics on main calculation parameters (e.g., calculation period, discount rate, real interest rate, energy prices, annual increase in energy prices, CO_2 emissions costs)	Annex 1, point 4 [6]
4	Sensitivity analysis	• Details on the sensitivity analyses carried out and related outcomes	Annex 1, point 5 [10]
5	Derivation of a cost-optimal level of energy performance for each reference building	• Identification of the cost-optimal levels for the different (new, existing) building types • Definition of a cost-optimal range selecting the optimal variants • Analysis of the requirements resulting in a lower use of primary energy (left border of the cost-optimal range) and of the economic perspectives (financial and macroeconomic) used to derive the cost-optimal levels and related differences	Annex 1, point 6 [6]
6	Plan to reduce the gap	• Comparison between the current/future/NZEB requirements and the identified cost-optimal levels • Identification of the gaps between current/future/NZEBs energy requirements and the cost-optimal levels for new and existing buildings • Analysis of the plan and appropriate steps to reduce the gaps	Art. 5(3) [7]; Art. 6(2) [6]

The methodology included the analysis and comparison of key input parameters (discount rate, energy prices, annual increase in energy prices) both for residential and non-residential buildings. Then, the cost-optimal levels were derived in terms of primary energy consumptions and global costs. Gaps between current requirements and cost-optimal levels were also checked, both in terms of primary energy and U-value of building elements. A comparison with previously cost-optimal reports was also made. Finally, an analysis of the conformity with the common methodology as well as of the plausibility of the input parameters and calculated cost-optimal levels was carried out to verify the calculation compliance with the cost-optimal framework based on the policy requirements [6,7]. The most recent cost-optimal reports are publicly available [45]. This review takes into account 30 cost-optimal reports, which were delivered by: Austria (AT), Belgium-Brussels (BEbru), Belgium-Flanders (BEfla), Belgium-Wallonia (BEwal), Cyprus (CY), Czech Republic (CZ), Germany (DE), Denmark (DK), Estonia (EE), Greece (EL), Spain (ES), Finland (FI), France (FR), Hungary (HU), Ireland (IE), Italy (IT), Lithuania (LT), Latvia (LV), Luxembourg (LU), Malta (MT), the Netherlands (NL), Poland (PL), Portugal (PT), Romania (RO), Sweden (SE), Slovenia (SI), Slovakia (SK), and the United Kingdom (UK).

3. Results

The results obtained are discussed in the following sections, as follows:

- Section 3.1: analysis and comparison of key input parameters (discount rate, energy prices for both residential and non-residential buildings, annual increase in energy prices for both residential and non-residential buildings).
- Section 3.2: analysis of cost-optimal levels (primary energy consumption and global costs).
- Section 3.3: comparison with first cost-optimal levels (2013) and consolidated version.
- Section 3.4: gaps between current requirements and cost-optimal levels.

3.1. Input Parameters Analysis and Comparison

The applied primary energy factors (PEF), the discount rate, the energy prices, and their development over the calculation period have significant impact on the outputs of the cost-optimal calculations. Under the EPBD framework, Member States can adapt them to their context. Figure 3 shows and compares their choices, and provides some key statistics.

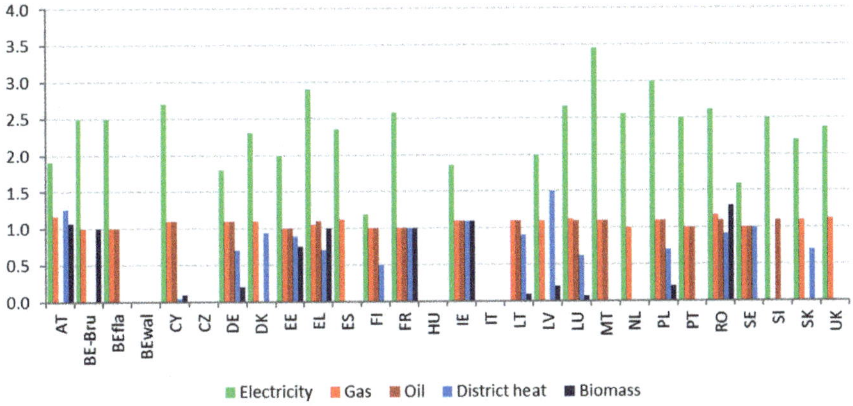

	Electricity	Gas	Oil	District Heat	Biomass
Min	1.20	1.00	1.00	0.05	0.07
Max	4.03	1.22	1.10	1.50	1.30
Average	2.39	1.08	1.06	0.85	0.67
Median	2.50	1.10	1.10	0.91	1.00

Figure 3. Primary energy factors used by Member States and associated statistics.

Regarding the Primary Energy Factors applied for the main energy fuels (Figure 3), the main differences refer to electricity, with a maximum of 3.5 (Malta) and a minimum of 1.2 (Finland). We also observed good variation for biomass, which some Member States (Cyprus, Germany, Lithuania, Latvia, Luxembourg, and Poland) considered almost fully renewable (PEF close to zero), and some others (Austria, Estonia, Greece, France, Ireland, Romania) assimilated into fossil fuels (PEF close to 1). Discount rates (DR) are reported in Figure 4.

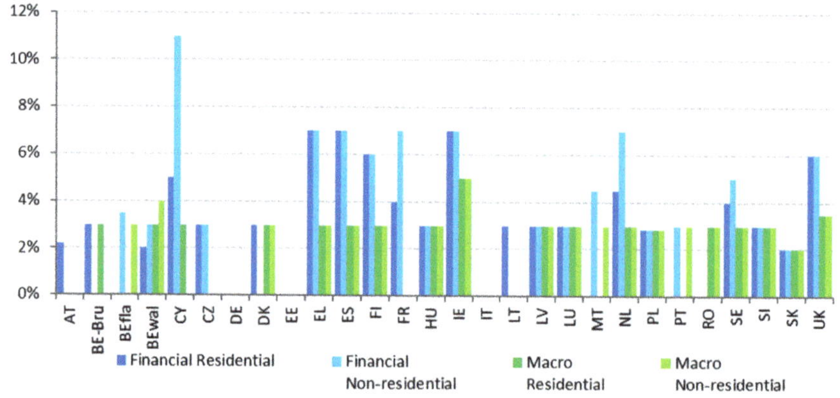

	Financial Residential	Financial Non-Residential	Macro-residential	Macro-non-Residential
Min	2%	2%	2%	2%
Max	7%	11%	5%	5%
Average	4%	5%	3%	3%
Median	3%	5%	3%	3%

Figure 4. Discount rate used by Member States and associated statistics.

As shown in Figure 4, almost all Member States (with a few exceptions) made the same choices for the macro-economic perspective (DR = 3%), following the indication of the official Guidelines [11]. The picture for the financial perspective is more heterogeneous, with an average of 4% for the residential and 5% of the non-residential sector.

The energy prices and the annual increase in energy price applied by Member States for residential and non-residential building types are shown in Figures 5 and 6. Minimum, maximum, average, and median values are reported for both residential and non-residential sectors in Tables 2 and 3.

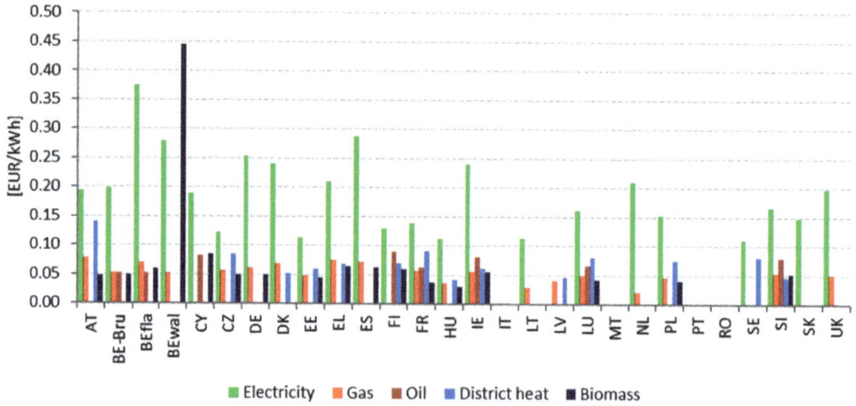

Figure 5. Energy prices used by Member States for residential sector.

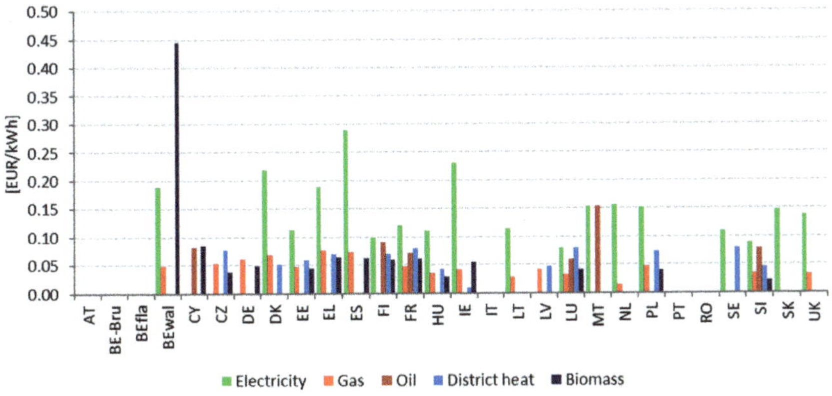

Figure 6. Annual increase in energy prices used by Member States for non-residential sector.

Table 2. Statistics about the energy prices selected for residential (Res) and non-residential (N-Res) buildings.

	Electricity		Gas		Oil		District Heat		Biomass	
	Res	N-Res	Res	N-Res	Res	N-Res	Res	N-Res	Res	N-Res
Min	0.110	0.080	0.021	0.014	0.053	0.059	0.043	0.010	0.030	0.023
Max	0.375	0.311	0.078	0.075	0.090	0.154	0.140	0.081	0.445	0.445
Average	0.191	0.156	0.054	0.046	0.070	0.089	0.071	0.061	0.075	0.079
Median	0.190	0.143	0.054	0.047	0.072	0.081	0.070	0.069	0.050	0.052

Table 3. Statistics about the annual increase in energy prices used for residential (Res) and non-residential (N-Res) buildings.

	Electricity		Gas		Oil		District Heat		Biomass	
	Res	N-Res	Res	N-Res	Res	N-Res	Res	N-Res	Res	N-Res
Min	0.2%	0.2%	1.0%	1.0%	1.2%	1.2%	0.6%	1.0%	0.7%	0.7%
Max	5.4%	2.8%	3.6%	2.8%	3.0%	5.3%	2.8%	2.8%	2.8%	2.8%
Average	1.7%	1.4%	1.9%	1.8%	2.4%	2.9%	1.6%	1.8%	1.7%	1.8%
Median	1.6%	1.5%	1.8%	1.8%	2.5%	2.6%	1.5%	1.9%	2.0%	2.0%

The figures demonstrate that the divergences are not as relevant for energy prices. Only a few cases (e.g., Flanders, Belgium for electricity and Wallonia, Belgium for biomass) referred to values not close to the averages.

The analysis of the annual increase in energy prices (Figure 6, Table 3) showed that the average of the Member States' values for electricity was closer than gas to the indication of the European Commission (For the estimation of fuels and electricity price development trends, the Commission reference data to Member States: https://ec.europa.eu/energy/en/data-analysis/market-analysis, https://ec.europa.eu/energy/en/news/reference-scenario-energy, accessed on 10 June 2022) (respectively 1.4% and 2.8%). However the deviation is quite high for all energy vectors, especially for electricity in the residential sector. For instance, here, Brussels applied only 0.2% against 5.4% for the Netherlands.

3.2. Cost-Optimal Level Assessment

From the reports analyzed, it was not possible to extract all the cost-optimal levels for all Member States, as half of them did not derive them in a clear and complete way. Plotting the values collected in the primary energy/global cost domain (Figures 7 and 8), an overall picture was obtained for new and existing buildings. The values pertain to single family houses (SFH), multi-family houses (MFH), and offices and other non-residential (N-R) buildings. The latter is the least-covered category, for which Member States are free to select the reference type. In some cases, these results refer to a particular type (often an educational building), in others to an average among different building types (e.g., school, commercial and hospital). For this reason, the comparison between countries cannot be fully consistent.

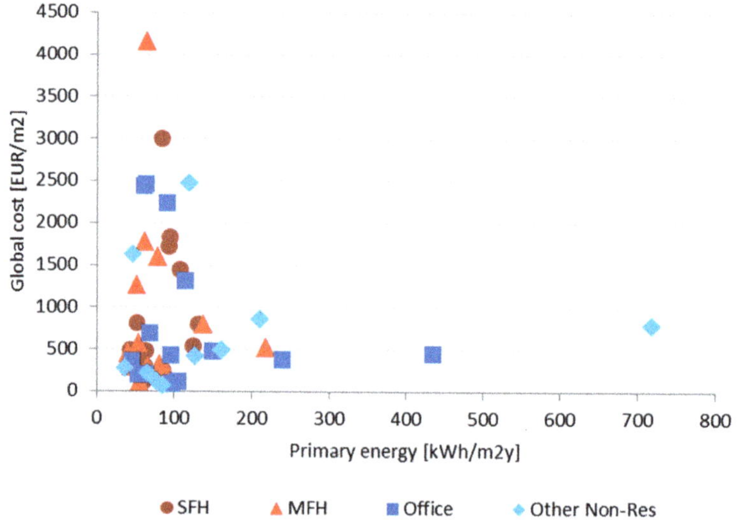

Figure 7. Cost-optimal levels for new buildings.

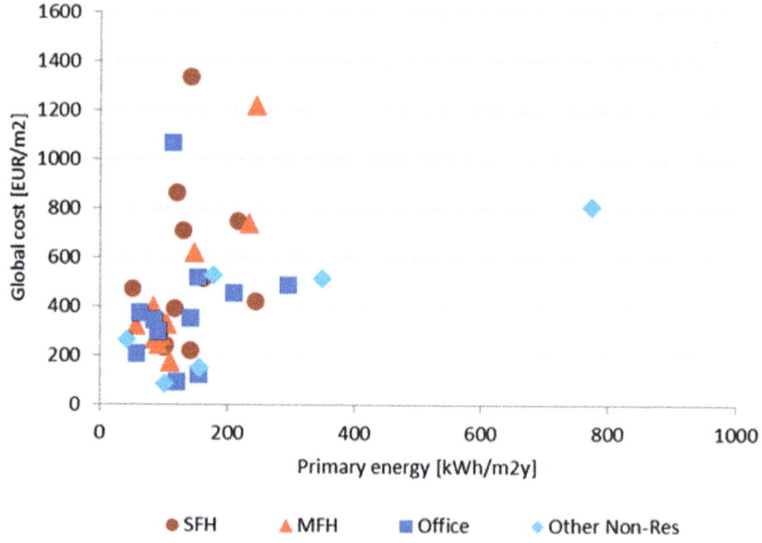

Figure 8. Cost-optimal levels for existing buildings.

For new buildings, the majority of cost-optimal points fell between 50 and 100 kWh/m^{2}y, with an average of 80 kWh/m^{2}y for the residential sector and 140 kWh/m^{2}y for the non-residential sector. Associated global costs were often lower than 1500 EUR/m^2, with an average of 925 EUR/m^2 for the residential and 800 EUR/m^2 for the non-residential sector.

For existing buildings, the majority of cost-optimal levels fell between 75 and 175 kWh/m^{2}y, with an average of 130 kWh/m^{2}y for the residential and 180 kWh/m^{2}y for the non-residential sector. For existing buildings, the global costs were generally lower than 600 EUR/m^2, with an average of 500 EUR/m^2 for the residential and 385 EUR/m^2 for the non-residential sector.

An additional analysis grouped countries per climate zone (Tables 4 and 5) [50]. We applied an official approach [51], including:

- Mediterranean zone: Cyprus, Greece, Spain, Italy, Malta, and Portugal;
- Oceanic zone: Belgium, France, Ireland, Luxembourg, the Netherlands, and the UK;
- Continental zone: Austria, Czech Republic, Estonia, Germany, Hungary, Lithuania, Latvia, Poland, Romania, Slovenia, and Slovakia;
- Nordic zone: Denmark, Finland, and Sweden.

Table 4. Average cost-optimal levels (PE: Primary Energy and GC: Global Cost) for new buildings per climatic condition.

Climate Zone	New SFH		New MFH		New Office		New Other n-R	
	PE [kWh/m^{2}y]	GC [EUR/m^2]	PE [kWh/m^{2}y]	GC [EUR/m^2]	PE [kWh/m^{2}y]	GC [EUR/m^2]	PE [kWh/m^{2}y]	GC [EUR/m^2]
Mediter.	81	887	105	698	221	648	423	607
Oceanic	86	760	66	746	94	1214	140	992
Continental	81	419	93	356	80	157	67	173
Nordic	77	1882	62	2076	66	1681	120	2481

Table 5. Average cost-optimal levels (PE: Primary Energy and GC: Global Cost) for existing buildings per climatic condition.

Climate Zone	Existing SFH		Existing MFH		Existing Office		Existing Other n-R	
	PE [kWh/m^{2}y]	GC [EUR/m^2]	PE [kWh/m^{2}y]	GC [EUR/m^2]	PE [kWh/m^{2}y]	GC [EUR/m^2]	PE [kWh/m^{2}y]	GC [EUR/m^2]
Mediter.	161	500	148	467	175	396	775 *	808 *
Oceanic	124	670	142	628	160	682	264	522
Continental	97	329	100	237	112	143	102	166
Nordic	183	643	77	303	78	336	122	236

* Available only for Malta (average of values obtained for schools, hotels, restaurants, shops, elderly centers and sport complex).

From this comparison, it is interesting to observe that in in all cases the primary energy consumption associated with cost-optimal levels is lower in the cold zones. Global costs are lower in the Continental zone, which includes the States of Eastern Europe.

3.3. Comparison with First Cost-Optimal Levels (2013)

Table 6 shows a comparison between the primary energy consumption associated with the cost-optimal levels of 2013 and 2020, when both these values were available. A reduction in primary energy consumption associated with the cost-optimal levels (negative percentage) can be observed for almost all building types: about −20% for the residential (new and existing) and new office buildings.

Table 6. Comparison between cost-optimal levels obtained Member States in 2020 and 2013.

MS	New SFH	New MFH	New Office	Existing SFH	Existing MFH	Existing Office
AT	−53%	−47%		−55%		
BEwal	−27%	−44%				
DE	−60%	−36%	−36%			
DK	−13%	−26%	−11%	−23%	−21%	−43%
EE	−46%	−39%	−44%	11%	−24%	−11%
ES	−12%	6%	35%	−9%	−17%	16%
FI	−38%	−32%	−44%	−17%	−38%	−24%
FR	−18%	−10%	−2%	66%	52%	108%
HU	−1%	4%	26%	21%	−22%	−19%
IE	56%	−4%	−10%			
LT	15%	16%	18%	9%	17%	12%
LU	0%	−1%	1%	0%	0%	−7%
PL	−26%	−21%		−1%	−8%	
RO	−48%	−60%	−5%	−46%	−25%	11%
SE	57%	35%	−28%			
SI	−100%	−100%	−100%	−100%	−100%	−100%
SK	−71%	−68%	−38%	−83%	−68%	−44%
UK	−32%	−41%	−17%			
Average	−23%	−23%	−17%	−17%	−21%	−9%

Table 7 displays a consolidated version of cost-optimal energy levels that considers 2013 levels when 2020 ones are missing, for all eight building types previously discussed. Only two Member States (Bulgaria and the Netherlands) did not provide a clear cost-optimal requirement.

Table 7. Consolidated version of cost-optimal levels in kWh/m^{2}y (considering both 2020 and 2013 results).

MS	New SFH	New MFH	New Office	New Other n-R	Existing SFH	Existing MFH	Existing Office	Existing Other n-R
AT	41	41	291	n/a	44	44	n/a	n/a
BEbru	64	55	n/a	n/a	119	149	n/a	n/a
BEfla	n/a	n/a	63	46	n/a	n/a	71	51
BEwal	82	63	64	48	143	246	115	n/a
BG	n/a	n/a	n/a	n/a	n/a	n/a	n/a	n/a
CY	156	217	280	n/a	160	178	252	n/a
CZ	84	95	210	n/a	144	78	193	n/a
DE	38	44	110	110	n/a	n/a	n/a	n/a
DK	52	40	46	n/a	121	58	63	n/a
EE	87	103	93	n/a	250	130	160	n/a
EL	109	52	114	n/a	163	107	143	n/a
ES	53	45	96	n/a	78	70	86	n/a
FI	92	82	86	117	233	102	93	130
FR	76	60	109	112	178	166	141	208
HR	84	105	62	n/a	79	144	55	n/a
HU	132	138	106	86	141	113	156	158
IE	52	66	69	211	104	88	210	260
IT	66	32	46	n/a	n/a	n/a	n/a	n/a
LT	141	154	127	118	156	168	140	131
LV	74	75	93	n/a	114	84	93	n/a
LU	125	82	150	180	131	85	150	192
MT	32	15	436	844	23	15	n/a	908
NL	n/a	n/a	n/a	n/a	n/a	n/a	n/a	n/a
PL	89	83	97	79	89	83	97	79
PT	40	29	n/a	128	61	48	n/a	n/a
RO	70	40	50	64	105	63	90	135
SE	85	65	62	n/a	89	115	99	n/a
SI	86	92	64	66	96	94	121	104
SK	29	29	71	n/a	24	27	46	n/a
UK	96	77	105	201	n/a	n/a	n/a	n/a

3.4. Gaps between Current Requirements and Cost-Optimal Levels

The evaluation of the gaps between cost-optimal levels and current requirements is a relevant step in the calculation, since it should provide useful indications for the update of existing energy performance regulations. National minimum energy performance requirements should not be higher than 15% compared to the outcome of the cost-optimal levels. A plan should be defined to reduce the gaps that cannot be strongly justified.

The gap quantification allows the comparison of results among countries, reference building types, and requirements for renovation of building or components. Therefore, it is important to note that the results are not fully comparable among Member States, since they are free to choose the macroeconomic or financial perspective for deriving the cost-optimal levels and to apply different national standards to calculate the energy performance of buildings. Other differences (e.g., related to investment costs) should reflect national market conditions, and are thus not a limit to, but an integral part of the comparison.

The comparison between the levels assessed in 2013 and 2020 can also be extended to the shortfall against minimum energy performance requirements in force. Tables 8 and 9 collect the available information, providing the gap evolution when data are available from both the 2013 and 2020 reports. A negative percentage means that the analyzed energy requirement is lower than the cost-optimal level. We report a single gap value for cases that were assessed only in a calculation round (by indicating the reference year), and both gap values when available. The Member States for which it was not possible to derive clear figures for any building types are not included in the tables.

Table 8. Consolidated version of shortfalls against minimum energy performance requirements for new buildings, with comparisons between the 2013 and 2020 values when available. A negative gap means that the requirement is lower than the cost-optimal level.

MS	New SFH	New MFH	New Office	New Other N-R
AT	48% (2013)	55% (2013)	6% (2013)	n/a
BEbru	−30% (2020)	−18% (2020)	n/a	n/a
BEwal	from 15% to 40%	from 15% to 83%	n/a	n/a
CY	33% (2013)	−2% (2013)	−7% (2013)	n/a
DE	0% (2013)	0% (2013)	0% (2013)	n/a
DK	from −10% to −22%	from −11% to −15%	from 33% to −11%	n/a
EE	from 0% to 46%	from −12% to 31%	from −3% to 42%	n/a
ES	from 8% to 14%	from 7% to 21%	7% (2013)	n/a
FI	from 15% to 14%	from 0% to 9%	from 3% to 13%	14% (2020)
FR	from −46% to −24%	from −6% to 2%	from −13% to 3%	from −13% to −10%
HR	49% (2013)	n/a	n/a	n/a
HU	from 67% to −32%	from 26% to −38%	from 60% to −6%	−1% (2020)
IE	from 51% to −5%	from −11% to −35%	from 299% to 9%	−9% (2020)
IT	31% (2013)	80% (2013)	−22% (2013)	n/a
LT	from 20% to −10%	from 11% to −4%	from 21% to 1%	2% (2020)
LU	from 0% to −21%	from 0% to −18%	from 16% to 0%	−21% (2020)
MT	172% (2013)	680% (2013)	61% (2020)	from 63% to 30%
NL	5% (2020)	5% (2020)	−3% (2020)	−3% (2020)
PL	from 17% to 7%	from 43% to 3%	−38% (2020)	−24% (2020)
PT	−10% (2013)	−12% (2013)	n/a	25% (2020)
RO	from 102% to 120%	from 33% to 195%	from 16% to 20%	77% (2020)
SE	from 35% to 6%	from 52% to 24%	from −6% to 23%	n/a
SI	14% (2013)	14% (2013)	20% (2013)	8% (2020)
SK	116% (2013)	43% (2013)	35% (2013)	n/a
UK	from −21% to −3%	from −21% to 19%	−5% (2013)	−12% (2020)

Table 9. Consolidated version of shortfalls against minimum energy performance requirements for existing buildings, with comparison between 2013 and 2020 values when available.

MS	Existing SFH	Existing MFH	Existing Office	Existing Other N-R
AT	54% (2013)	45% (2013)	n/a	n/a
BEbru	−54% (2020)	−61% (2020)	n/a	n/a
CY	n/a	n/a	−7% (2013)	n/a
DK	from 1% to 10%	from 3% to 63%	from −3% to 39%	n/a
EE	from −7% to 16%	from 6% to 38%	from 17% to 31%	n/a
ES	−29% (2013)	−27% (2013)	−6% (2013)	n/a
FI	from 19% to 8%	from 1% to 4%	from 11% to 0%	10% (2020)
FR	from 10% to 1%	from 10% to −5%	from 49% to −34%	from 49% to 6%
HR	59% (2013)	n/a	n/a	n/a
HU	from 71% to −6%	from −15% to 4%	from −31% to −10%	−30% (2020)
IE	23% (2020)	42% (2020)	−28% (2020)	82% (2020)
LT	from 54% to 7%	from 56% to 5%	from 80% to 10%	10% (2020)
LU	from 8% to 8%	from 7% to 8%	from 5% to 14%	−16% (2020)
MT	270% (2013)	680% (2013)	n/a	from 74% to 27%
PL	from 79% to 7%	from 92% to 3%	−38% (2020)	−24% (2020)
PT	1% (2013)	−2% (2013)	n/a	n/a
RO	from 65% to 46%	from 58% to 115%	from 40% to −33%	−16% (2020)
SE	2% (2020)	−35% (2020)	−24% (2020)	n/a
SI	15% (2013)	19% (2013)	−37% (2013)	−17% (2020)
SK	48% (2013)	49% (2013)	85% (2013)	n/a

Focusing on Member States for which data from both 2013 and 2020 are available, it can be observed that in some cases the gaps decreased, and in others they increased with the second calculation round. This progress can be assessed as quite positive, especially for existing buildings. A key implication of this aspect for a future revision of the methodology is accounting for the developments of the construction market as well as the technological innovation of building products, considering the evolution of energy prices, primary energy factors, and materials. Another important implication is the identification of the most suitable renovation packages that Member States should consider to achieve ZEBs.

4. Discussion

The cost-optimal methodology appears to be currently entrenched at government and scientific levels. It represents a well-established tool that includes both energy and economic evaluations. Progress is clearly assessed in Member States in relation to the analyzed categories. In support of future reviews of the cost-optimal methodology implementation, the following proposals are elaborated.

- A common template to collect data

A future common template would be useful to collect key input data and outputs. A large number of input data may impact final results (e.g., implemented technology costs, normally collected from databases or construction projects), while a common template would allow easier and comparable methodology implementation, and consequently coordinated assessment.

- Increase the number of variants

Considering a higher number of variants could be useful, as the current number (10 packages/variants plus the reference case) may not guarantee a comprehensive coverage of applicable technologies. A number of 20 packages/variants appears suitable, as some Member States and other studies [52–55] have already shown this figure as feasible.

- Guidance for historical buildings

Indications could be provided for historical buildings. Normally the selected existing building types date from the years 1960s–1980s, and a few Member States also considered historical buildings. Guidance for selecting energy efficiency measures in heritage buildings could be useful [56].

- Indications on indoor air quality and comfort

Currently, a few Member States considered comfort and indoor air quality within their cost-optimal framework, probably due to the complexity of their inclusion in the calculations. Summer comfort is particularly overlooked, and an active air conditioning system is not often included among the selected efficiency measures. One possible procedure would refer to the operative indoor temperature provided by Standard EN 16798 [57], which replaced EN 15251. According to this set-point, the energy demand for cooling should be calculated for all building variants, and an active cooling system should be considered for variants with an energy demand higher than a certain limit (e.g., 3 or 5 kWh/month).

- Introduction of a base refurbishment level

The cost-optimal framework could take into account additional variables, such as the increase in the building value due to better energy performance, as well as further co-benefits deriving from the application of energy efficiency measures and the use of renewable sources, such as improved comfort, noise reduction, and indoor air quality.

The introduction of a "base refurbishment level" for existing buildings would take these factors into account. It would represent the lower renovation level to be compared with more efficient ones. This base level should comprise building façade and roof rehabilitation, old windows and heating/cooling system replacement, active cooling installation. Its definition would allow the obtainment of harmonized building variants in terms of functionality, aesthetics, and liveability, would recognize which costs could be omitted, and would fix the end-life of the original building components (normally older than 40 years). An example of how the base refurbishment level can be applied with good results is given in the Entranze project [20].

- Additional co-benefits of better performing buildings

Member States are free to extend the methodology to full life-cycle costs, although none of them have. As recently discussed [58], the global cost calculation could follow a revised formula, which includes new savings related to the environmental impacts (C_{EN}), economic impacts (C_{EC}, i.e., the change in GDP and the cost savings due to improved energy security), and social impacts (C_{SO}), as shown in equation 2:

$$C_G(\tau) = C_I - C_{EN} - C_{EC} - C_{SO} + \sum_j \left[\sum_{i=1\ldots\tau} (C_{a,i}(j) \times R_d(i) + C_{C,i}(j)) - V_{f,\tau}(j)\right] \quad (2)$$

where:

$C_G(\tau)$ is the life-cycle cost including environmental, economic, and social indicators [€];

C_I is the sum of initial investment costs;

$C_{a,i}(j)$ is the annual cost for component j at the year i;

$R_d(i)$ is the discount term for year i;

$C_{c,i}(j)$ means carbon cost per measure or set of measures j during year i (not included in the financial perspective methodology);

$V_{f,\tau}(j)$ is the final value of component j at the end of the calculation period.

The inclusion of this approach in the cost-optimal calculation would be relevant to the wider benefits of energy efficiency and policy decisions that shape the market transition.

- Derivation of the cost-optimal level

In relation to the cost-optimal level identification, two aspects deserve more attention:

- the application of the "cost-optimal range" is too often ignored, despite affecting the final results;
- the choices related to the averaging operation applied to the cost-optimal levels (derived for different building types) to quantify the shortfall against current requirements should always be thoroughly discussed and motivated.

The first point also relates to the wider benefits of energy efficiency previously discussed. If this effect is not explicitly considered in the optimization method, it becomes more important to select the optimum level in a non-conservative way. That is, by applying a tolerance on global costs (3%–5%) and choosing the lowest energy level (in the so called "left border of the cost-optimal range").

- Enhancing sensitivity

The selection of input parameters highly influences results. Consequently, sensitivity analysis is of a key importance in controlling their impact and the robustness of related outcomes. This allows quantification of the uncertainty of the data, especially those derived from databases. Two approaches are possible: changing parameters one by one (as done by Member States) or assigning a probabilistic function (e.g., Monte Carlo simulation) to input parameters and obtaining more precise distributions of outcomes. This approach takes into account more fully the heterogeneity of the different data sources, providing results based on intervals. As example of this approach, a study from Becchio et al. [59] considered, for each identified parameter, a minimum, average, and maximum value in a triangular distribution.

- Cost-optimality at district level

It would be useful to define a cost-optimal approach focused to the district level, where environmental and social impacts are considered in terms of the co-benefits linked to the environment and the involved community [60]. This could also encourage the development of decarbonization strategies for reducing environmental impacts at local level [61].

Finally, it is worth underlining that, in compliance with Article 6 of [8] (ex-Article 5 of the EPBD), the calculation of cost-optimal levels falls within the core of the Green Deal [3], as costs of greenhouse gas allowances as well as environmental and health aspects of energy use will have to be considered in order to derive the lowest costs [62]. This is also in line with the update of the NZEB definitions, the introduction of energy requirements and incentive mechanisms for renovation [63], as stated in the Renovation Wave [4], and the environmental targets for the building sector. The cost-optimal level shall lie within the range of performance levels where the cost benefit analysis calculated over the estimated economic lifecycle is positive.

5. Conclusions

Reducing the energy demand in buildings is a prerequisite for meeting Europe's energy efficiency and GHG-emission reduction targets. The cost-optimal and NZEB implementation can support this process and provide useful references to develop policy and incentive measures to retrofit the existing building stock.

This paper gives a comprehensive assessment of the cost-optimal methodology in Member States based on latest reports. This review appears necessary in order to assess the implementation progress and outline weak and strong points as well as future developments. The review outputs highlight how cost-optimality is currently a well-established concept in Member States, and an overall positive development can be acknowledged. Based on the collected data, some statistics can be presented. The average cost-optimal level is assessed at 80 kWh/m^{2}y for the new residential sector and 140 kWh/m^{2}y for the new non-residential sector, while it is 130 kWh/m^{2}yr for the existing residential sector and 180 kWh/m^{2}y for existing non-residential sector. For almost all building types, the primary energy consumptions associated with cost-optimal levels were lower in cold regions, while global costs are normally lower in countries with a warm and mild climate.

The positive assessment achieved by Member States in implementing the cost-optimal methodology implies its potential for adaptation toward the gradual introduction of zero-emission buildings (ZEBs), for both new and existing buildings and at building element level.

The comparison with cost-optimal levels of the previous calculation round (2013) shows that for almost all building types, lower values were obtained in 2020. Looking at the consolidated cost-optimal levels, only two Member States did not clearly provide any cost-optimal requirement. Regarding conformity to the EU policy framework, some gaps can be highlighted:

- almost two-thirds of the Member States failed to cover one or more objectives, such as a required building type or the cost-optimal levels for building elements installed in existing buildings.
- almost half of the Member States applied derivation methodologies that did not fully conform or did not provide sufficient information to allow assessment of this calculation step.
- many Member States did not provide convincing explanations about the existing gaps or clear plans to reduce them, or did not even discuss the shortfalls against current requirements. Only in a few cases was a comparison with NZEB requirements available.

However, the situation looks positive with regard to the plausibility of input parameters and results obtained. As a consequence, the identified areas for improvement of the cost-optimal methodology include modalities of adapting to ZEBs to better enlighten the environmental and externalities aspects. The implications of such adaptations will have to be tested in terms of calculation robustness, confidence level, and sensitivity analysis.

The future cost-optimal framework will enable the calculation of both energy and emission performance, taking into account environmental and health externalities, as well as the emissions trading system (ETS) extension and carbon prices. The update of the calculations of cost-optimal levels will have to be monitored while also considering:

- the update of the NZEB definitions for new and renovated buildings;
- the introduction of energy requirements and incentive mechanisms for existing buildings in line with the Renovation Wave Strategy;
- the environmental targets for 2030 and the carbon neutrality goal for 2050 for the building sector as a whole.

This review also details suggestions on the points that could be considered in the future cost-optimal methodology that will be reviewed by 2026. Among the implications of the results of this review are the introduction of cost-optimality at district level and historical buildings; the use of a common template to collect data; harmonization of results and streamlining of the administrative aspect. A base refurbishment level could be useful to take into account the increased value of energy efficiency implementation in building retrofit. Enhanced importance could be placed on comfort and indoor air quality, as well as on the inclusion of wider energy efficiency benefits, such as a reduced import dependency and a positive impact on the economy (e.g., job creation). In this way, the global cost concept could be more in line with a complete life-cycle assessment that would take into account all environmental impacts throughout the life-cycle. In particular, the introduction of socioeconomic benefits could fulfil ambitious requirements and unlock higher energy and monetary savings. From this perspective, the outcomes of this cost-optimal review assume a strategic value, since they allow a comprehensive comparison among Member States for assessing progress and related strengths and weaknesses. Its future development and implementation will be crucial for the renovation of existing buildings and the achievement of a zero-carbon society in 2050.

Supplementary Materials: Cost-optimal reports analyzed in this paper are available at: https://ec.europa.eu/energy/topics/energy-efficiency/energy-performance-of-buildings/energy-performance-buildings-directive/eu-countries-2018-cost-optimal-reports_en?redir=1.

Author Contributions: Conceptualization, P.Z., D.D., R.A. and P.B.; methodology, P.Z., D.D. and R.A.; software, P.Z., D.D. and R.A.; formal analysis, P.Z., D.D. and R.A.; investigation, P.Z., D.D. and R.A.; data curation, P.Z., D.D. and R.A.; writing—original draft preparation, P.Z. and D.D.; writing—review and editing, D.D.; visualization, P.Z. and D.D.; project administration, P.B. All authors have read and agreed to the published version of the manuscript.

Funding: This research received no external funding.

Data Availability Statement: Additional data and datasets related to this paper will be available in: "Datasets and analysis of the cost-optimal methodology implementation in European Member States in compliance with the Energy Performance of Buildings Directive", Paolo Zangheri, Delia D'Agostino, Roberto Armani, and Paolo Bertoldi, submitted to Data (MDPI).

Acknowledgments: For the second round of cost-optimal calculations, the Member States submitted their reports to the Commission, and the Joint Research Centre (JRC) assumed the role of analyzing and assessing these documents in order to verify the States' conformity to the common methodology and the plausibility of the results obtained. This activity falls within the tasks of the Commission, as formalized by the Energy Performance of Buildings Directive (EPBD), in the assessment of the progress of the Member States in reaching cost-optimal levels of minimum energy performance requirements. We are grateful for the useful and constructive comments we received from our reviewers, in particular through the iterative process with the European Commission's Directorate-General for Energy (DG ENER).

Conflicts of Interest: The authors declare no conflict of interest.

List of Abbreviations and Definitions

EPBD	Energy Performance of Buildings Directive
EED	Energy Efficiency Directive
MFH	Multi-family house
n-R	Non-residential
NZEB	Nearly zero-energy building
EE	Energy efficiency
RES	Renewable energy systems
SFH	Single-family house

References

1. Economidou, M.; Todeschi, V.; Bertoldi, P.; D'Agostino, D.; Zangheri, P.; Castellazzi, L. Review of 50 years of EU energy efficiency policies for buildings. *Energy Build.* **2020**, *225*, 110322. [CrossRef]
2. EC, A Clean Planet for All, COM. 773 Final, A Clean Planet for All a European Strategic Long-Term Vision for a Prosperous, Modern, Competitive and Climate Neutral Economy. 2018. Available online: https://eur-lex.europa.eu/legal-content/EN/TXT/PDF/?uri=CELEX:52018DC0773&from=EN (accessed on 4 May 2022).
3. EC, the European Green Deal, COM (2019) 640 Final, Communication from the Commission to the European Parliament, the European Council, the Council, the European Economic and Social Committee and the Committee of the Regions. Available online: https://ec.europa.eu/info/sites/info/files/european-green-deal-communication_en.pdf (accessed on 8 June 2022).
4. EC, A Renovation Wave for Europe—Greening our Buildings, Creating Jobs, Improving, COM (2020) 662 Final. Available online: https://eur-lex.europa.eu/legal-content/EN/TXT/?qid=1603122220757&uri=CELEX:52020DC0662 (accessed on 29 June 2022).
5. EC 2013, Green Paper, A 2030 Framework for Climate and Energy Policies, COM (2013) 169 Final. Available online: https://eur-lex.europa.eu/LexUriServ/LexUriServ.do?uri=COM:2013:0169:FIN:en:PDF (accessed on 5 July 2022).
6. Regulation (EU) 2018/1999 of the European Parliament and of the Council of 11 December 2018 on the Governance of the Energy Union and Climate Action. Available online: https://eur-lex.europa.eu/legal-content/EN/TXT/?uri=uriserv:OJ.L_.2018.328.01.0001.01.ENG&toc=OJ:L:2018:328:FULL (accessed on 17 May 2022).
7. European Parliament. EPBD Recast, Directive 2010/31/EU of the European Parliament and of the Council of 19 May 2010 on the Energy Performance of Buildings (Recast). *Official Journal of the European Union*, 18 June 2010; L 153/13.
8. Revision of the Directive of the European Parliament and of the Council on the Energy Performance of Buildings (Recast), COM (2021) 802 Final. Available online: https://ec.europa.eu/energy/sites/default/files/proposal-recast-energy-performance-buildings-directive.pdf (accessed on 19 May 2022).
9. EC. Commission Staff Working Document, Good Practice in Energy Efficiency, Accompanying the Document: Energy efficiency for a Sustainable, Safer and More Competitive Europe (Communication from the Commission to the European Parliament and the Council). 27 April 2017. Available online: https://eur-lex.europa.eu/legal-content/EN/TXT/?uri=celex%3A52014DC0520 (accessed on 1 August 2022).
10. European Parliament. Commission Delegated Regulation (EU) No 244/2012 of 16 January 2012 Supplementing Directive 2010/31/EU of the European Parliament and of the Council on the Energy Performance of Buildings by Establishing a Comparative Methodology Framework for Calculating Cost-Optimal Levels of Minimum Energy Performance Requirements for Buildings and Building elements. *Off. J. Eur. Union* **2012**, *136*, 18–36. Available online: https://eur-lex.europa.eu/LexUriServ/LexUriServ.do?uri=OJ:L:2012:081:0018:0036:EN:PDF (accessed on 15 September 2022).

1. European Parliament. Guidelines accompanying Commission Delegated Regulation (EU) No. 244/2012 of 16 January 2012 supplementing Directive 2010/31/EU of the European Parliament and of the Council on the energy performance of buildings by establishing a comparative methodology framework for calculating cost-optimal levels of minimum energy performance requirements for buildings and building elements. *Off. J. Eur. Union* **2012**. Available online: https://op.europa.eu/en/publication-detail/-/publication/b6d03fa4-b936-43f1-9120-10b027c04e3a/language-en (accessed on 15 September 2022).
2. D'Agostino, D.; Tzeiranaki, S.T.; Zangheri, P.; Bertoldi, P. Assessing Nearly Zero Energy Buildings (NZ EBs) development in Europe. *Energy Strat. Rev.* **2021**, *36*, 100680. [CrossRef]
3. D'Agostino, D.; Parker, D.; Epifani, I.; Crawley, D.; Lawrie, L. How will future climate impact the design and performance of nearly zero energy buildings (NZEBs)? *Energy* **2021**, *240*, 122479. [CrossRef]
4. D'Agostino, D.; Mazzarella, L. What is a Nearly zero energy building? Overview, implementation and comparison of definitions. *J. Build. Eng.* **2018**, *21*, 200–212. [CrossRef]
5. Pikas, E.; Thalfeldt, M.; Kurnitski, J. Cost optimal and nearly zero energy building solutions for office buildings. *Energy Build.* **2014**, *74*, 30–42. [CrossRef]
6. D'Agostino, D.; Mazzarella, L. Data on energy consumption and Nearly zero energy buildings (NZEBs) in Europe. *Data Brief* **2018**, *21*, 2470–2474. [CrossRef]
7. Kurnitski, J.; Kuusk, K.; Tark, T.; Uutar, A.; Kalamees, T.; Pikas, E. Energy and investment intensity of integrated renovation and 2030 cost optimal savings. *Energy Build.* **2014**, *75*, 51–59. [CrossRef]
8. Buso, T.; Becchio, C.; Corgnati, S.P. NZEB, cost- and comfort-optimal retrofit solutions for an Italian Reference Hotel. *Energy Procedia* **2017**, *140*, 217–230. [CrossRef]
9. Ashrafian, T.; Yilmaz, A.Z.; Corgnati, S.P.; Moazzen, N. Methodology to define cost-optimal level of architectural measures for energy efficient retrofits of existing detached residential buildings in Turkey. *Energy Build.* **2016**, *120*, 58–77. [CrossRef]
20. Zangheri, P.; Armani, R.; Pietrobon, M.; Pagliano, L. Identification of cost-optimal and NZEB refurbishment levels for representative climates and building typologies across Europe. *Energy Effic.* **2017**, *11*, 337–369. [CrossRef]
21. CA EPBD, Concerted Action EPBD: Implementing the Energy Performance of Buildings Directive (EPBD). Information of the Joint Initiative of EU Member States and the European Commission. Available online: http://www.epbd-ca.eu/themes/cost-optimum (accessed on 19 July 2022).
22. Corgnati, S.P.; Fabrizio, E.; Filippi, M.; Monetti, V. Reference buildings for cost optimal analysis: Method of definition and application. *Appl. Energy* **2013**, *102*, 983–993. [CrossRef]
23. Longo, S.; Montana, S.L.F.; Sanseverino, E.R. A review on optimization and cost-optimal methodologies in low-energy buildings design and environmental considerations. *Sustain. Cities Soc.* **2018**, *45*, 87–104. [CrossRef]
24. Ferrara, M.; Monetti, V.; Fabrizio, E. Cost-Optimal Analysis for Nearly Zero Energy Buildings Design and Optimization: A Critical Review. *Energies* **2018**, *11*, 1478. [CrossRef]
25. Østergård, T.; Jensen, R.L.; Maagaard, S.E. Building simulations supporting decision making in early design—A review. *Renew. Sustain. Energy Rev.* **2016**, *61*, 187–201. [CrossRef]
26. Becchio, C.; Bottero, M.C.; Corgnati, S.P.; Dell'Anna, F. Decision making for sustainable urban energy planning: An integrated evaluation framework of alternative solutions for a NZED (Net Zero-Energy District) in Turin. *Land Use Policy* **2018**, *78*, 803–817. [CrossRef]
27. Becchio, C.; Ferrando, D.G.; Fregonara, E.; Milani, N.; Quercia, C.; Serra, V. The cost-optimal methodology for the energy retrofit of an ex-industrial building located in Northern Italy. *Energy Build.* **2016**, *127*, 590–602. [CrossRef]
28. Zacà, I.; D'Agostino, D.; Congedo, P.M.; Baglivo, C. Data of cost-optimality and technical solutions for high energy performance buildings in warm climate. *Data Brief* **2015**, *4*, 222–225. [CrossRef]
29. D'Agostino, D.; Parker, D.; Melià, P.; Dotelli, G. Optimizing photovoltaic electric generation and roof insulation in existing residential buildings. *Energy Build.* **2022**, *255*, 111652. [CrossRef]
30. Becchio, C.; Ferrando, D.; Fregonara, E.; Milani, N.; Quercia, C.; Serra, V. The Cost Optimal Methodology for Evaluating the Energy Retrofit of an ex-industrial Building in Turin. *Energy Procedia* **2015**, *78*, 1039–1044. [CrossRef]
31. Baglivo, C.; Congedo, P.M.; D'Agostino, D.; Zacà, I. Cost-optimal analysis and technical comparison between standard and high efficient mono-residential buildings in a warm climate. *Energy* **2015**, *83*, 560–575. [CrossRef]
32. Corrado, V.; Ballarini, I.; Paduos, S. Assessment of Cost-optimal Energy Performance Requirements for the Italian Residential Building Stock. *Energy Procedia* **2014**, *45*, 443–452. [CrossRef]
33. D'Agostino, D.; Parker, D. A framework for the cost-optimal design of nearly zero energy buildings (NZEBs) in representative climates across Europe. *Energy* **2018**, *149*, 814–829. [CrossRef]
34. D'Agostino, D.; Parker, D. Data on cost-optimal Nearly Zero Energy Buildings (NZEBs) across Europe. *Data Brief* **2018**, *17*, 1168–1174. [CrossRef] [PubMed]
35. Ferrara, M.; Fabrizio, E.; Virgone, J.; Filippi, M. A simulation-based optimization method for cost-optimal analysis of nearly Zero Energy Buildings. *Energy Build.* **2014**, *84*, 442–457. [CrossRef]
36. Fokaides, P.A.; Papadopoulos, A.M. Cost-optimal insulation thickness in dry and mesothermal climates: Existing models and their improvement. *Energy Build.* **2014**, *68*, 203–212. [CrossRef]
37. Aste, N.; Adhikari, R.S.; Manfren, M. Cost optimal analysis of heat pump technology adoption in residential reference buildings. *Renew. Energy* **2013**, *60*, 615–624. [CrossRef]

38. Teodoriu, G.; Balan, M.; Șerbănoiu, I.; Verdeș, M. Cost-optimal analysis of performance relation thermal insulation—Hydronic heating system applied to Romanian residential buildings. *Procedia Technol.* **2014**, *12*, 583–590. [CrossRef]
39. Pikas, E.; Kurnitski, J.; Liias, R.; Thalfeldt, M. Quantification of economic benefits of renovation of apartment buildings as a basis for cost optimal 2030 energy efficiency strategies. *Energy Build.* **2015**, *86*, 151–160. [CrossRef]
40. Schimschar, S.; Blok, K.; Boermans, T.; Hermelink, A. Germany's path towards nearly zero-energy buildings—Enabling the greenhouse gas mitigation potential in the building stock. *Energy Policy* **2011**, *39*, 3346–3360. [CrossRef]
41. Ballarini, I.; Corgnati, S.P.; Corrado, V. Use of reference buildings to assess the energy saving potentials of the residential building stock: The experience of TABULA project. *Energy Policy* **2014**, *68*, 273–284. [CrossRef]
42. Kurnitski, J.; Saari, A.; Kalamees, T.; Vuolle, M.; Niemelä, J.; Tark, T. Cost optimal and nearly zero energy performance requirements for buildings in Estonia. *Est. J. Eng.* **2013**, *19*, 183–202. [CrossRef]
43. D'Agostino, D.; Zacà, I.; Baglivo, C.; Congedo, P.M. Economic and thermal evaluation of different uses of an existing structure in a warm climate. *Energies* **2017**, *10*, 658. [CrossRef]
44. D'Agostino, D.; Cuniberti, B.; Bertoldi, P. Data on European non-residential buildings. *Data Brief* **2017**, *14*, 759–762. [CrossRef]
45. EC. EU Countries' Cost-Optimal Reports. 2018. Available online: https://ec.europa.eu/energy/topics/energy-efficiency/energy-performance-of-buildings/energy-performance-buildings-directive/eu-countries-2018-cost-optimal-reports_en?redir=1 (accessed on 28 April 2022).
46. ECOFYS. *Assessment of Cost Optimal Calculations in the Context of the EPBD (ENER/C3/2013-414)*; Final Report; Ecofys: Brussles, Belgium, 2015.
47. EC. Report from the Commission to the European Parliament and the Council—Progress by Member States in Reaching Cost-Optimal Levels of Minimum Energy Performance Requirements. COM(2016) 464 final of 29 July 2016. Available online: https://eur-lex.europa.eu/legal-content/EN/TXT/?uri=CELEX%3A52016DC0464 (accessed on 1 August 2022).
48. *EN 15459:2007*; Energy Performance of Buildings—Economic Evaluation Procedure for Energy Systems in Buildings. CEN: 2007; CEN: Tallinn, Estonia, 2007.
49. BPIE (Buildings Performance Institute Europe). *Implementing the Cost-Optimal Methodology in EU Countries*; pp. 1–82, ISBN 9789491143083. Available online: http://bpie.eu/costoptimalmethodology.html (accessed on 26 May 2022).
50. D'Agostino, D.; Parker, D.; Epifani, I.; Crawley, D.; Lawrie, L. Datasets on Energy Simulations of Standard and Optimized Buildings under Current and Future Weather Conditions across Europe. *Data* **2022**, *7*, 66. [CrossRef]
51. Recommendation (EU) 2016/1318 Commission Recommendation (EU) 2016/1318 of 29 July 2016 on guidelines for the promotion of nearly zero-energy buildings and best practices to ensure that, by 2020, all new buildings are nearly zero-energy buildings. *Off. J. Eur. Union* **2016**, L 208/46. Available online: https://eur-lex.europa.eu/legal-content/EN/TXT/?uri=CELEX%3A32016H1318 (accessed on 15 September 2022).
52. Hamdy, M.; Hasan, A.; Siren, K. A multi-stage optimization method for cost-optimal and nearly-zero-energy building solutions in line with the EPBD-recast 2010. *Energy Build.* **2013**, *56*, 189–203. [CrossRef]
53. de Vasconcelos, A.B.; Pinheiro, M.D.; Manso, A.; Cabaço, A. EPBD cost-optimal methodology: Application to the thermal rehabilitation of the building envelope of a Portuguese residential reference building. *Energy Build.* **2016**, *111*, 12–25. [CrossRef]
54. Ortiz, J.; Fonseca i Casas, A.; Salom, J.; Garrido Soriano, N.; Fonseca i Casas, P. Cost-effective analysis for selecting energy efficiency measures for refurbishment of residential buildings in Catalonia. *Energy Build.* **2016**, *128*, 442–457. [CrossRef]
55. Kona, A.; Economidou, M.; Zangheri, P. *Analysis of the Annual Reports 2017 under the Energy Efficiency Directive—Summary Report*; EUR 28922 EN; Publications Office of the European Union: Luxembourg, 2017; ISBN 978-92-79-77040-1. [CrossRef]
56. Ascione, F.; Cheche, N.; De Masi, R.F.; Minichiello, F.; Vanoli, G.P. Design the refurbishment of historic buildings with the cost-optimal methodology: The case study of a XV century Italian building. *Energy Build.* **2015**, *99*, 162–176. [CrossRef]
57. *EN 16798-1:201*; Energy Performance of Buildings—Ventilation for Buildings—Part 1: Indoor Environmental Input Parameters for Design and Assessment of Energy Performance of Buildings Addressing Indoor Air Quality, Thermal Environment, Lighting and Acoustics. Available online: http://store.uni.com/catalogo/uni-en-16798-1-2019 (accessed on 15 June 2022).
58. Shnapp, S.; Paci, D.; Bertoldi, P. *Untapping Multiple Benefits: Hidden Values in Environmental and Building Policies*; EUR 30280 EN, Publications Office of the European Union: Luxembourg, 2020; ISBN 978-92-76-19983-0. [CrossRef]
59. Becchio, C.; Bertoncini, M.; Boggio, A.; Bottero, M.; Corgnati, S.P.; Dell'Anna, F. *The Impact of Users' Lifestyle in Zero-Energy and Emission Buildings: An Application of Cost-Benefit Analysis*; Springer Nature: Berlin/Heidelberg, Germany, 2019. [CrossRef]
60. Congedo, P.M.; Baglivo, C.; Zacà, I.; D'Agostino, D. High performance solutions and data for nZEBs offices located in warm climates. *Data Brief* **2015**, *5*, 502–505. [CrossRef]
61. D'Agostino, D.; Parker, D.; Melià, P.; Dotelli, G. Data on roof renovation and photovoltaic energy production including energy storage in existing residential buildings. *Data Brief* **2022**, *41*, 107874. [CrossRef] [PubMed]
62. D'Agostino, D.; Cuniberti, B.; Maschio, I. Criteria and structure of a harmonised data collection for NZEBs retrofit buildings in Europe. *Energy Procedia* **2017**, *140*, 170–181. [CrossRef]
63. D'Agostino, D.; Tzeiranaki, S.T.; Zangheri, P.; Bertoldi, P. Data on nearly zero energy buildings (NZEBs) projects and best practices in Europe. *Data Brief* **2021**, *39*, 107641. [CrossRef] [PubMed]

buildings

MDPI

Article

Integration of Photovoltaic Systems for Energy Self-Sufficient Low-Rise Multi-Family Residential Buildings in Republic of Korea

Byung Chang Kwag, Gil Tae Kim * and In Tae Hwang

Center for Housing Environment Research and Innovation, Korea Land and Housing Research Institute, Sejong 34047, Republic of Korea; byungchang.kwag@lh.or.kr (B.C.K.); hit1009@lh.or.kr (I.T.H.)
* Correspondence: gtkim1@lh.or.kr; Tel.: +82-44-902-9136; Fax: +82-44-902-9130

Abstract: Globally, building energy consumption has been rising, emphasizing the need to reduce energy usage in the building sector to lower national energy consumption and carbon dioxide emissions. This study analyzes the applicability of photovoltaic (PV) systems in enhancing the energy self-sufficiency of small-scale, low-rise apartment buildings. The analysis is based on a case study using Republic of Korea's Zero-Energy Building Certification System. By employing the ECO2 simulation program, this research investigates the impact of PV system capacity and efficiency on the energy self-sufficiency rate (ESSR). A series of parametric analyses were carried out for various combinations of building-attached photovoltaic (BAPV) roofs and building-integrated photovoltaic (BIPV) facades, considering the initial cost of BIPV facades. The simulations demonstrate that achieving the target ESSR requires a combination of BAPV roofs and BIPV facades, due to limited roof areas for PV systems. Additionally, this study reveals that BIPV facades can be cost-effective when their unit price, relative to BAPV roofs, is below 62%. Based on the ECO2 simulations, a linear regression formula is proposed to predict the ESSR for the case study building. Verification analysis shows that the proposed formula predicts an ESSR of 74.1%, closely aligned with the official ESSR of 76.9% certified by the Korean government. Although this study focuses on the case of a specific apartment building and lacks actual field data, it provides valuable insights for future applications of PV systems to enhance energy self-sufficiency in small-scale, low-rise apartment buildings in Republic of Korea.

Keywords: BAPV roof; BIPV facade; nearly zero-energy building; low-rise residential building; energy self-sufficiency rate

1. Introduction

Globally, the building sector constitutes a substantial portion of national energy consumption. According to the International Energy Agency (IEA), the building sector accounts for over 40% of energy consumption in North America, more than 37.3% in Europe, and above 24% in Asia. Specifically, residential buildings represent over 17% in North America, 25% in Europe, and 19% in Asia, in terms of the total energy consumption [1]. The U.S. Energy Information Administration (EIA) similarly reports that the building sector represents approximately 21% of total energy consumption in the United States, with residential buildings accounting for about 11% [2]. The Korea Energy Economics Institute (KEEI) indicates that the building sector comprises about 22% of national energy consumption, with residential buildings making up about 11% [3]. According to the KEEI, the important fact is that energy consumption in the building sector is continuously increasing [3].

The government in Republic of Korea introduced the Zero-Energy Building Certification (ZEBC) System in 2017, aiming to reduce energy consumption in the building sector. Since 2020, the ZEBC requires public buildings to achieve an energy self-sufficiency rate (ESSR) of at least 20%. Starting in 2023, this mandate has been extended to public multi-family residential buildings with 30 or more dwelling units [4]. The ESSR is the ratio between the source energy generation and the source energy consumption of a building. It serves as the basis for determining the Zero-Energy Building Grade; Grade 5 requires an ESSR from 20% to less than 40%, Grade 4 requires an ESSR from 40% to less than 60%, Grade 3 requires an ESSR from 60% to less than 80%, Grade 2 requires an ESSR from 80% to less than 100%, and Grade 1 requires an ESSR from 100% or higher [4].

However, as mentioned earlier, in Republic of Korea, multi-family residential buildings with 30 or more dwelling units are required to comply with the Zero-Energy Building Certification [4]. This highlights a lack of standards or systems for achieving zero energy in small-scale, low-rise, multi-family residential buildings in Republic of Korea. Consequently, there is also a scarcity of research focused on achieving energy self-sufficiency for these low-rise residential buildings.

According to the Korea Statistical Information Service (KOSIS), as of 2022, multi-family housing accounted for 79% of the total number of dwellings in Republic of Korea [5]. Data from the Korea Energy Agency indicates that since the implementation of the Zero-Energy Building Certification in 2017, only 20 out of 975 certified buildings are residential buildings, representing just 2.1% of the total number of certified buildings [4]. Among the certified residential buildings, 11 buildings are multi-family residential buildings, showing no significant difference in proportion compared to single-family homes. However, when comparing the ZEB grades, 45% of single-family homes achieved a ZEB Grade 3 or higher, while only 33% of multi-family residential buildings reached this level. Therefore, given the high prevalence of multi-family residential buildings in Republic of Korea, it is crucial to enhance the energy self-sufficiency of these buildings to achieve significant energy savings in the building sector.

Research aimed at improving the energy performance and energy self-sufficiency of housing has been actively conducted worldwide, including in Republic of Korea. Some previous studies have performed research using case studies on various net-zero energy buildings, demonstrating the necessity of applying diverse energy reduction technologies and renewable energy sources to achieve net-zero energy buildings [6–9]. Notably, these studies identified that photovoltaic (PV) systems have been employed in most cases. This underscores that PV systems are considered the primary renewable energy solution for achieving net-zero energy buildings. Furthermore, various pieces of global research on implementing net-zero energy buildings have consistently shown that both PV (including building-attached photovoltaic or BAPV and building-integrated photovoltaic or BIPV) systems and solar thermal systems are considered as energy sources in many of these types of studies [10–22].

In Republic of Korea, various studies have also been conducted to implement net-zero energy residential buildings, as summarized in Table 1 [23–48]. Unlike international studies, Korean research has predominantly focused on energy reduction technologies and the analysis of building energy policies and their effectiveness, rather than the implementation of zero energy multi-family residential buildings. This suggests that there is a lack of research on application strategies and capacity estimation methods for PV systems, which are essential for achieving nearly or net-zero energy residential buildings in Korean studies. Furthermore, Korean research has primarily focused on high-rise apartments, reflecting the predominant residential culture in Republic of Korea. Consequently, there is a notable deficiency in the research aimed at improving energy performance and achieving nearly or net-zero energy in small-scale, low-rise, multi-family residential buildings, with fewer than 30 dwelling units, in Republic of Korea.

In general, achieving net-zero energy buildings with PV systems can be easier for single-family homes compared to multi-family residential buildings. This is because the energy generated by PV systems installed on the same roof area only needs to cover the energy load of a single household in a single-family home. In contrast, in multi-family housing, PV systems must cover the energy load of multiple households. Consequently, achieving energy self-sufficiency solely through rooftop PV systems could be challenging for multi-family residential buildings.

Table 1. Summary of previous research on energy-efficient buildings in Republic of Korea.

Author	Method	Building Type	Indicator	Variables	Type of Renewable Energy	Conclusion
Bo Rang Park et al. [23]	Simulation	Residential building	Energy use intensity	• Infiltration • Lighting power density • Heating system efficiency • Plug-load density	-	• Achieving carbon neutrality requires a comprehensive approach beyond regulating individual technology elements.
Jihye Choi et al. [24]	Survey	Residential building	Choice probability	• Respondents • Subsidy • Building energy level		• Cost-related factors are the most significant in terms of public acceptance of zero-energy residential building.
Changyoon Ji et al. [25]	Statistics	Non-residential building	Energy use intensity	• Building Energy Efficiency Certification (BEEC)	-	• Improving BEEC is suggested to better account for all energy uses and enhance its effectiveness in reducing greenhouse gas emissions.
Changyoon Ji et al. [26]	Statistics	Residential building	Greenhouse gas (GHG) emissions	• Building type • Year it was built • Climate region • Thermal conductivity of the envelope	-	• Strengthening BEEC has reduced greenhouse gas emissions from heating energy use in new buildings, but it has not significantly impacted on baseload and cooling energy emissions.
KyungSoo Kim et al. [27]	Survey Statistics Simulation	Residential Building	Energy consumption	• Building Type • Building Size • Built Year • Window-Wall-Ratio • Building Orientation • Aspect Ratio	-	• The developed standard building model showed a high level of efficiency and reliability in predicting energy consumption, with a total energy consumption error rate of 12.67%.
Yujun Jung et al. [28]	Simulation	Residential building	Energy and economic metrics; life cycle performance	• Air tightness • Occupants • Window–wall ratio	• Hydropower • Solar power	• The electricity-based energy scenario has better energy and environmental performance compared to the hydrogen-based scenario, achieving energy savings of 62% and 53%, respectively.

Table 1. *Cont.*

Author	Method	Building Type	Indicator	Variables	Type of Renewable Energy	Conclusion
Changyoon Ji et al. [29]	Statistics	Residential building	Energy use intensity	• Climate region • Floor area • Building energy certificate	-	• The BEEC effectively reduces heating energy consumption, but its impact on electricity use is insignificant.
Munkhbat Undram et al. [30]	Simulation	Residential building	ESSR	• Photovoltaic (PV) area • Building area	• Solar power	• District heating apartments can achieve Zero-Energy Building (ZEB) Grade 5 with rooftop PV installations alone, but individual apartments require additional measures to reduce primary energy consumption.
Seongjo Wang et al. [31]	Simulation	Non-residential building	ESSR	• Building size • Cooling/heating capacity • Renewable energy capacity	• Solar power • Fuel cell • Geothermal	• ESSR prediction formulas are proposed to design optimal renewable energy systems for zero-energy buildings.
Hyomun Lee et al. [32]	Field test	Residential building	Energy generation	• Season	• Solar power • Solar thermal	• The house consumed 7368.95 kWh/yr and generated 11,439.68 kWh/yr through roof-integrated photovoltaics (RIPVs), achieving a 35.58% annual energy surplus.
Sungwoong Yang et al. [33]	Survey simulation	Residential building	Energy consumption	• Air tightness • Lighting power density • Thermal conductivity of the envelope • Heating system efficiency	-	• Applying energy-saving technologies can lead to substantial energy savings and improved energy efficiency in low-rise residential buildings.
Yeweon Kim et al. [34]	Review	Residential and non-residential buildings		-	-	• Ground-source heat pumps generally showed higher self-sufficiency rates than water-source heat pumps.

Table 1. *Cont.*

Author	Method	Building Type	Indicator	Variables	Type of Renewable Energy	Conclusion
Byung Chang Kwag et al. [35]	Simulation	Residential building	Heat loss form factor; energy load; energy use intensity	• Building size • Envelope thermal conductivity • Household location • Lighting power density • Heating system efficiency	-	• Improving the boiler efficiency is the most effective measure for reducing energy consumption, while enhancing the building envelope has the least impact.
Joohyun Lee et al. [36]	Simulation	Residential building	Energy load; energy use intensity	• Passive house • Conventional house	-	• It is important to consider the local climate and construction practices in achieving energy efficiency of a passive house.
Hye Soo Suh et al. [37]	Simulation	Community building	Energy consumption	• Thermal conductivity of the envelope • Air tightness • Lighting power density • Ventilation efficiency • Cooling/heating system efficiency	• Solar power • Geothermal • Solar thermal	• Combining passive and active design strategies with renewable energy systems can significantly improve energy performance and achieve nearly zero-energy targets for community buildings.
Jeonghun Song et al. [38]	Simulation	Residential and non-residential buildings	Building energy performance	• Building type • Renewable energy type	• Solar power • Fuel cell • Geothermal • Solar thermal	• Integrating renewable energy sources like photovoltaics, ground-source heat pumps, and fuel cells can reduce electricity and gas supply from the grid by 17% and 3%, respectively.
Jiyoung Eum et al. [39]	Simulation	Residential building	Total net present cost	• Energy storage system (ESS) capacity/price • PV capacity/price • Grid price	• Solar power	• ESS prices should decrease significantly to be economically viable, especially with government support and the implementation of time-of-use (TOU) rates.

Table 1. *Cont.*

Author	Method	Building Type	Indicator	Variables	Type of Renewable Energy	Conclusion
Tae-Hyoung Kim et al. [40]	Statistics	Residential and non-residential buildings	Greenhouse gas emissions	• Climate region • Energy source • Year	-	• This study suggests that the building sector has significant potential for energy efficiency improvements and greenhouse gas emission reductions.
Sung-Yul Kim et al. [41]		Residential building	Optimum solar power generation	• Building design • Subsidy • Initial cost • Operation cost	• Solar power	• Economic analyses show that connecting solar power generation capacity from a zero-energy perspective is more effective than maximizing the capacity based on the building structure.
Kwon Sook Park et al. [42]	Statistics	Residential building	Energy consumption	• Number of family members • Building type • Climate condition	-	• Improving energy performance in existing buildings, strengthening building design criteria, and revitalizing Korea's energy-saving policies are essential to reduce energy demand and GHG emissions in the residential buildings sector.
Jeongyoon Oh et al. [43]	Review	Residential and non-residential buildings		• Passive strategies • Active strategies	-	• This study proposes integrating and optimizing the passive and active strategies in the early phase of a building's life cycle and real-time monitoring during the usage phase to achieve net-zero energy buildings.
Chang Heon Cheong [44]	Simulation	Residential building	Energy reduction	• Locations of PV panels • Building story	• Solar power	• To achieve net-zero energy in apartment buildings, an advanced energy mix is necessary. The policy for zero-energy buildings may only be feasible for detached houses.

Table 1. *Cont.*

Author	Method	Building Type	Indicator	Variables	Type of Renewable Energy	Conclusion
Jin-Hee Kim et al. [45]	Statistics	Residential and non-residential buildings	Energy generation	• Energy demand • PV installation type • PV cell type	• Solar power • Solar thermal	• PV systems are essential for achieving net-zero energy in buildings, but their application requires careful consideration of the economic viability, architectural integration, and surrounding environment.
Duk Joon Park et al. [46]	Review	Residential and non-residential buildings		• Number of energy certified buildings • Energy performance	-	• The BEEC has influenced gas and district heating consumption in residential buildings and shown a decreasing trend in primary energy consumption for non-residential buildings since its implementation.
Lim Jae-Han et al. [47]	Simulation	Residential building	Energy load	• Thermal conductivity of the envelope • Air tightness • Ventilation efficiency • Heating system efficiency	-	• Enhancing insulation, window performance, ventilation, and air tightness can significantly reduce energy consumption in apartment buildings.
Kyoung-ho Lee et al. [48]	Field test	Residential building	Energy generation	• Thermal demands • Outdoor air temperature • Solar irradiance	• Solar power • Solar thermal	• It is important to integrate solar energy systems in buildings to achieve energy savings and improve the overall energy efficiency. • The total annual solar fraction of the solar heating system was 69.7%.

This study aims to analyze how PV systems can be applied to achieve energy self-sufficiency in a small-scale, low-rise, multi-family residential building in Republic of Korea, particularly during the schematic design process. For this research, an actual low-rise residential building with 28 dwelling units that achieved ZEB Grade 3 will be used as a case study.

2. Methodology

The main purpose of this research is to analyze the integration of PV systems into low-rise residential buildings to achieve energy self-sufficiency in Republic of Korea, particularly during the schematic design process. This study analyzes the energy performance of buildings and conducts a sensitivity analysis of the impact of PV systems using building energy simulation, as shown in Figure 1.

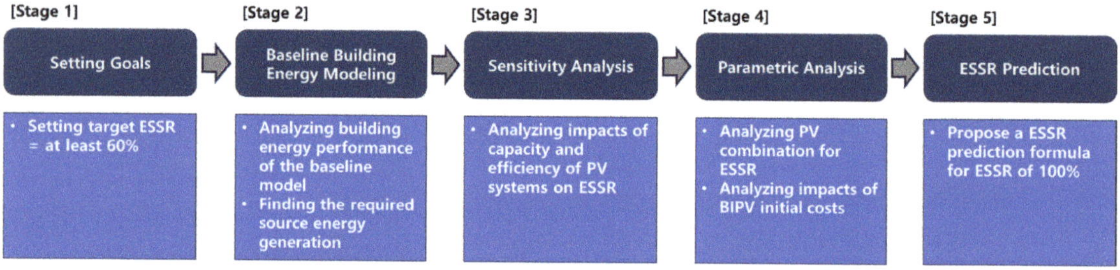

Figure 1. Flowchart of the research.

The target ESSR of 60% was set based on the requirements of the city government of Sejong City for public rental housing. Subsequently, using the Korean building energy code and other criteria, the baseline building was modelled in the ECO2 building energy simulation and the building energy performance was evaluated to calculate the required source energy generation needed to achieve the target ESSR.

To understand the contributions of the BAPV roof and BIPV facade towards achieving the target ESSR, a sensitivity analysis of the PV system's capacity and efficiency was conducted using the ECO2 simulation. Additionally, this research carried out parametric analysis to analyze the different combinations of BAPV roofs and BIPV facades, as well as the impact of the initial cost of a BIPV facade on the different combinations of PV systems. Based on the ECO2 simulation results and regression analysis, this research proposed a calculation formula for predicting the ESSR.

The energy performance analysis of the baseline buildings was conducted using ECO2, the official building energy analysis program in Republic of Korea, as illustrated in Figure 2. Internationally, various building energy simulation tools, such as EnergyPlus, TRNSYS, and DOE-2 based EQUEST, are commonly used to assess and evaluate building energy performance. However, in Republic of Korea, the ECO2 simulation program serves as the official tool for analyzing building energy loads and consumption in the context of the Zero-Energy Building Certification. The ECO2 program calculates building energy loads and consumption, aiming to evaluate the overall energy performance [30,34,35,49,50]. Previous studies have shown that the ECO2 simulation program yields energy usage trends consistent with the reference data [51–53]. In this research, the ECO2 simulation program is employed to analyze building energy performance and assess the impact of PV systems on achieving energy self-sufficiency rates.

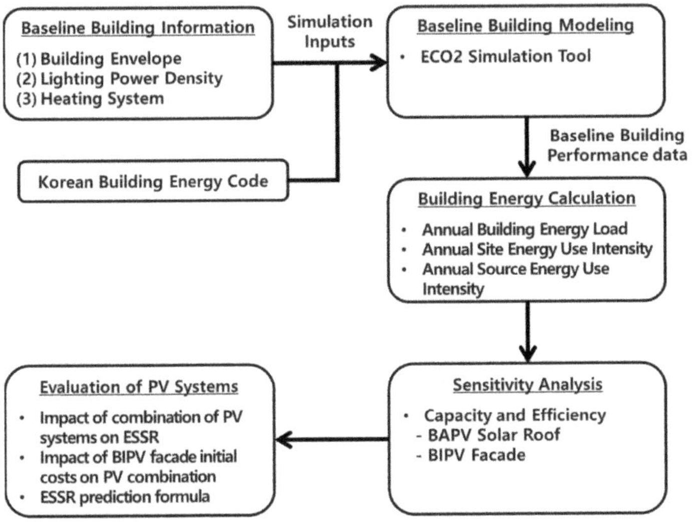

Figure 2. Flowchart of ECO2 building energy simulations.

3. Building Energy Modeling

3.1. Baseline Building Energy Modeling

The case study building is a small-scale five-story multi-family public housing building with 28 dwelling units located in Sejong, Republic of Korea. Excluding the first floor, which is a public space, including a lobby and monitoring room, the second to fifth floors consist of seven dwelling units per floor, as shown in Figure 3. This public apartment aims to achieve an energy self-sufficiency rate of over 60% and obtain at least ZEB Grade 3 certification. Each dwelling unit is planned to have an area of 39 m^2. The description of the case study building is summarized in Table 2.

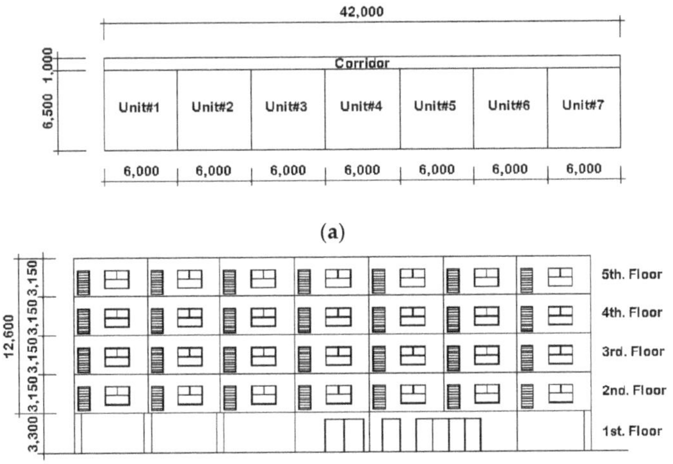

Figure 3. Schematic drawings of the case study building: (**a**) floor plan, (**b**) elevation (unit: mm).

Table 2. Description of the case study building.

Category	Specification
Floor to Ceiling	2.3 m
Floor to Floor	3.2 m
Dwelling Unit Area	Total 1092 m^2 39 m^2/unit × 7 units/floor × 4 floor
Roof Area	306 m^2
Front Exterior Wall Area	530 m^2
Front Exterior Wall Window Area of a Dwelling Unit	(Living Space) Window 1: 3.8 m^2/Window 2: 2.9 m^2 (Mechanical Room) Louver Window: 1.6 m^2
Number of Floors	Five-Story Building First floor: Parking Lots, Lobby, Monitoring Room Second~Fifth Floor: Dwelling Units (Seven Units Per Floor)
Main Building Material	Reinforced Concrete
Target ESSR	Over 60% (ZEB Grade 3)

The ECO2 building energy simulation inputs are summarized in Table 3. These inputs are primarily based on reference data from the Energy Saving Design Standards for Buildings of the Korean Government [49]. According to the Energy Saving Design Standards for Buildings of the Ministry of Land, Infrastructure, and Transport of Republic of Korea, residential buildings can exclude the cooling system from the evaluation criteria for the Zero-Energy Building Certification [54]. Therefore, the baseline building in this study does not consider the cooling system and has excluded space cooling from the ZEB certification evaluation criteria.

Table 3. Baseline modeling inputs.

Element	Category	Specification
Exterior Wall	Thermal Transmittance (U-value)	0.167 W/(m^{2}K)
Window	Thermal Transmittance (U-value)	1.00 W/(m^{2}K)
Slab on Grade Floor	Thermal Transmittance (U-value)	0.17 W/(m^{2}K)
Roof	Thermal Transmittance (U-value)	0.15 W/(m^{2}K)
Space Heating	Type	Radiant Floor Heating
	Water Temperature	Supply 80 °C/Return 60 °C
	Efficiency [%]	91% (Decentralized Natural Gas Condensing Boiler)
Domestic Hot Water	Type	Natural Gas Condensing Boiler
	Water Temperature	Supply 80 °C/Return 60 °C
Ventilation	Type	Energy Recovery Ventilator
	Capacity (Air Change per Hour)	0.5
	Efficiency	Heating 70% Cooling 45%
Lighting	Lighting Power Density	7 W/m^2
Infiltration	Air Change per Hour	6

3.2. Parameters for the Sensitivity Analysis

The study applied BAPV on the roof and BIPV on the facade of the baseline building and conducted a sensitivity analysis to assess the impact of the BAPV roof and BIPV facade on the building's ESSR. The parameters investigated in this study included the installation area and efficiency of the BAPV roof and BIPV facade, as outlined in Table 4.

Table 4. Description of the options for the sensitivity analysis.

Parameter		Options
BAPV Roof	Area [m^2]	30, 60, 90, 120, 150, 180, 210, 240, 270, 300
	Efficiency [%]	10, 12, 14, 16, 18, 20, 22, 24
BIPV Facade	Area [m^2]	16, 32, 48, 64, 80, 96, 112, 128, 144, 160
	Efficiency [%]	10, 12, 14, 16, 18, 20, 22, 24

4. Building Energy Simulation Results

4.1. Energy Performance of the Baseline Model

Table 5 presents the results of the ECO2 simulation for the baseline model. Heating accounted for approximately 54.2% of the total building energy load, followed by domestic hot water at 32.3%, and lighting at 13.5%. When considering site energy consumption, calculated taking into account the equipment efficiency and operating hours, heating energy consumption represented 60.2% of the total, domestic hot water 21.0%, lighting 7.9%, and ventilation 10.8%. In terms of source energy consumption, space heating represented 48.0%, domestic hot water 16.3%, lighting 15.1%, and ventilation 20.6%. Overall, heating dominated the building energy performance, highlighting its significant contribution to energy consumption.

Table 5. Simulation results of the baseline model.

Category	Space Heat	Space Cool	Hot Water	Lighting	Ventilation	Total
Building Energy Load [kWh/m^2· yr]	51.5	0.0	30.7	12.8	0.0	95.0
Site Energy Consumption [kWh/m^2· yr]	97.4	0.0	34.0	12.8	17.5	161.7
Source Energy Consumption [kWh/m^2· yr]	111.9	0.0	37.9	35.1	48.1	233.0

4.2. Results of the Sensitivity Analysis

4.2.1. Impact of PV System Capacities

This study analyzed the impact of capacity changes in the BAPV roof and BIPV facade on the ESSR for the case study building. In the ECO2 simulation, the BAPV roof efficiency and installation angle was fixed at 20% and 30 degrees, respectively, while the BIPV facade efficiency was assumed to be 15%. Notably, when conducting simulations for the BAPV roof or BIPV facade, each type of PV system was simulated independently. In other words, the BIPV facade was not applied during the simulations for the BAPV roof. The simulation results, as shown in Figures 4 and 5, revealed that the ESSR increases linearly as the installation area of both systems increases and the installed capacity increases.

The simulation analysis results indicate that achieving the target ESSR of 60% solely with either a BAPV roof or a BIPV facade is challenging due to installation area limitations. Specifically, a BAPV roof would need to cover more than 280 m^2, while a BIPV facade would require at least 560 m^2, according to the linear regression equation. However, local government regulations in Republic of Korea restrict solar panel installation to a maximum of 70% of the flat roof area [55,56]. Given the total roof area of 300 m^2 in the case study building, a BAPV roof can be installed on up to 210 m^2 of the horizontal flat roof surface. Applying an installation angle of 30°, the maximum installable area for the BAPV roof is 243 m^2. Therefore, achieving the target ESSR necessitates combining both a BAPV roof and a BIPV facade.

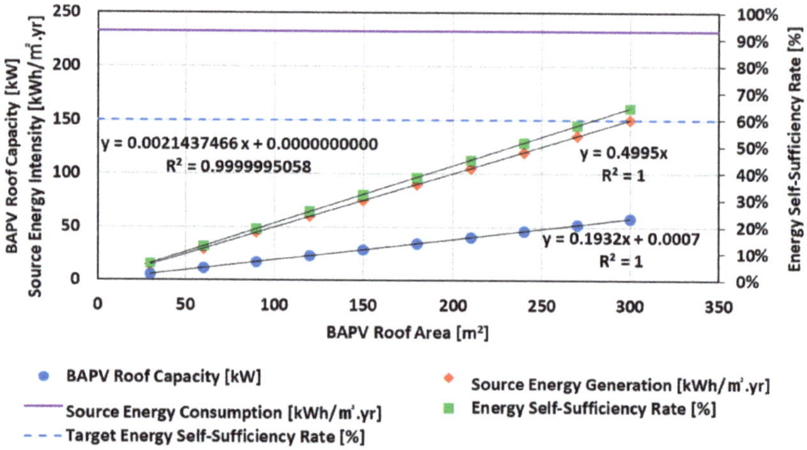

Figure 4. Source energy intensity for variations in the BAPV roof area.

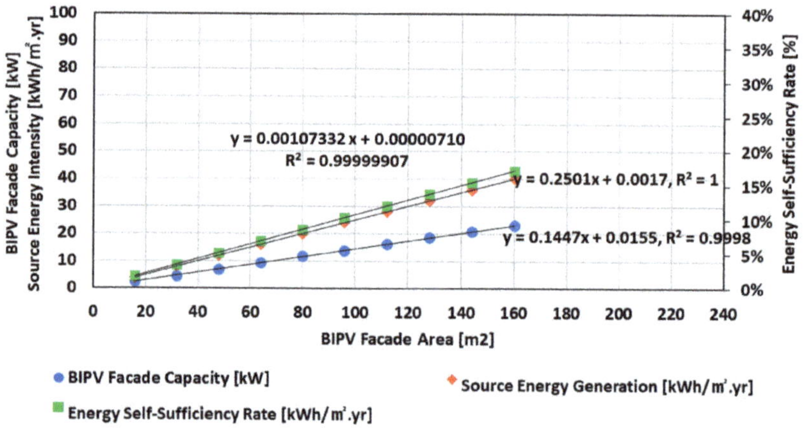

Figure 5. Source energy intensity for variations in the BIPV facade area.

4.2.2. Impact of PV System Efficiencies

This study conducted simulations to analyze the impact of changes in the efficiency of the PV system on the ESSR of the case study building. When simulating the BAPV roof, it was assumed that it would be installed on 300 m^2 of the roof, with no installation of a BIPV facade. For the BIPV facade, it was assumed that it would cover 160 m^2, corresponding to about 30% of the front wall area of 530 m^2, without a BAPV roof. As a result of the analysis, it was observed that the change in the building's ESSR due to efficiency changes followed a linear trend, as shown in Figures 6 and 7. Specifically, for the BAPV roof, an efficiency saving of at least 16% was necessary to achieve the target ESSR for the case study building. However, in the case of the BIPV facade, regardless of how high the efficiency was, the target ESSR of 60% could not be reached.

The limitation arises from the BIPV facade installation area being restricted to 30% of the front wall.

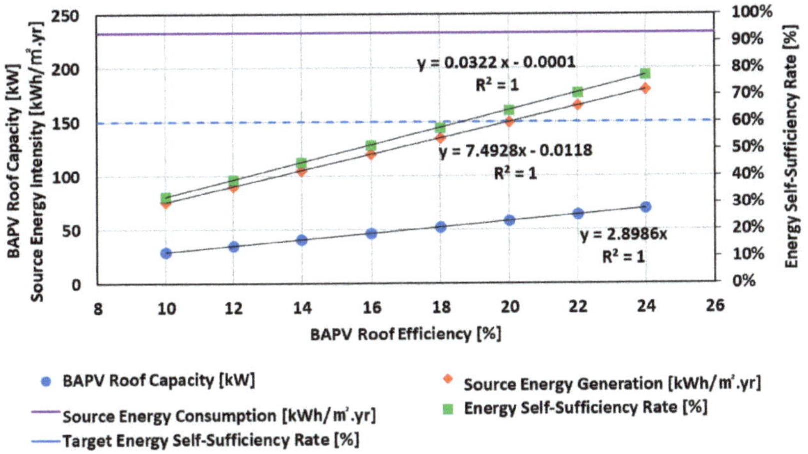

Figure 6. Source energy intensity for variations in the BAPV roof efficiency.

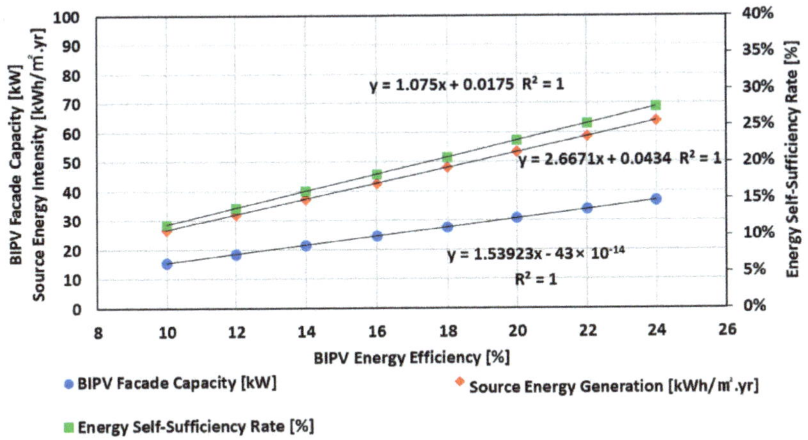

Figure 7. Source energy intensity for variations in the BIPV facade efficiency.

5. Analysis

5.1. Impact of Combining PV Systems on Achieving the Target ESSR

Since the source energy demand of the baseline model is 233 kWh/(m^2·yr), achieving the target ESSR of 60% requires a source energy generation of at least 139.8 kWh/(m^2·yr). This study analyzed various combinations of BAPV roofs and BIPV facades to meet the target ESSR, as summarized in Table 6. For this analysis, the PV efficiencies of 20% and 15% were used for the BAPV roof and BIPV facade, respectively. As shown in Table 6, an area of 280 m^2 for the BAPV roof alone is sufficient to achieve the target ESSR of 60%. However, due to regulatory constraints, which allows only up to 270 m^2 for the BAPV roof installation in the case study building, achieving the target ESSR solely with a BAPV roof is not feasible. Therefore, a combination of a BAPV roof and a BIPV facade is necessary to achieve the target ESSR of 60% for the case study building.

Table 6. Simulation results for various combinations of BAPV roofs and BIPV facades.

CASE	Surface Area [m²]		PV Capacity [kW]		Source Energy Generation [kWh/m²·yr]			ESSR [%]
	BAPV Roof	BIPV Facade	BAPV Roof	BIPV Facade	BAPV Roof	BIPV Facade	Total	
Case 1	30	500	6	78	15.0	124.8	139.8	60.0%
Case 2	60	440	12	68	30.0	109.8	139.8	60.0%
Case 3	90	380	17	59	44.9	94.9	139.8	60.0%
Case 4	120	320	23	50	59.9	79.9	139.8	60.0%
Case 5	150	260	29	40	74.9	64.9	139.8	60.0%
Case 6	180	200	35	31	89.9	49.9	139.8	60.0%
Case 7	210	140	41	22	104.8	35.0	139.8	60.0%
Case 8	223	114	43	18	111.4	28.5	139.8	60.0%
Case 9	240	80	46	12	119.9	19.9	139.8	60.0%
Case 10	270	20	52	3	134.9	4.9	139.8	60.0%
Case 11	280	0	54	0	139.8	0.0	139.8	60.0%
Case 12	300	0	58	0	149.9	0.0	149.9	64.3%

5.2. Impact of BIPV Facade Initial Costs

This study examined the combination of a BAPV roof and a BIPV facade to achieve the target ESSR, considering the installation costs. As shown in Table 7, the baseline installation cost for PV systems was obtained from the Korea Energy Agency's 2021 report [57]. Since there are no recent official cost estimates for BIPV facade panels from the Korea Energy Agency (KEA), this research used the cost difference ratio between BAPV roofs and BIPV facades from the 2011 and 2012 KEA reports [58,59]. Specifically, the cost of BIPV facade panels was assumed to be approximately 2.0 times that of BAPV roof panels, estimated at KRW 2724 per kW.

Table 7. Standard price per 1 kW for PV systems released by KEA.

Type of PV System		Year		
		2011	2012	2021
BAPV Roof	1000 KRW/kW	5650	4972	1816
	USD/kW	4238	3729	1362
BIPV Facade	1000 KRW/kW	13,055	9553	-
	USD/kW	9791	7165	-

(Exchange rate: KRW 1000 = USD 0.75, Google Finance, 8 March 2024).

To evaluate the effect of the initial cost of a BIPV facade, this study conducted ECO2 simulations for various combinations of BAPV roof and BIPV facade installations needed to achieve the target ESSR of 60%. Different initial costs for a BIPV facade were then applied, as shown in Table 8. Throughout the analysis, the installation areas for BAPV roof and BIPV facade panels, as well as the cost of a BAPV roof, remained constant, while the cost of the BIPV facade was varied.

Table 9 presents the initial cost associated with different combinations of BAPV roofs and BIPV facades for various implementation surfaces. The minimum initial cost is highlighted in yellow. Notably, for Option 1 and Option 2, a higher utilization of the BIPV facade results in a lower initial cost for implementing PV systems. In contrast, other options achieve the minimum initial cost by using only a BAPV roof. These results indicate that applying a BIPV facade is cost-effective when its initial price, relative to a BAPV roof, is below 62%. Therefore, to promote the extensive adoption of BIPV facades, reducing the initial installation cost through financial support systems, such as government subsidies, is crucial.

Table 8. Unit cost options for a BIPV facade.

Unit Cost Option	Option 1	Option 2	Option 3	Option 4	Option 5	Option 6	Option 7
BAPV Roof [USD]	1362	1362	1362	1362	1362	1362	1362
BIPV Facade [USD]	681	845	953	1362	2043	2724	4086
Ratio	0.5	0.62	0.7	1.0	1.5	2.0	3.0

Table 9. The resulting initial cost for various combinations and cost options, with the minimum total value highlighted in yellow.

Case	BAPV Roof [USD]	Option 1		Option 2		Option 3		Option 4		Option 5		Option 6		Option 7	
		BIPV Facade [USD]	Total Cost [USD]	BIPV Facade [USD]	Total Cost [USD]	BIPV Facade [USD]	Total Cost [USD]	BIPV Facade [USD]	Total Cost [USD]	BIPV Facade [USD]	Total Cost [USD]	BIPV Facade [USD]	Total Cost [USD]	BIPV Facade [USD]	Total Cost [USD]
Case 1	7900	52,924	60,824	65,631	73,530	74,082	81,982	105,848	113,748	158,773	166,672	211,697	219,597	317,545	325,445
Case 2	15,786	46,569	62,354	57,750	73,535	65,186	80,972	93,138	108,923	139,707	155,492	186,276	202,061	279,413	295,199
Case 3	23,685	40,245	63,930	49,908	73,593	56,334	80,019	80,490	104,175	120,735	144,421	160,981	184,666	241,471	265,156
Case 4	31,585	33,869	65,454	42,000	73,585	47,409	78,994	67,737	99,322	101,606	133,191	135,475	167,060	203,212	234,797
Case 5	39,484	27,513	66,998	34,119	73,603	38,513	77,997	55,027	94,511	82,540	122,025	110,054	149,538	165,080	204,565
Case 6	47,370	21,158	68,528	26,238	73,608	29,617	76,987	42,316	89,686	63,474	110,845	84,632	132,003	126,948	174,319
Case 7	55,270	14,824	70,094	18,383	73,653	20,750	76,020	29,648	84,918	44,471	99,741	59,295	114,565	88,943	144,213
Case 8	58,687	12,070	70,757	14,968	73,655	16,896	75,582	24,140	82,827	36,210	94,897	48,281	106,967	72,421	131,107
Case 9	63,170	8458	71,627	10,489	73,658	11,839	75,009	16,916	80,085	25,374	88,543	33,832	97,001	50,747	113,917
Case 10	71,056	2071	73,127	2568	73,624	2899	73,954	4142	75,198	6213	77,269	8284	79,340	12,426	83,482
Case 11	73,687	0	73,687	0	73,687	0	73,687	0	73,687	0	73,687	0	73,687	0	73,687

5.3. ESSR Prediction Models for PV Systems

5.3.1. ESSR Prediction Model Regression Analysis

Table 10 shows variations in the PV capacity, source energy consumption, and ESSR for changes in the BIPV facade area, based on different BAPV roof areas. As shown in Table 10, since the trend of increasing the ESSR with an increase in the BIPV facade area is linear, the required installation area of the BIPV facade for the case study building to achieve net-zero energy can be estimated through linear regression.

Figure 8 shows that the trend of the ESSR variation with respect to changes in the BIPV facade area remains consistent regardless of the BAPV roof area, indicating only a shift in the y intercept. Figure 9 further demonstrates that the y intercept linearly correlates with the BAPV roof area. By combining these linear regression relationships, Equation (2) could be derived, suggesting that it is feasible to predict the ESSR based on variations in both the BAPV roof and BIPV facade areas for a specific building.

$$ESSR[\%] = 0.001071 \times A_{BIPV} + 0.002142 \times A_{BAPV} + 0.000199 \quad (1)$$

However, this equation only considers the influence of the installation areas of the BAPV roof and BIPV facade, failing to adequately reflect the direct correlation with the source energy consumption for ESSR calculations. In other words, Equation (1) may show high predictive accuracy for the case study building used to derive Equation (1), however it can lack accuracy when building loads and source energy consumption vary. Therefore, to address such limitations and account for source energy consumption, this study used the correlation between the PV capacity and building source energy consumption to assess the achievable ESSR through PV system implementation.

Table 10. Simulation results for various installation areas of the BAPV roof and BIPV facade.

Category	BAPV Roof Area [m²]	BIPV Facade Area [m²]								
		100	150	200	250	300	350	400	450	500
Total PV Capacity [kW]	150	45	52	60	68	76	83	91	99	107
	180	51	58	66	74	82	89	97	105	113
	210	57	64	72	80	88	95	103	111	119
	240	62	69	77	85	93	100	108	116	124
Total Source Energy Generation [kWh/m². yr]	150	99.9	112.3	124.8	137.3	149.8	162.3	174.8	187.2	199.7
	180	114.9	127.3	139.8	152.3	164.8	177.3	189.8	202.2	214.7
	210	129.8	142.2	154.7	167.2	179.7	192.2	204.7	217.1	229.6
	240	144.9	157.3	169.8	182.3	194.8	207.3	219.8	232.2	244.7
ESSR [%]	150	42.9	48.2	53.6	58.9	64.3	69.7	75.0	80.4	85.7
	180	49.3	54.7	60.0	65.4	70.7	76.1	81.4	86.8	92.2
	210	55.7	61.1	66.4	71.8	77.1	82.5	87.9	93.2	98.6
	240	62.2	67.5	72.9	78.2	83.6	88.9	94.3	99.7	105.0

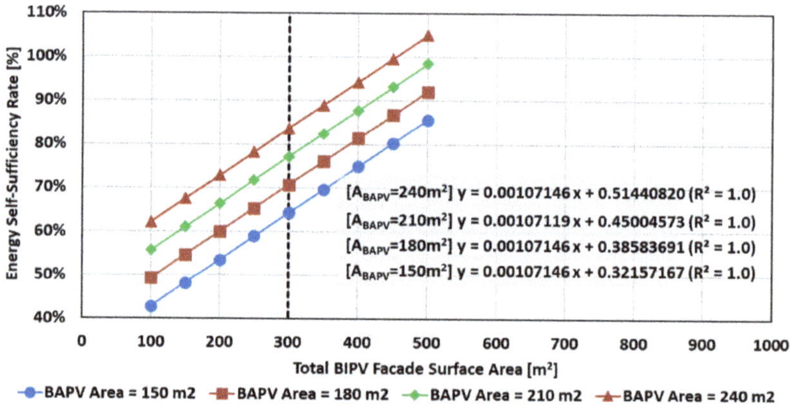

Figure 8. ESSR variations according to the BIPV facade area and BAPV roof area.

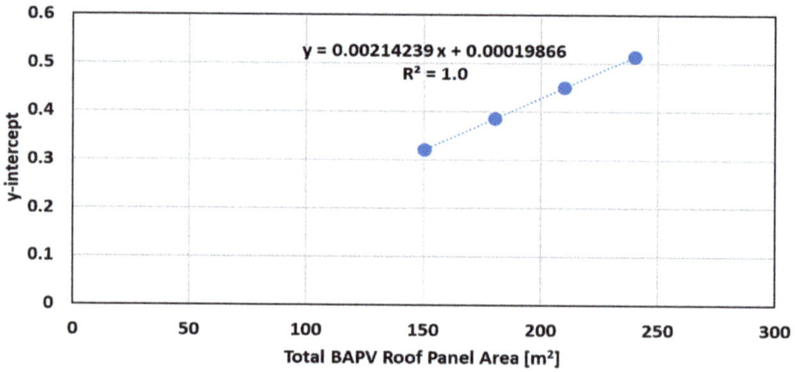

Figure 9. Relationship between the BAPV roof area and the y intercept in Figure 8.

As shown in Figure 10, the ESSR variation relative to the ratio of the PV capacity to the source energy consumption follows a linear trend. The slope of the change, which varies based on the BAPV roof area, remains consistent, while the y intercept differs. Figure 11 shows that the y intercept linearly correlates with changes in the BAPV roof area. By combining these linear regression equations, depicted in Figures 10 and 11, this study derived Equation (2), which can predict the building's ESSR.

$$ESSR[\%] = 1.6066 \times \frac{Total\ Solar\ Power\ Capacity}{Annual\ Primary\ Energy\ Consumption} + 0.000833 \times A_{BAPV} - 0.005180 \quad (2)$$

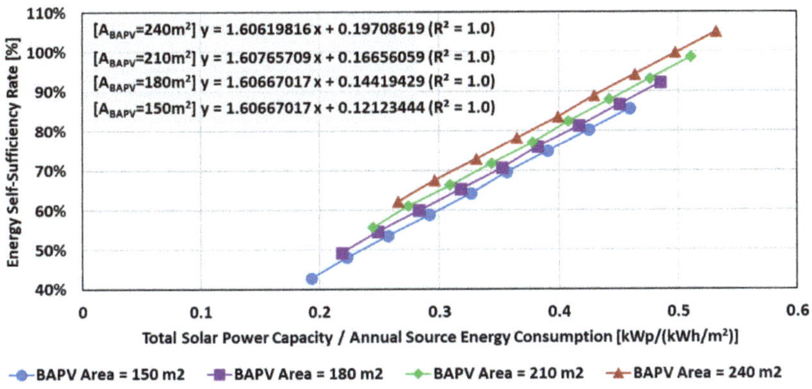

Figure 10. ESSR variations for the ratio between total PV capacity and source energy consumption.

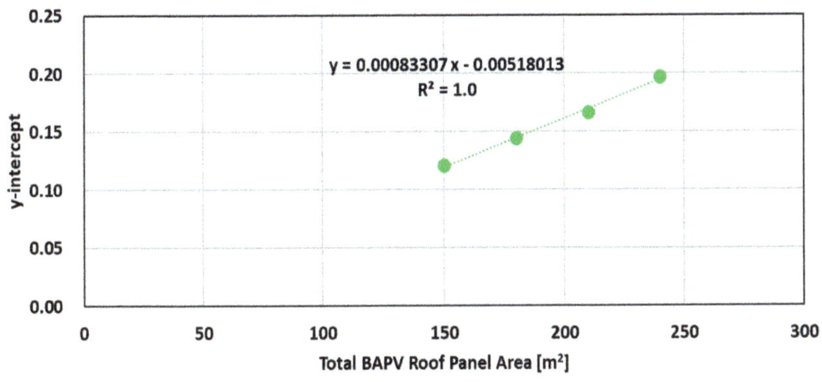

Figure 11. Relationship between BAPV roof area and y intercept in Figure 10.

5.3.2. Verification of the ESSR Prediction Models

The case building studied in this research was built in August 2023, as shown in Figure 12, along with the floor plan shown in Figure 13. While the main design remained consistent with the conceptual design, the dwelling unit area was expanded beyond the initial plans. Notably, mechanical ventilation was not implemented in the final design. The photographs and specifications of the completed building are presented in Figure 12 and Table 11.

Based on the specifications of the building shown in Table 11, the official results of the Zero-Energy Building Certification of Republic of Korea are presented in Table 12, indicating the final ESSR of 76.9%. This exceeds the target ESSR of 60%. The variation is attributed to changes in the boundary conditions, including envelope U-values, the infiltration rate, and lighting power density, as well as adjustments to the floor area compared to the initial plans.

(**a**) (**b**) (**c**)

Figure 12. Images of final design of the case study building: (**a**) elevation, (**b**) BIPV facade, (**c**) BAPV roof.

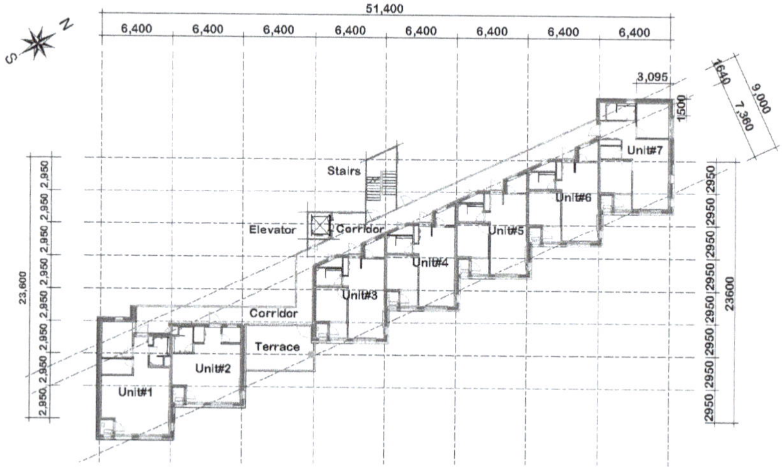

Figure 13. Floor plan of the final design of the case study building (unit: mm).

Table 11. Details of the final design specifications of the case study building.

Energy Efficiency Measure			Options				
Dwelling Unit Information		Floor Area of Units [m^2]	49	48	43	39	36
		Number of Dwelling Units	2	3	3	16	4
Passive System	Building Envelop: Opaque Element	Thermal Transmittance (U-value)	0.137 W/(m^2K)				
	Building Envelop: Window	Thermal Transmittance (U-value)	0.693 W/(m^2K)				
	Infiltration	Air Change per Hour	5.50				
Active System	Interior Lighting	Lighting Power Density	4.67 W/m^2				
	Space Cooling	Type	Space Cooling Not Installed				
	Space Heating	Type	Radiant Floor Heating				
		Efficiency	91.1% (Natural Gas Boiler)				
	Domestic Hot Water	Type	Natural Gas Boiler				
		Water Temperature	Supply 80 °C/Return 60 °C				
	Ventilation	Type	Mechanical Ventilator Not Installed				
Energy Generation System	BAPV Roof	Area	240 m^2 (52 kW = 500 Wp × 104 panels				
		Efficiency	21.6%				
	BIPV Facade	Area	138 m^2 (17.2 kW = 123 Wp * 140 panels)				
		Efficiency	12.3%				

Table 12. The official simulation results of the Zero-Energy Building Certification for the case study building.

Category	Space Heat	Space Cool	Hot Water	Lighting	Ventilation	Total	Energy Generation
Building Energy Load [kWh/m^2.yr]	63.6	0	30.7	8.5	0	102.8	0
Site Energy Consumption [kWh/m^2.yr]	126.2	0	32.7	8.5	0	167.4	56.9
Source Energy Consumption [kWh/m^2.yr]	145.9	0	34.3	23.4	0	203.6	156.6

The ESSR predictions were made using Equations (1) and (2), considering the final building specifications, source energy consumption, and PV capacity. The calculated ESSR prediction from Equation (1) was 66.2%, whereas Equation (2) predicted an ESSR of 74.1%. This discrepancy highlights that Equation (2) provides a more accurate prediction.

The reason lies in that Equation (1) focuses solely on variations in the installation area of PV systems, without directly considering the building's source energy consumption. While Equation (1) suited the sample data used for its derivation, it exhibited significant errors when the building's energy usage changed. In contrast, Equation (2) accounts for both PV capacity and source energy consumption, resulting in a more reliable ESSR prediction.

6. Conclusions

This study analyzed various factors related to implementing PV systems for achieving nearly zero energy in small-scale, low-rise, multi-family residential buildings. The analysis was based on a case study building located in Republic of Korea, a newly constructed five-story multi-family residential building with 28 dwelling units. The research investigated the impact of PV capacity and efficiency variations, BIPV facade costs, and the combined effects of a BAPV roof and BIPV facade on the building's ESSR. Based on these analyses, the study proposed equations to predict the ESSR when implementing PV systems in small-scale, low-rise, multi-family residential buildings in Republic of Korea.

During the conceptual design process, the case study building set a target ESSR of 60%, excluding space cooling and plug loads. To achieve this target ESSR, this study analyzed the impact of PV capacity and efficiency. The simulation results showed that achieving the target ESSR using only a BAPV roof or a BIPV facade would be challenging due to the limited roof area for BAPV systems in multi-family residential buildings and the lower efficiency of BIPV facades compared to BAPV roofs. Therefore, a combination of a BAPV roof and BIPV facade emerges as a viable strategy to achieve the desired ESSR.

Due to the initial cost of a BIPV facade being twice as high as that of BAPV roof, it was more cost-effective to use more BAPV roof panels than BIPV facade panels. To enhance the cost-effectiveness of the BIPV facade, the sensitivity analysis revealed that its initial cost should be 62% lower than that of a BAPV roof. This difference primarily stems from the lower energy performance of BIPV facades compared to BAPV roofs.

Based on the simulation results for various combinations of BAPV roofs and BIPV facades, this study derived equations using linear regression analysis to predict the ESSR for a small-scale, low-rise, multi-family residential building. By utilizing the final design details of the case study residential building, the proposed equation, which considers both the PV capacity and building source energy consumption, was validated to predict the ESSR, which was closely aligned with the actual ESSR.

However, it is important to acknowledge that this study has limitations. Conducted on a specific case study building, further analysis is necessary to determine whether the results can be generalized to multi-family residential buildings of varying scales and locations. Additionally, since this study primarily relied on simulation analysis, future validation using real building energy performance data and PV system-generated energy data is essential.

Despite these limitations, this study differs from previous research by investigating the application of PV systems as a strategy to achieve the ESSR in small-scale, low-rise, multi-family residential buildings in Republic of Korea. This research could be useful for future studies exploring strategies for achieving zero-energy, multi-family residential buildings using PV systems.

Author Contributions: Conceptualization, B.C.K.; Methodology, B.C.K.; Formal analysis, B.C.K.; Investigation, B.C.K.; Data curation, B.C.K.; Writing—original draft, B.C.K.; Writing—review & editing, G.T.K. and I.T.H.; Visualization, B.C.K. and I.T.H.; Supervision, G.T.K.; Project administration, G.T.K.; Funding acquisition, G.T.K. All authors have read and agreed to the published version of the manuscript.

Funding: This research was funded by City Government of Sejong [Sejong New Deal Policy for Urban Regeneration].

Data Availability Statement: The data presented in the study are included in the article material; further inquiries can be directed to the corresponding author.

Conflicts of Interest: The authors declare no conflicts of interest.

References

1. International Energy Agency. Efficiency & Demand. Available online: https://www.iea.org/countries (accessed on 22 June 2024).
2. U.S. Energy Information Administration. Annual Energy Review 2023. Available online: https://www.eia.gov/totalenergy/data/annual/index.php (accessed on 22 June 2024).
3. Korea Energy Economics Institute. *2023 Yearbook of Energy Statistics—Data up to 2022*; Korea Energy Economics Institute: Ulsan, Republic of Korea, 2023; ISSN 1226-606X.
4. Korea Energy Agency. Zero Energy Building Certification. Available online: https://zeb.energy.or.kr/BC/BC00/BC00_01_001.do (accessed on 22 June 2024).
5. Korean Statistical Information Service. Housing Census. Available online: https://kosis.kr/eng/statisticsList/statisticsListIndex.do?parentId=I1.1&menuId=M_01_01&vwcd=MT_ETITLE&parmTabId=M_01_01#content-group (accessed on 22 June 2024).
6. Wu, W.; Skye, H.M. Residential net-zero energy buildings: Review and perspective. *Renew. Sustain. Energy Rev.* **2021**, *142*, 110859. [CrossRef] [PubMed]
7. Feng, W.; Zhang, Q.; Ji, H.; Wang, R.; Zhou, N.; Ye, Q.; Hao, B.; Li, Y.; Luo, D.; Lau, S.S.Y. A review of net zero energy buildings in hot and humid climates: Experience learned from 34 case study buildings. *Renew. Sustain. Energy Rev.* **2019**, *114*, 109303. [CrossRef]
8. Wells, L.; Rismanchi, B.; Aye, L. A review of Net Zero Energy Buildings with reflections on the Australian context. *Energy Build.* **2018**, *158*, 616–628. [CrossRef]
9. Ullah, K.; Prodanovic, V.; Pignatta, G.; Deletic, A.; Santamouris, M. Technological advancements towards the net-zero energy communities: A review on 23 case studies around the globe. *Sol. Energy* **2021**, *224*, 1107–1126. [CrossRef]
10. Omar, A.I.; Khattab, N.M.; Abdel Aleem, S.H.E. Optimal strategy for transition into net-zero energy in educational buildings: A case study in El-Shorouk City, Egypt. *Sustain. Energy Technol. Assess.* **2022**, *49*, 101701. [CrossRef]
11. Alawode, A.; Rajagopalan, P. Feasibility of net zero energy high rise apartment buildings in Australia. *Sol. Energy* **2022**, *231*, 158–174. [CrossRef]
12. Ni, S.; Zhu, N.; Zhang, Z.; Hou, Y.; Li, S. The operational performance of net zero energy wooden structure building in the severe cold zone: A case study in Hailar of China. *Energy Build.* **2021**, *257*, 111788. [CrossRef]
13. Nematchoua, M.K.; Nishimwe, A.M.; Reiter, S. Towards nearly zero-energy residential neighbourhoods in the European Union: A case study. *Renew. Sustain. Energy Rev.* **2021**, *135*, 110198. [CrossRef]
14. Zhang, X.; Wang, A.; Tian, Z.; Li, Y.; Zhu, S.; Shi, X.; Jin, X.; Zhou, X.; Wei, S. Methodology for developing economically efficient strategies for net zero energy buildings: A case study of a prototype building in the Yangtze River Delta, China. *J. Clean. Prod.* **2021**, *320*, 128849. [CrossRef]
15. Panagiotidou, M.; Aye, L.; Rismanchi, B. Optimisation of multi-residential building retrofit, cost-optimal and net-zero emission targets. *Energy Build.* **2021**, *252*, 111385. [CrossRef]
16. Ali, H.H.; Al-Rub, F.A.A.; Shboul, B. Evaluation of Near-net-zero-energy Building Strategies: A Case Study on Residential Buildings in Jordan. *Int. J. Energy Econ. Policy* **2020**, *10*, 325–336. [CrossRef]
17. Lan, L.; Wood, K.L.; Yuen, C. A holistic design approach for residential net-zero energy buildings: A case study in Singapore. *Sustain. Cities Soc.* **2019**, *50*, 101672. [CrossRef]
18. Alajmi, A.; Rodríguez, S.; Sailor, D. Transforming a passive house into a net-zero energy house: A case study in the Pacific Northwest of the U.S. *Energy Convers. Manag.* **2018**, *172*, 39–49. [CrossRef]

19. Ascione, F.; Bianco, N.; Böttcher, O.; Kaltenbrunner, R.; Vanoli, G.P. Net zero-energy buildings in Germany: Design, model calibration and lessons learned from a case-study in Berlin. *Energy Build.* **2016**, *133*, 688–710. [CrossRef]
20. Zhou, Z.; Feng, L.; Zhang, S.; Wang, C.; Chen, G.; Du, T.; Li, Y.; Zuo, J. The operational performance of "net zero energy building": A study in China. *Appl. Energy* **2016**, *177*, 716–728. [CrossRef]
21. Tsalikis, G.; Martinopoulos, G. Solar energy systems potential for nearly net zero energy residential buildings. *Sol. Energy* **2015**, *115*, 743–756. [CrossRef]
22. Cellura, M.; Guarinoa, F.; Longoa, S.; Mistretta, M. Energy life-cycle approach in Net zero energy buildings balance: Operation and embodied energy of an Italian case study. *Energy Build.* **2015**, *72*, 371–381. [CrossRef]
23. Park, B.R.; Chung, M.H. Analysis of the additional energy-saving potential of residential buildings after mandatory zero-energy buildings to achieve carbon neutrality in South Korea. *Build. Environ.* **2023**, *228*, 109908. [CrossRef]
24. Choi, J.; Kee, D.; Lee, J.; Kim, J.J. Understanding heterogeneous consumer preferences for residential zero-energy buildings (ZEBs) in South Korea: A latent class approach. *Sustain. Cities Soc.* **2023**, *97*, 104747. [CrossRef]
25. Ji, C.; Hong, T.; Kim, H.; Yeom, S. Effect of building energy efficiency certificate on reducing energy consumption of non-residential buildings in South Korea. *Energy Build.* **2022**, *255*, 111701. [CrossRef]
26. Ji, C.; Hong, T.; Kim, H. Statistical analysis of greenhouse gas emissions of South Korean residential buildings. *Renew. Sustain. Energy Rev.* **2022**, *156*, 111981. [CrossRef]
27. Kim, K.S.; Yoo, D.C.; Choi, C.; Jang, H.I. Verification of Energy Usage Based on Standard Building Model Development of Low-Rise Residential Buildings in South Korea. *Adv. Civ. Eng.* **2021**, *2021*, 6679108. [CrossRef]
28. Jung, Y.; Heo, Y.; Cho, H.; Kang, Y.T.; Kim, Y.; Lee, H. A plan to build a net zero energy building in hydrogen and electricity-based energy scenario in South Korea. *J. Clean. Prod.* **2023**, *397*, 136537. [CrossRef]
29. Ji, C.; Choi, M.; Hong, T.; Yeom, S.; Kim, H. Evaluation of the effect of a building energy efficiency certificate in reducing energy consumption in Korean apartments. *Energy Build.* **2021**, *248*, 111168. [CrossRef]
30. Munkhbat, U.; Byun, S.Y.; Lee, D.H. A Study on the Prediction of Achieving of Zero Energy Building Certification in an Apartment House according to Early Adoption of Zero-energy Buildings. *J. KIAEBS* **2021**, *15*, 313–325.
31. Wang, S.; Tae, S.; Jang, H. Prediction of the Energy Self-Sufficiency Rate of Major New Renewable Energy Types Based on Zero-Energy Building Certification Cases in South Korea. *Sustainability* **2021**, *13*, 11552. [CrossRef]
32. Lee, H.; Choi, M.; Lee, R.; Kim, D.; Yoon, J. Energy performance evaluation of a plus energy house based on operational data for two years: A case study of an all-electric plus energy house in Korea. *Energy Build.* **2021**, *252*, 111394. [CrossRef]
33. Yang, S.; Cho, H.M.; Yun, B.Y.; Hong, T.; Kim, S. Energy usage and cost analysis of passive thermal retrofits for low-rise residential buildings in Seoul. Renew. *Sustain. Energy Rev.* **2021**, *151*, 111617. [CrossRef]
34. Kim, Y.; Yu, K.H. Study on the Certification Policy of Zero-Energy Buildings in Korea. *Sustainability* **2020**, *12*, 5172. [CrossRef]
35. Kwag, B.C.; Han, S.; Kim, G.T.; Kim, B.; Kim, J.Y. Analysis of the Effects of Strengthening Building Energy Policy on Multifamily Residential Buildings in South Korea. *Sustainability* **2020**, *12*, 3566. [CrossRef]
36. Lee, J.; Shepley, M.M.; Choi, J. Exploring the localization process of low energy residential buildings: A case study of Korean passive houses. *J. Build. Eng.* **2020**, *30*, 101290. [CrossRef]
37. Suh, H.S.; Kim, D.D. Energy performance assessment towards nearly zero energy community buildings in South Korea. *Sustain. Cities Soc.* **2019**, *44*, 488–498. [CrossRef]
38. Song, J.; Oh, S.-D.; Song, S.J. Effect of increased building-integrated renewable energy on building energy portfolio and energy flows in an urban district of Korea. *Energy* **2019**, *189*, 116132. [CrossRef]
39. Eum, J.; Kim, Y. Applicability analysis of residential energy storage system (ESS) using Homer in Korea. In Proceedings of the Grand Renewable Energy 2018 Proceedings, Pacifico Yokohama, Yokohama, Japan, 17–22 June 2018.
40. Kim, T.H.; Jeong, Y.S. Analysis of Energy-Related Greenhouse Gas Emission in the Korea's Building Sector: Use National Energy Statistics. *Energies* **2018**, *11*, 855. [CrossRef]
41. Kim, S.-Y.; Choi, H.-S.; Eum, J.-H. Energy-Independent Architectural Models for Residential Complex Plans through Solar Energy in Daegu Metropolitan City, South Korea. *Sustainability* **2018**, *10*, 482. [CrossRef]
42. Park, K.S.; Kim, M.J. Energy Demand Reduction in the Residential Building Sector: A Case Study of Korea. *Energies* **2017**, *10*, 1506. [CrossRef]
43. Oh, J.; Hong, T.; Kim, H.; An, J.; Jeong, K.; Koo, C. Advanced Strategies for Net-Zero Energy Building: Focused on the Early Phase and Usage Phase of a Building's Life Cycle. *Sustainability* **2017**, *9*, 2272. [CrossRef]
44. Cheong, C.H. Appropriate Size of a Net Zero-Energy Multi-Family Housing using PV system. *J. KIAEBS* **2017**, *11*, 113–120.
45. Kim, J.H.; Kim, H.R.; Kim, J.T. Analysis of Photovoltaic Applications in Zero Energy Building Cases of IEA SHC/EBC Task 40/Annex 52. *Sustainability* **2015**, *7*, 8782–8800. [CrossRef]
46. Park, D.J.; Yu, K.H.; Yoon, Y.S.; Kim, K.H.; Kim, S.S. Analysis of a Building Energy Efficiency Certification System in Korea. *Sustainability* **2015**, *7*, 16086–16107. [CrossRef]
47. Lim, J.H.; Kim, S.I.; Song, S.Y. Improvement of Design Criteria in Heating and Cooling Equipment According to the Consolidation of Design Standard for Energy Saving in Apartment Buildings of Korea. *J. Korean Sol. Energy Soc.* **2014**, *34*, 89–97. [CrossRef]
48. Lee, K.-H.; Lee, J.-K.; Yoon, E.-S.; Joo, M.-C.; Lee, S.-M.; Baek, N.-C. Annual measured performance of building-integrated solar energy systems in demonstration low-energy solar house. *J. Renew. Sustain. Energy* **2014**, *6*, 042013. [CrossRef]

49. Ministry of Land, Infrastructure and Transport. Energy Saving Design Standards for Buildings. Available online: http://www.law.go.kr/ (accessed on 22 June 2024).
50. Oh, T.-K.; Lee, D.; Park, M.; Cha, G.; Park, S. Three-Dimensional Visualization Solution to Building-Energy Diagnosis for Energy Feedback. *Energies* **2018**, *11*, 1736. [CrossRef]
51. Kim, C.S. The Influence of Unit Plan Shapes to the Energy Eciency of Collective Housing Simulated by ECO2 Software. *Korea Inst. Ecol. Archit. Environ. J. (KIEAE J.)* **2015**, *15*, 89–94.
52. Lee, A.R.; Kim, J.G.; Kim, J.H.; Jeong, H.G.; Jang, C.Y.; Song, K.D. Comparing the actual heating energy with calculated energy by the amended standard building energy rating system for apartment buildings. *Korea Inst. Ecol. Archit. Environ. J. (KIEAE J.)* **2015**, *15*, 103–107.
53. Kwon, J.H.; Jang, H.I.; Kim, M.K.; Park, H.S.; Suh, S.J. Comparison on the Annual Energy Demand by Building Energy Assessment Tool and Dynamic Energy Simulation. In Proceedings of the 2012 Summer Conference of the Society of Air-Conditioning and Refrigerating Engineers of Korea, Pyeongchang, Republic of Korea, 27–29 June 2012; pp. 363–366.
54. Kim, S.H.; Kwak, Y.H.; Kim, C.S. The Analysis on Energy Performance of Collective Housing using ECO2 and DesignBuilder Softwares. *Korea Inst. Ecol. Archit. Environ. J. (KIEAE J.)* **2018**, *18*, 47–54.
55. Seoul Metropolitan Government. Standards for Installation and Management of Solar Power Facilities in Seoul. Available online: https://www.seoul.go.kr (accessed on 22 June 2024).
56. National Agency for Administrative City Construction. Solar Power Facility Installation Guidelines. Available online: https://naacc.go.kr (accessed on 22 June 2024).
57. Korea Energy Agency. 2021 Standard Unit Price for Renewable Energy Sources. Available online: https://www.knrec.or.kr/biz/pds/pds/view.do?no=303 (accessed on 22 June 2024).
58. Korea Energy Agency. 2012 Standard Unit Price for Renewable Energy Sources. Available online: https://www.knrec.or.kr/biz/pds/businoti/view.do?no=598 (accessed on 22 June 2024).
59. Korea Energy Agency. 2011 Standard Unit Price for Renewable Energy Sources. Available online: https://www.knrec.or.kr/biz/pds/businoti/view.do?no=395 (accessed on 22 June 2024).

buildings MDPI

Article

The Design of a Wellbeing Center with an Advanced Parametric Roof in Jabel Hafeet Mountain Located in Al Ain City, United Arab Emirates

Marwa Alsuwaidi, Noura Almahri, Latifa Alazeezi, Taif Alkhzaimi, Lindita Bande *, Anwar Ahmad, Omar Al Khatib, Fayez Fikry and Young Ki Kim

Architectural Engineering Department, College of Engineering, United Arab Emirates University, Al Ain P.O. Box 15551, United Arab Emirates; 201703016@uaeu.ac.ae (N.A.); 201700972@uaeu.ac.ae (L.A.); 201704924@uaeu.ac.ae (T.A.); anwar.ahmad@uaeu.ac.ae (A.A.); omar.alkhatib@uaeu.ac.ae (O.A.K.); fayez.fikry@uaeu.ac.ae (F.F.); youngki.kim@uaeu.ac.ae (Y.K.K.)
* Correspondence: lindita.bande@uaeu.ac.ae

Abstract: Building construction in the UAE is a growing industry. While Abu Dhabi and Dubai have a majority of mid-rise and high-rise buildings, AL Ain has mostly low-rise and mid-rise buildings. Additionally, the construction industry in Al Ain has seen an increase in the recent years. The aim of this study is to design a wellness center with a parametric roof in the challenging topography of Jebel Hafeet Mountain in AL Ain. The building must be socially and environmentally sustainable and must align with the local culture. The methodology used includes the following steps: performing an analysis of the selected site and climate; building the design on challenging topography; creating a parametric roof design; conducting an energy and radiation analysis. Wellness is the act of practicing healthy habits to attain better physical and mental health outcomes. The center will be for females only to provide privacy and to achieve social sustainability (based on the local culture). It will be in Al Ain (Jabel Hafeet Mountain) because unlike the other cities in the UAE, Al Ain does not contain any wellness centers. The design of the parametric roof aims to adapt to the mountain and decrease solar radiation on the roof. Both of the above aim to bring innovation to the city, country, and region for the purpose of building a sustainable design that is adapted to the local culture and environment. The results show a significant energy and solar radiation after the use of such structure. These findings are relevant to local authorities and the private construction industry.

Keywords: wellbeing center; energy simulation; parametric structure; python language

1. Introduction

1.1. Wellness Center Design

Wellbeing embraces six mutually co-dependent dimensions: physical, mental, spiritual, emotional, social, and environmental. If any one of these is neglected over time, it will affect one's health, wellbeing, and quality of life. So, to reduce stress levels, the possibility of suffering from mental illness, and the risk of illness, and to strengthen community relations, this study focuses on designing a wellness center.

A study conducted in 2020 showed that 57% of people in the UAE suffer from at least one mental health disorder; the two most common ones, at 56.4% and 31.5% of the population, are anxiety and depression [1]. Another survey conducted by Cigna showed that 89% of women in the UAE are stressed compared to 85% globally [2].

Research studies show that people who take care of their wellbeing are healthier, happier, and more productive, miss less work, and have lower healthcare costs. Wellness centers are places that allow people to take care of their wellbeing by offering health services for the body and mind. Skincare services and body services such as fitness classes, personal training, and nutrition consulting are usually offered at wellness centers [3]. The government of Dubai issued a wellbeing certificate to support the construction of sustainable building, with the attention focused on ensuring benefits to the people. This certification encourages local authorities, contractors, and other professionals to improve wellbeing conditions through building design [4,5].

1.2. Local Standards and Passive Strategies

Abu Dhabi Emirate has several initiatives and standards regarding energy efficiency in buildings, with the goal of reducing total consumption in cities under this Emirate. As a result, the new ESTIDAMA rating system is being used in new constructions. The Urban Planning Council introduced this building code, which is part of the Abu Dhabi 2030 sustainability plan [6]. The Arabic world of sustainability is a vision for a new sustainable way of life in the UAE and Arabic region. Estidama's goal is to protect and enhance Abu Dhabi's physical and cultural identity while also improving residents' quality of life in four equal pillars of sustainability: the environment, the economy, the society, and the culture. Estidama's green building strategy incorporates sustainability factors into all stages of building construction, with a focus on the design, construction, operation, and maintenance phases [7]. According to these regulations, new construction must meet Estidama's mandatory requirements. Estidama specifies the primary building characteristics for new construction, including the materials, the area, the orientation, and a variety of other passive strategies [8]. Passive strategies refer to any techniques or design features utilized that decrease building temperatures without using electricity. The use of passive methods for heating or cooling has gained popularity in recent years, particularly in the last ten years, as part of the trend toward green building. In summertime, efficient building envelopes maximize air movement while controlling heat from the sun. In hot, arid climates like in the United Arab Emirates (UAE), several passive cooling strategies can be recommended [8]. Evaporative cooling, shading (including parametric shading), insulation, light-colored coatings with high reflection, green roofing, and other technologies are examples [9,10].

1.3. Energy in Buildings

In the last 70 years, Abu Dhabi has gone through a fast urban development. The city is growing to become the country's center. According to a recent study conducted in the city of Abu Dhabi, the building sector consumes 80% of all energy, with 70% of that energy being used to cool the buildings [11]. About 90% of the electricity used in the United Arab Emirates is consumed by the nation's existing building stock [12]. Heating, ventilation, and air conditioning (HVAC) equipment generates the most electrical load due to the harsh weather in the United Arab Emirates. It accounts for 40% of the total electrical load year-round and up to 60% of the peak electrical load during the summer [13]. Exposure to high levels of solar radiation causes increased air temperature inside the buildings [14]. The building envelope has a major impact on how much heat is absorbed by buildings [15]. There are three methods for lowering heat gain through the envelop: reduce the amount of glazing, create shade, and lessen the solar gain characteristics of the building envelope materials [14].

On the other hand, a study was conducted in two phases with the purpose of improving a corridor's visual and environmental conditions, including solar radiation, exposure, and daylighting, at the University of Sharjah's campus in the United Arab Emirates, and the results show a significant improvement in UDLI (100–2000) when compared to the corridor's current status, with an increase of more than 54%, as well as a significant decrease in vertical and horizontal radiation exposure [16].

One of the most important examples of applying a parametric structure to the buildings in Abu Dhabi is Al Bahar Towers, which are two high-rise office buildings in Abu Dhabi, the UAE. The shading tool for the façade was created by using parametric geometry modeled after the traditional Islamic lattice and translated using bio-inspiration, regional design, guide grid, and performance-oriented technology principles. The triangular elements of the façade are supported by a system of steel rods and hydraulic pistons and are made of a lightweight composite material (PTFE and fiberglass coat). A computer algorithm controls the system, adjusting the angle of the triangular elements based on the position of the sun and the desired level of shading. The façade not only provides a sustainable solution for reducing energy consumption, but it also creates a visually appealing environment [17].

1.4. Parametric Architecture

The principles of parametric design related to the use of several parameters in the design process represent everything that can be designed [18]. The process of developing a certain complex design is defined as parametric design. This representation is based on variable-controlled relationships between objects. Changing the variables results in different models. The solution is then chosen based on criteria such as performance, ease of construction, budget constraints, user needs, esthetics, or a combination of these [19]. Topological forms described by parametric functions provide a wide range of structure creation options. A geometrical form is shaped in parametric design by parameters and equations that describe it, rather than by declaration of its shape and structure. There are two types of parametric design: conceptual parametric design and constructive parametric design. Constructive parametric design refers to more information used for 3D object determination. Many variables influence the architectural/engineering object being designed in this context, and designing can be viewed as a multicriteria decision making process [20].

However, the digital environment provides a fresh perspective on design. On the one hand, it allows for a great deal of freedom in shaping free forms; on the other hand, it allows designers to use computer software for project optimization and simulation. Following the first generation of digital design processes and the application of new digital design tools, novel structures and connections have emerged, as have new processes. Design tools were created to align the digital form with reality on the construction site. A variety of software programs and plug-ins are used to generate parametric structures, including Dynamo in Revit 2020 [21], CATIA 2014, which stands for Computer-Aided Three-Dimensional Interactive Application [22], Marionette in Vectorworks 2022 [23], and Grasshopper for Rhinoceroses 2021 [24]. Comparing the eligibility of the previously mentioned parametric design tools, Grasshopper is the most eligible to achieve this goals of this study compared to the other tools due to its open-source nature, its low-cost software, the validated simulation plug-ins, and the many learning materials available for free which help in developing the skills easily. It is possible to design sustainable communities with parametric design processes using advanced tools such as Grasshopper [25].

Grasshopper (2021) is "a plug-in software that uses a visual programming language to allow architects to easily isolate the driving parameters of their design while iteratively tuning them. It lowers the entry level for using complex parametric tools [24]. Visual programming (VP) systems were developed to aid designers in the development of scripts for the construction of parametric models that consider various parameters such as climatic conditions, energy consumption, and so on. VP systems are defined as "any system that allows the user to specify a program in a two-(or more)-dimensional fashion" [26].

1.5. Parametric Roof Design

Every culture and society that has been established in a particular climate zone for a considerable amount of time has its optimal archetypal cognitive model represented by the roof [27]. The roof of a building is its covering that provides protection from the elements, including wind, rain, snow, sunlight, and temperature extremes. Technical, financial, or esthetic factors have all influenced the construction of roofs, which can be flat, pitched, vaulted, domed, or a combination of these shapes [28]. The roof design process in a hot, arid climate should take into consideration strength and stability, energy conservation and ventilation, the control of solar heat and air temperature, daylight and the control of glare, durability, and ease of maintenance [29,30]. Design, thermal performance, and structural integrity can now be optimized and regulated thanks to recent advancements in parametric programming. Still, physical form-finding is not routinely incorporated into manufacturing processes in the industry. Deciphering the grid meshes in a structural solution for free-formed structures (such as grid shells), which are often constructed with bars and nodal connectors, has required a great deal of engineering work. The current trend stems from the discovery of bio-mimetic principles within a distinct computational design framework. This part of the generative rule set is used by the parametric model. In addition to providing structural analysis, the generated model can provide the data needed for mechanical manufacturing that uses a numerical control system [31].

A parametric roof is a roof design strategy that is designed based on computational methods and algorithms to create a structure that responds to specific design criteria and performance goals. Componential methods enable architects and engineers to control various design parameters such as shape, orientation, and materials to create a unique and optimized design. Parametric roofs are often used in buildings where the roof has a complex geometry, and where there is a need to control the solar radiation and energy flows. The use of parametric roofs has become increasingly popular in recent years, as it allows for more efficient and sustainable designs that are tailored to specific site conditions and functional requirements. In one of the recent studies, the specific design criteria for the roof design are based on the Catalan geometry analysis. By simulating several shapes, the optimum design is defined by the area that the roof covers, the shape of the roof, the structure, and the roof shell behavior under self-load. There were no pre-defined performance goals or threshold [32]. A second study on roof optimization for emergency units has a threshold of a better solution than the 600 × 400 cm plastic one. Based on the parameters of climate, shape, flexibility of the structure, connection of the structure, load, and footprint area, a new shape and structure emerged. This new shape can be easily manufactured by local producers [33]. One of the examples of a parametric roof in the UAE is the Louver Museum of Abu Dhabi; its massive dome structure serves environmental, esthetic, and structural purposes. Environmentally, it provides shade and cooler temperatures to the outdoor public spaces that connect the museum village's multiple buildings. The parameters that define the shape openings of the structure serve as an external open-air canopy, filter natural sunlight and create a dramatic and constantly changing lighting effect beneath the dome, and improve the walkability and the thermal comfort [34].

In one study that investigates designing parametric roofs, the author follows a digitalized approach by using Grasshopper to develop the roof on three levels. The results suggest that Grasshopper provides the potential to explore a wider range of well-informed design solutions and the potential to find novel and intriguing solutions to a design problem from the perspective of design [35]. Another study has investigated using parametric modeling to achieve performance-oriented design. The case study's design topic is a wide span roof, which is a component of a bigger, ongoing project in Bologna, Italy. The study emphasizes daylighting and the passive mitigation of summertime overheating. A parametric model has been created for these tasks. The study discussed three project scales: Large-scale parametric variations in the roof's overall shape are studied in connection to ventilation-based cooling; in this case, the parametric model enables the creation of various roof configurations, including variations in the structural morphology and structural tessellation. The integration of openable modules is examined at a medium scale in connection with air extraction for cooling; in this regard, the parametric model enables the exploration of openings based on size and distribution variations. In order to minimize direct solar gain while maintaining the flow of indirect natural light, a number of options are investigated for the cladding system at a small scale. The parametric model has also been used to look into the arrangement of self-shading modules and how they fit into the structure [36]. Another paper presents an investigation of a parametric system to produce variable designs for inexpensive roof units with central support, known as "tulips". The units are designed to cover outdoor areas and are made up of digitally manufactured plates, commercial wood struts, and a fabric cover, which allows for mass production and customization of the units to meet various structural, functional, and climatic requirements. To create models and full-scale prototypes, the development has integrated parametric programming, genetic algorithms, topological analysis, and digital fabrication. This procedure offers a method for rationalizing a design that facilitates rapid and flexible elaboration through digital implementation and mathematical analysis [37].

2. Methodology

This research paper has a linear methodology, which is a systematic approach that breaks down the design process into individual steps. These steps are often sequential and involve site selection, climate analysis, the development and testing of sustainable building concepts, the modeling and testing of parametric roof concepts, and the simulation of building energy with and without roofs. The case study is located in Al Ain city. This city has a hot, arid climate with low levels of humidity in comparison to Dubai and Abu Dhabi (coastal cities). Local data such as the Al Ain Airport file and the Jebel Hafeet Mountain topography were used in various stages of this study. The case study is located in Jebel Hafeet mountain. The modeling and simulation of the base case refers to the model without the parametric roof. The research continues with the addition of the parametric roof and energy simulation based on this scenario. Figure 1 shows schematically the process followed in this study by defining the two main objectives and the steps used to achieve them. The addition of the parametric roof is a unique approach that aims to connect the building shape to the mountain topography; moreover, the structure shall reduce the energy consumption of the building (among other passive solutions).

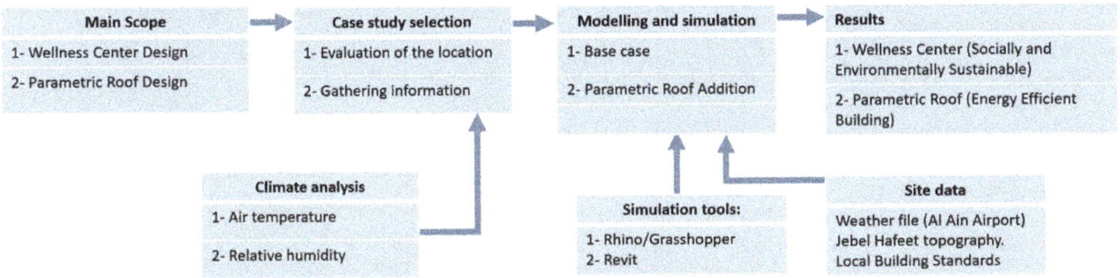

Figure 1. Schematic methodology followed in this study.

2.1. Site Selection and Climate Analysis

This site is situated on top of Jabel Hafeet, providing a higher level of privacy. The location boasts a unique view of the sunset and Al Ain downtown, which plays a significant role in promoting mental wellbeing. There are also available parking areas on the site. The site map with the entrances, sun path, and direction of the main view were carefully analyzed. Figure 2 shows the average monthly temperatures and precipitation related to the city of Al Ain and Table 1 shows the main points of the analysis.

Table 1. Existing building context (analysis/decisions).

	Analysis	Decisions
1. Temperature	The path of the sun in the summer is higher than its path in winter. Hot summers and cool winters. The warmest month of the year is August, with an average high temperature of 37 °C.	Use vertical shading devices (louvres) on the west side. Use horizontal shading devices on the south side. Use heat insulators. Use PV panels

Table 1. *Cont.*

	Analysis	Decisions
2. Wind	The calmer time of year lasts from August to January. The windier part of the year lasts from January to August.	Use cross ventilation.
3. Illumination	The length of the day in Al Ain varies over the course of the year. In 2022, the shortest day was December 22, with 10 h, 38 min of daylight; the longest day was June 21, with 13 h, 38 min of daylight.	Use vertical shading devices (louvres) on the west side. Use horizontal shading devices on the south side. Use heat insulators. Use PV panels
4. Humidity	Al Ain experiences significant seasonal variation in the perceived humidity. The muggier period of the year lasts from June to October, during which time the comfort level is muggy.	Use automatic controls to keep all types of HVAC equipment up and running to provide better comfort for occupants.
5. Noise and Views	The site is far from the city noise. There is a view of mountain on the southwest side.	The main views will be on the southwest to provide privacy and a pleasant view.
6. Accessibility	The site is accessible from one main road. There is a close parking area that can be used as a drop-off area.	Pedestrian and secondary roads must be built around the site. Provide emergency service and visitors entrances. Provide direction signs.

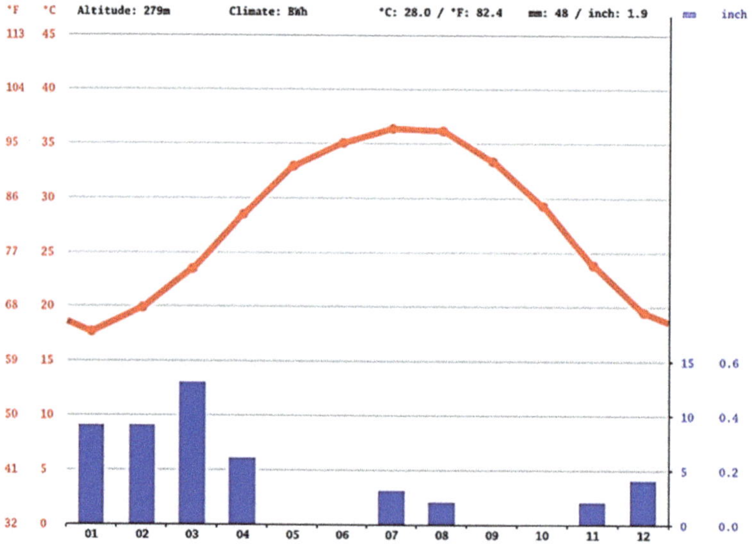

Figure 2. Average temperature and precipitation graph (the unit system on the vertical axis is in degree Fahrenheit and Celsius) [38].

2.2. *Sustainable Wellbeing Center Design Process*

2.2.1. Socially Sustainable Wellbeing Center

There are noticeable cultural differences between Al Ain and other cities in the UAE such as Dubai and Abu Dhabi, and it is imperative to respect the local customs and beliefs of the community. The wellness center provides a private and secure environment that is specifically designed for women in order to promote social sustainability and provide a safe space for women. The wellness center also provides a variety of services, including massages, spa treatments, and swimming, so women can relax and unwind. Women can also benefit from a range of activities, such as yoga and meditation, to improve their physical and mental health.

2.2.2. Environmentally Sustainable Wellbeing Center

Jebel Hafeet forms a perfect place for such a center because it offers the best view during sunset, as well as privacy. The center will contain a pool, gym, spa, sauna, restaurant, library, and lecture halls. The center will satisfy the needs of Al Ain city and follow the sustainable principles of design. The construction must be carried out without damaging the natural environment.

The first concept alternative is shown in Figure 3 below. The site plan of the building is shown in Figure 4. The design was inspired by the shape of the rocks of Jebel Hafeet. In this design, the building consists of three buildings. Each building has a different height and contains spaces that are related. Figure 5 shows the functional diagram of the building. The building will contain a reception area, offices, lockers, prayer rooms, toilets, cafeteria, yoga rooms, spa rooms, a gym and swimming pool, a restaurant, and service areas.

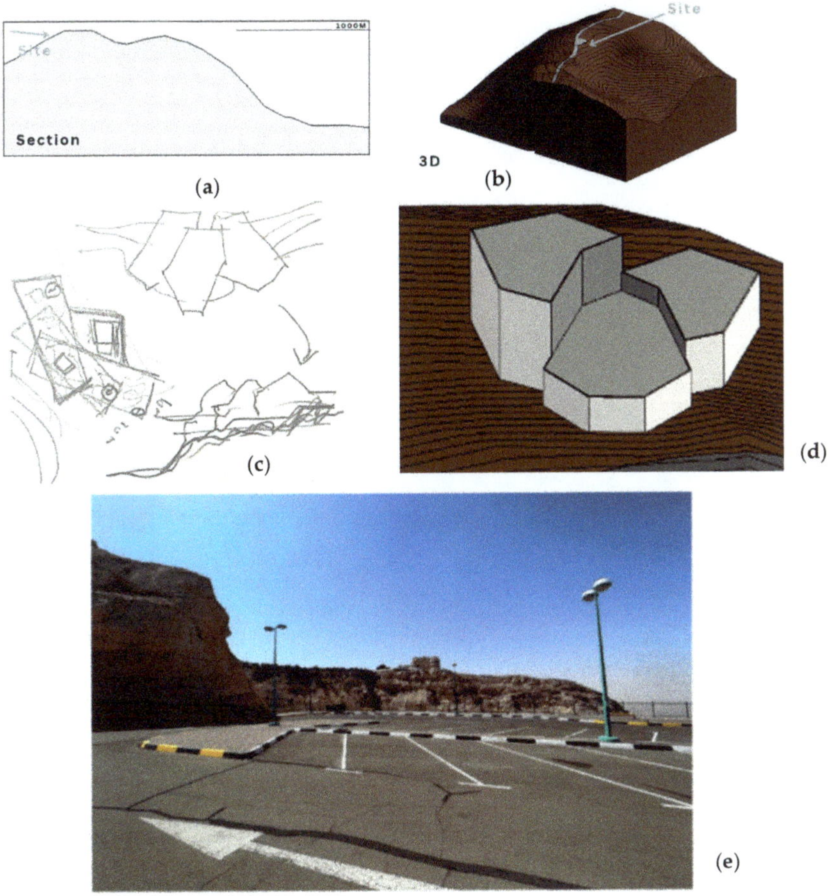

Figure 3. (**a**) Topographic location in 2D (section) and (**b**) 3D; (**c**) sketch; (**d**) conceptual 3D image of initial concept; (**e**) site picture from parking lot.

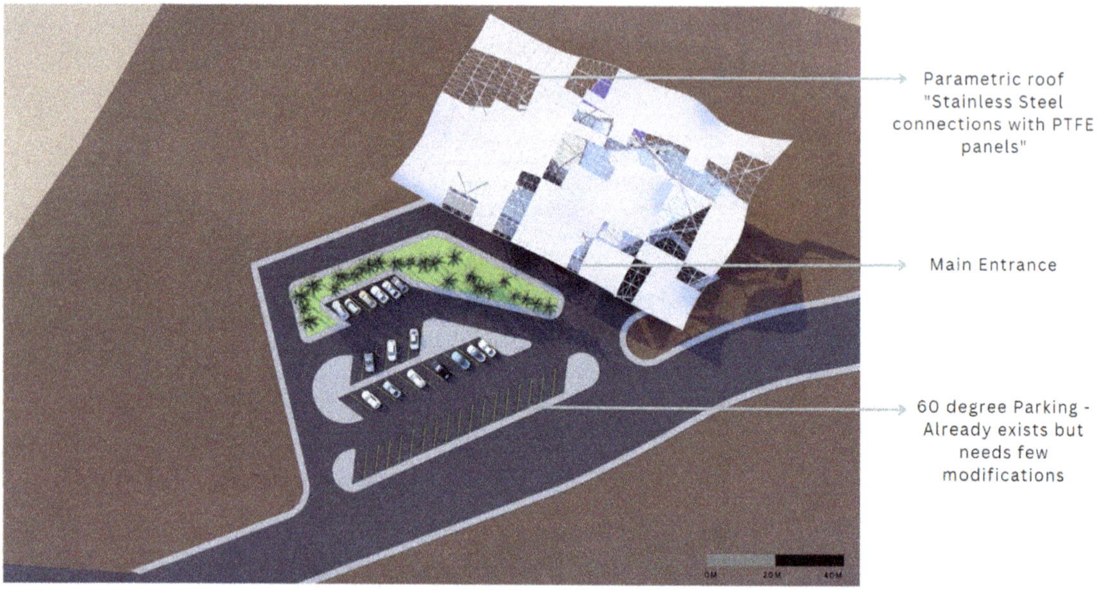

Figure 4. The site plan of the building with the integrated parametric roof.

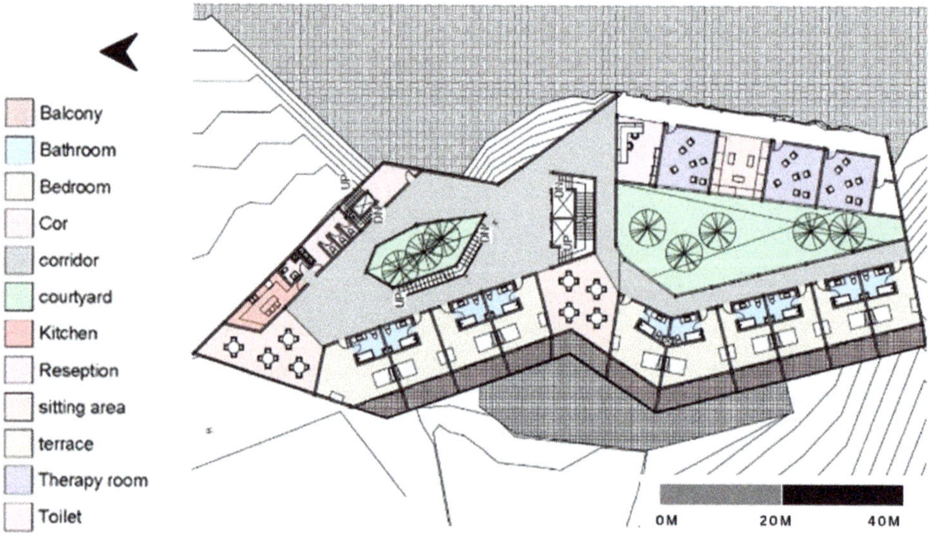

Figure 5. The functional distribution of the areas in the building for level −1.

To develop the selected concept, we started by drawing the contour lines of the site and creating sections to better understand the topography. From there, we drew the external walls of the 3 floors, following the contour lines. We then started dividing each floor into different spaces, taking into consideration the specific needs and functions of each area. Once we had a clear idea of the internal layout, we added the parametric roof and began optimizing it to fit the design. The aim is to create a building that not only offers a functional and efficient space for wellness activities, but also fits seamlessly into the natural landscape and creates a sense of tranquility and relaxation for its users. Figures 3–7 below show the concept development process. As stated before, the project is a wellness center in Jebel Hafeet, Al Ain. The center is for females only and will be called The Rose Quartz, upon the name of a quintessential crystal/stone which represents female empowerment. Rose quartz crystals can also help to boost women's inner confidence and encourage them to embrace their femininity. The center has a total area of 1200 m^2 and can occupy around 200 females per day. Visitors can have a stayover if they want to. The center is aiming to offer a place where females from all backgrounds can visit to relieve their stress.

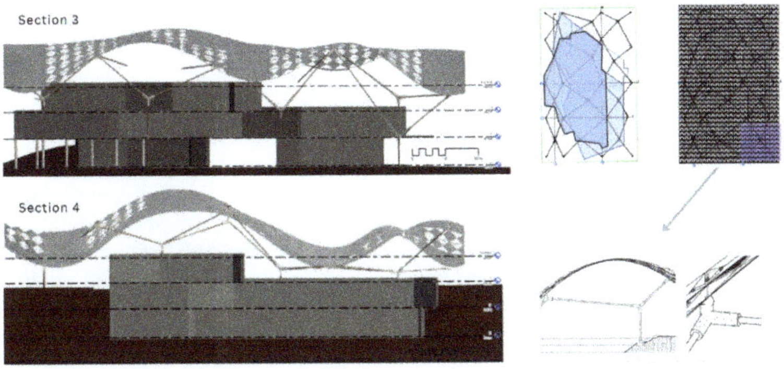

Figure 6. The Construction sections show the roof construction details.

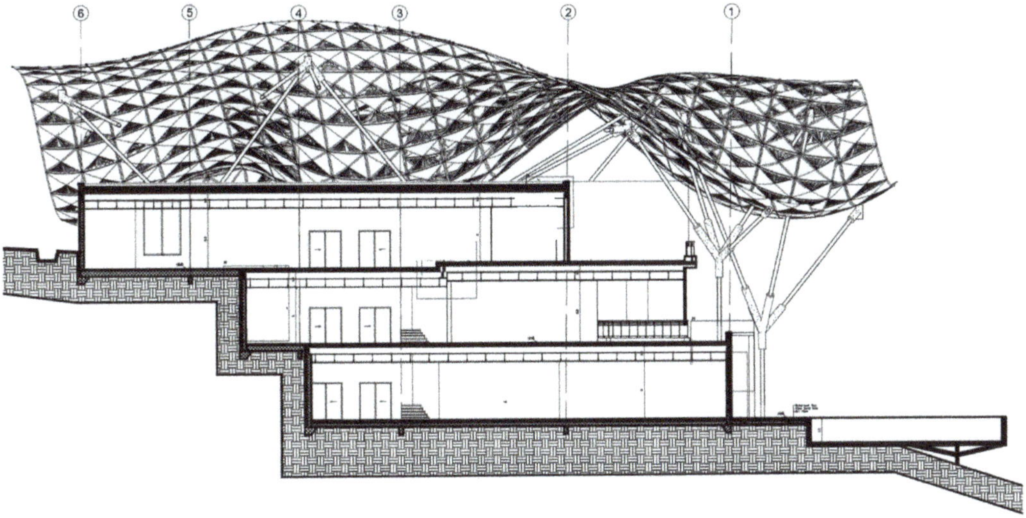

Figure 7. The construction section of the building and the parametric roof (with the aim of showing the roof integration with the current building).

2.3. Design of the Parametric Roof

The parametric roof was inspired from the red dunes around Al Ain as well as the road to the mountain. This mixture of curves makes a unique shape that inspires the design of such complex structure. The main tool used in this study is Rhinoceros. This tool enables the creation of complex shapes and analyzes in real-time the data from various scenarios. The tool has been updated over the years with several plug-ins, such as Grasshopper, Ladybug, Honeybee, Houdini, Daysim, and many more. Each shape introduced can be modified, and parameters can be added and optimized through the use of the plug-in. The scripting language is Python, and Grasshopper is the environment where the other plug-ins are connected. For energy simulations, Energy Plus is used as a plug-in. The process of connecting the scripts requires basic knowledge of the algorithm design.

2.4. Modeling and Simulations

2.4.1. Modeling of the Main Building

The main tools used in this study are Revit, Rhino, Grasshopper plug-ins, and STAAD (all 2023 versions). Each tool was used for a specific purpose. The modeling and simulations of the main building are carried out in Revit. All plans, sections, elevations, interiors, and exteriors are also developed in Revit and rendered in Lumion. Also, the energy simulation is performed in this software by bringing the selected roof scenario into the simulation [39].

2.4.2. Modeling of the Parametric Roof

The modeling of the solar radiation of the roof is conducted in Rhino/Grasshopper. A parametric roof will be used for several reasons. First, a parametric roof allows for greater flexibility and customization in terms of the shape and form of the roof. This is important for the project as we want to create a unique and visually striking building that stands out in its location. Additionally, a parametric roof can be designed to provide specific functional benefits, such as improved natural lighting and ventilation, which are important for a wellness center. Overall, the use of a parametric roof aligns with the design goal of creating a functional and esthetically pleasing wellness center that promotes wellness in all aspects. The parametric roof will be made of stainless-steel pipes welded together, holding the PTFE panels. The pipes will be supported by tree columns. The members of these columns will be connected to the pipes of the roof using steel spheres. Each sphere will be bolted to the tree column member below it. Some of the tree columns will be connected directly to the ground (and will have their own foundation pad), while others will be supported by the beams of the building. Figures 6 and 7 show the modeling and construction details of the parametric roof with the building [40].

To ensure the structural integrity and stability of the wellness center, it was necessary to evaluate the performance and sizing of all of the structural elements, including columns, beams, and sub-beams. To achieve this, we utilized the STAAD program, which is a comprehensive software tool that enables accurate analysis and design of complex structures. By inputting the building's specifications, including the dimensions, load requirements, and material properties, the program was able to generate a detailed model of the structure (shown in Figure 8) and provide us with the necessary data to make informed decisions regarding the size and placement of each element. The image is a visual representation of the structural complexity. The STAAD program is a powerful tool that allowed us to ensure that the wellness center would be safe, stable, and capable of withstanding the stresses and loads it is designed to support.

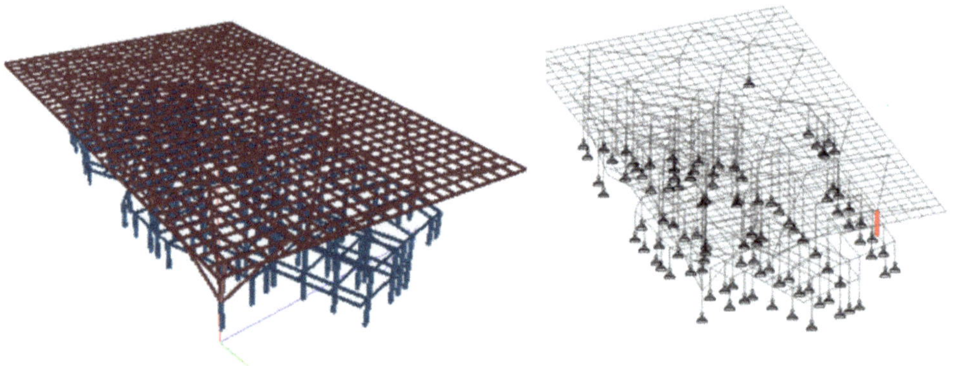

Figure 8. STAAD rendering and a sample output.

PTFE (polytetrafluoroethylene) with stainless-steel pipes is an ideal material for the parametric roof for several reasons. Firstly, PTFE is a highly durable, non-stick, and weather-resistant material that can withstand harsh weather conditions like extreme heat, UV rays, and heavy rainfall. Additionally, it is a lightweight material, which makes it easy to transport and install. The stainless-steel pipes used to hold the PTFE panels are strong, are corrosion-resistant, and can last for a long time. We will be using this material for the parametric roof by welding together stainless-steel pipes and covering them with PTFE panels to create a unique and visually stunning roof design. The roof will also be designed parametrically to optimize its performance and improve its overall efficiency.

2.4.3. Parametric Roof Optimization

For the design concept to achieve the desired design and objectives, a certain criterion must always be included. The optimal choice will be determined by a few parameters applied in Rhino and Grasshopper. The shading device must maximize the shading on the building roof, and the parametric roof must be used as shading for the building. Cross ventilation must increase the ventilation in the area between the building and the parametric roof which cools down the current roof. Natural light for the courtyard must be increased by adapting the panel sizes of the parametric roof based on the solar radiation analysis (Figure 9).

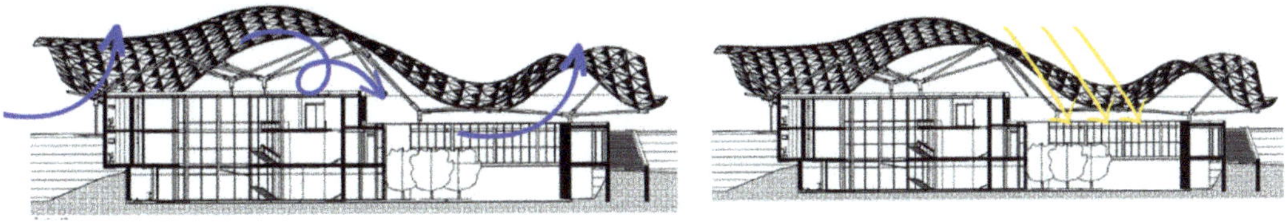

(a) (b)

Figure 9. Schematic images of (**a**) cross ventilation with the PR and (**b**) natural light with the PR.

2.4.4. Energy Simulation of the Building with and without the Roof

The energy simulations are performed in Revit. The second scenario of the roof was imported into the tool software and after adjusting the surface, the model could be simulated. In order to assess the energy efficiency of incorporating the parametric roof, we utilized Revit Insight to evaluate the energy consumption rate before implementing the parametric roof and wall insulation. We then recalculated the energy consumption rate after altering the U-values of the walls and roofs. The materials for the base case are as follows: for roofs and floors, metal decking covered with ordinary concrete which is an excellent option for our building's roofs and floors due to its durability, strength, and flexibility in design. It provides a solid base for different types of tiles, making it easy to adapt it to different spaces within the building. The metal decking will be covered with various types of tiles depending on the area it serves. For instance, the gym area will have rubber tiles to absorb shock and reduce noise, while the yoga area will have vinyl tiles that provide a non-slip surface and are easy to clean. The restaurant area will have ceramic tiles for their elegant look and easy maintenance, and the spa area will have mosaic tiles for their esthetic appeal and durability. By using metal decking, we can ensure that our building's roofs and floors will be structurally sound and esthetically pleasing. The U-values for the base case in the energy simulation are based on ASHRAE standards. Afterward, there is a second simulation that refers to the local standards of ESTIDAMA. And the third option integrates the parametric roof [41].

3. Results

3.1. Wellness Building Design

The first aim of this study was to design a sustainable wellness center by adapting the building to the mountain geography and by using a parametric roof related to the first point. Figures 10 and 11 show the outcome of this design. As it is shown in Figure 10, the internal space has natural light from the elevation and internal courtyard. The use of the courtyard improves ventilation. Additionally, the building adapts to the mountain shape, as seen in Figure 11.

Figure 10. A 3D image of the interior.

Figure 11. The elevations of the building.

3.2. Wellness Building Design with Integrated Parametric Roof

The target was to reduce energy consumption in the building by minimizing the solar radiation of the existing roof by adding a parametric roof (a comparison of various scenarios) and adapting the architecture of the parametric roof to the existing building. The solar radiation was studied in three different scenarios with different roof carvings to see which one would be applicable to the parametric roof. As shown in Figures 12 and 13, the red color shows a high level of radiation impact on the roof, while the blue color shows a lighter level of solar radiation. The second scenario (curved design) has a higher number of low radiation panels (45%) in the parametric roof itself. This analysis is performed in order to evaluate which panels can be removed for the roof to have more natural ventilation (based on the wind experienced in Al Ain). Figure 14 shows that the panels corresponding to the blue squares of direct solar radiation have a different design (with two openings out of a total of four).

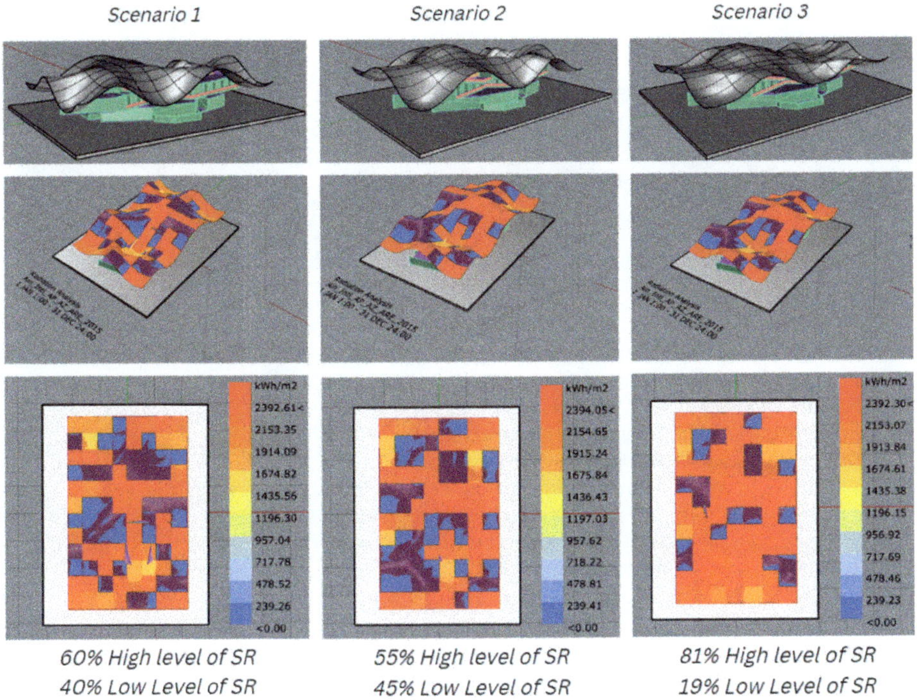

Figure 12. The solar radiation scenarios on the parametric roof.

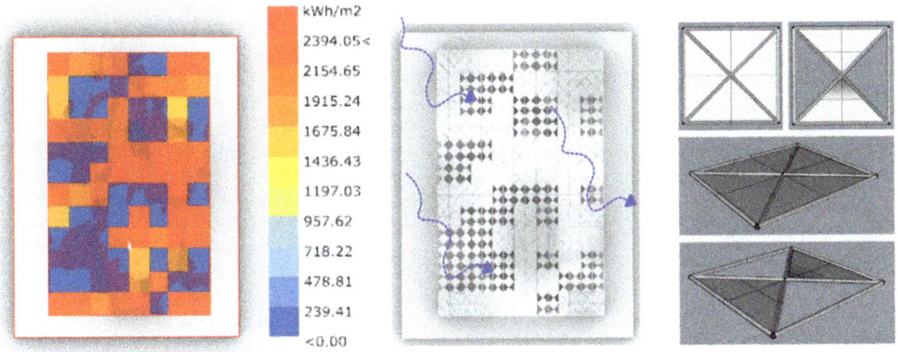

Figure 13. The solar radiation and wind study for the panels.

Figure 14. A 3D image of the building.

After studying the solar radiation and its impact on the roof, we decided to remove some panels to help eliminate the cost as well as to help the wind to go in and out of the parametric roof as we mentioned above. First, we selected the panels shown in scenario 2 with a blue color which have lower solar radiation. Then, after studying the wind wheel for the site in all seasons, "winter, spring, summer, and fall", it was found that the wind comes mostly from the northwest direction and the southeast direction.

To measure energy consumption reduction as a result of applying a parametric roof design and Estidama code, energy simulation was performed through Insight in Revit 2022. For this purpose, three models were created.

- Model A: for the base case model, the parametric roof was not applied and the ASHRAE 90.1 Standard [42] code was applied.
- Model B: in the base case, the Estidama building code was applied to the roof and exterior walls.
- Model C: a parametric roof using an optimization method was applied to model B.

First, the energy model attempted a simulation to determine changes in energy consumption patterns with and without the Estidama building code applied. For this purpose, the baseline model was created based on ASHRAE 90.1 (model A) and a proposed model applying the Estidama building code (model B). It can be seen from Table 2 that the Estidama code was applied to the roof and external walls. In order to confirm the effect of reducing energy consumption through a parametric roof with an optimized design for blocking solar radiation, the second two models were created as one with a parametrically designed roof through roof design optimization in Section 2.4.3, and one that completely excludes the roof. In this way, it is possible to confirm the impact of the roof of the proposed design, which was optimally designed with a double roof structure, on reducing energy consumption through solar radiation and cooling air circulation. Through simulations, the energy consumption of a model applying ASHRAE 90.1, a model applying the Estidama building code, and a model applying the Estidama building code and parametric roof design can be determined. For example, based on ASHRAE 90.1, an energy consumption of 304 kWh/m^2 is required, and in the case of the Estidama building code, an energy consumption of 155 kWh/m^2 is required.

Table 2. The energy consumption rates for the 3 cases.

	Base Case (ASHRAE Standard)	**Case A (Improved U Values Based on ESTIDAMA Standard)**	**Case B (Application of Parametric Roof)**
Image	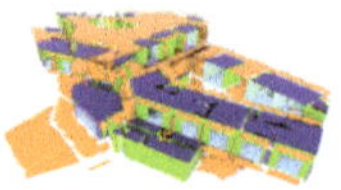	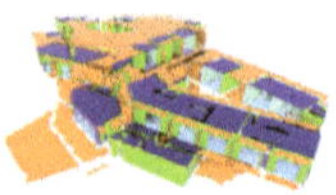	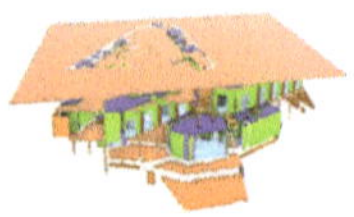
Energy Value	321 kw/m^2/year	174 kw/m^2/year	155 kw/m^2/year
Value in % compared to base case	100%	50%	34%
Value in % compared to case A	na	100%	66%

Based on a literature review, the below formula is used in REVIT for energy calculations of the models [43].

$$Q_{tot} = \sum \frac{H_{tot}\ (T_{in} - T_{out})\Delta t}{1000} \quad (1)$$

$$H_{tot} = \sum(U_{outside\ wall}\ A_{ouside\ wall}) + \sum\left(U_{roof}\ A_{roof}\right) + \sum\left(U_{ground\ floor}\ A_{ground\ floor}\right) + \sum(U_{window}\ A_{window}) + \sum(U_{door}\ A_{door}) \quad (2)$$

In order to assess the energy efficiency of incorporating the parametric roof, we utilized Revit Insight to evaluate the energy consumption rate before implementing the parametric roof and wall insulation. We then recalculated the energy consumption rate after altering the U-values of the walls and roofs. Finally, we added the parametric roof and re-evaluated the energy consumption rate. The results showed that the incorporation of the parametric roof reduced energy consumption rates by 34%, exceeding the initial goal of a 20% reduction. This significant reduction in energy consumption is a testament to the effectiveness of the design and demonstrates the potential benefits of implementing sustainable design strategies in building projects.

4. Discussion

The discussion shall be related to the two main objectives of this research paper.

4.1. Wellness Building Design

The first aim of this study is to design a socially and environmentally sustainable center. From the social perspective, adding such building typology is beneficial for the women of the city. There are noticeable cultural differences between Al Ain and other cities in the UAE such as Dubai and Abu Dhabi, and it is imperative to respect the local customs and beliefs of the community. The wellness center provides a private and secure environment that is specifically designed for women in order to promote social sustainability and provide a safe space for women. The wellness center also provides a variety of services, including massages, spa treatments, and swimming, so women can relax and unwind. Women can also benefit from a range of activities, such as yoga and meditation, to improve their physical and mental health. The wellness center's architectural form is designed to create an atmosphere of relaxation and tranquility. The interior design is carefully crafted to promote a sense of calm and peace, while the exterior design is focused on creating a welcoming and inviting atmosphere. The use of natural materials, such as steel and stone, helps to create a warm and inviting atmosphere. The center's layout is designed to provide an easy flow of movement and maximize the use of available space. In addition to enhancing the esthetics of the building, the shading device provides a comfortable lighting environment that is intended to be energy-efficient and provide a pleasant atmosphere. Its interior is spacious and bright, giving it a sense of openness to the outside and comfort. A serene and calming atmosphere is also created through the use of colors. Moreover, the wellness center provides psychological support to help women cope with the stress and anxiety associated with life's daily struggles. The center provides counseling services and a safe environment for individuals seeking support in managing their emotions and feelings. By providing these services, the wellness center strives to create a place of comfort, support, and healing for women.

However, further studies shall also be conducted related to the male population of the city. Wellbeing centers can be designed in the city center to be more accessible to all residents on a daily basis or even on an hourly basis. This is related to the social sustainable aspect.

In relation to the environment, a wellness center built on a mountain, such as Jebel Hafeet, can have a number of positive outcomes. The building adapts to the topography, thus minimizing the excavation of soil. Architecturally, the building has several levels adapting to the shape of the location. Furthermore, parts of the mountain are integrated in the interior design of the building. The application of the courtyards brings natural light to the internal spaces located on the side of the mountain.

4.2. Wellness Building Design with Integrated Parametric Roof

This section is related to the application of the parametric roof, which was the second aim of this study. This structure added to the main building reduces the cooling load by 34% by reducing the direct solar radiation. The application of the parametric roof in this building design was one of the aims related to building efficiency but also to better adaptation to the mountain topography in terms of the architectural aspect. The use of such structure was challenging in the modeling process due to many parameters being involved and the difficult location where the structure is placed. Building the structure might be challenging due to the difficult terrain. Additionally, the road of the mountain leading to the mountain peak is a two-way street. For such major works, the road can be blocked for weeks depending on the works scheduled.

Further study is needed of the local birds, such as the falcons. They have the mountain as their habitat, and therefore, understanding this bird population in relation to the various areas of the mountain could be beneficial to the application of the wellness center with the parametric roof.

4.3. Tools' Advantages and Limitations (Revit and Rhino)

To achieve the design goals and to create a building that is both energy-efficient and integrated with the surrounding environment, we have used a variety of software during the design process. The software that we have used and how they have helped us in the design process are explained in the following:

1. Revit: Revit 2023 was used to create detailed architectural plans and to perform energy simulations. This software allowed us to create 3D models of the building, which helped us to visualize the design and identify any potential problems before construction. We used the energy simulation feature to analyze the building's energy performance and to optimize the design for energy efficiency.
2. AutoCAD: AutoCAD 2023 was used to create detailed construction plans and sections. This software allowed us to create accurate drawings of the building, including floor plans, elevations, and sections. This was essential for communicating our design to the contractors and ensuring that the building was constructed correctly.
3. Rhino: Rhino 2023 was used to design the parametric roof of the building. This software allowed us to create complex shapes and to test different configurations, which helped us to optimize the design for solar shading and energy efficiency. [44]
4. Grasshopper 3D: A graphical algorithm editor plug-in for Rhino 3D 2023 is called Grasshopper. Using mathematical functions, it regulates the modeling procedures and generates both complicated and basic geometric structures. Anyone without a background in coding or programming could utilize it because it runs through coding. It is utilized for energy consumption analysis of buildings, lighting performance study, and parametric modeling. It offers excellent flexibility and precision while modeling and rendering.
5. Lumion: Lumion 2023 was used to create renders of the exterior and interior of the building and architectural elevations. This software allowed us to create high-quality images and videos of the building, which helped us to present the design to the client.
6. Staad: Staad 2023 was used to perform structural calculations. This software allowed us to analyze the building's structure and to ensure that it was safe and stable. We used it to validate the design and to make sure that it met the required standards and codes.

5. Conclusions

5.1. Wellness Building Design

The first aim of this study was to design a socially and environmentally sustainable wellness center. The wellness center was designed to meet the six dimensions of wellness and provide a sustainable and energy-efficient environment for users. The building's structural system and materials were selected based on the topography of the site, and advanced technologies were implemented in roof and façade structures to reduce the energy consumption by 34% and the solar radiation by 25%. The building construction and construction management processes associated with buildings on a mountain have been examined, and appropriate solutions have been implemented. The final design of the wellness center was developed to meet the client's demands and expectations as a whole. It is expected that the wellness center will contribute to the health and wellbeing of the community and promote a more sustainable and mindful way of life.

The wellness center design is socially sustainable as it refers to the local culture where genders are separated in this building typology. The building is designed to serve female residents only. The city of Al Ain has a different demographic from Dubai and Abu Dhabi, and therefore, designing such structure has a positive impact on the women's wellbeing. Additionally, the full structure is located on the only mountain in the city. The view of the horizon, the sunset, the nature of the building being well adapted to the geometry of the mountain, the careful selection of the materials, and the application of active and passive strategies make the project environmentally sustainable.

5.2. Wellness Building Design with Integrated Parametric Roof

The application of the parametric roof is innovative for the specific location/topography, the city, and the country. The application of this structure resulted in a reduction in the energy consumption by 34%. Based on the state-of-the-art parametric architecture, this generally relates to the façade structures and not roof design. The construction and operation of the center could result in the creation of jobs in the health and fitness industry, in the construction industry, and in other fields related to health and fitness. Tourists and visitors could be attracted to the center, which would positively impact the local economy by generating revenue from accommodation, food, and beverages. The construction of the center may result in an increase in property values in the surrounding area, which would benefit local property owners, considering the application of the advanced parametric roof as well.

The design of the parametric roof is innovative in relation to the location, city, country, and region. The parameters applied in the design of the roof include the shape of the mountain, the shape of the building, the orientation of the structure, the view, and the maximum solar energy reduction due to the roof of the wellbeing center. Such a structure can be beneficial in applications to governmental or private public buildings such as malls, hospitals, and hotels by reducing cooling costs. The innovative use of a parametric design in roof construction has ushered in a new era of architectural sophistication that directly addresses the challenges posed by solar radiation, particularly in reducing solar energy absorption by buildings. Parametric design uses advanced computer algorithms to analyze the characteristics of a roof and adjust its shape accordingly. This allows for more precise control over the amount of sunlight that is absorbed, as well as improved insulation and ventilation.

The primary advantage of a parametric roof is that it is capable of adapting complex and unusual geometries. These shapes are not merely esthetic choices but are calculated to optimize the roof's response to sunlight. For instance, certain roof shapes can cast shadows on the building's façade during the peak period of the day, thereby substantially reducing thermal gain and, consequently, the building's cooling needs.

Another aspect of parametric design is the use of a second dynamic building skin, such as a lattice truss skin, to create a nice curvy shape that will reflect the sun's rays in order to reduce solar radiation. The use of a lattice truss building skin, inspired by parametric design principles, represents a significant advance in the field of architectural strategies to reduce solar energy absorption by buildings. By using this innovative approach, it involves the creation of a secondary skin or façade that surrounds the exterior of a building and is characterized by an intricate lattice structure. A lattice truss can be meticulously optimized through the use of computational algorithms in order to control the amount of sunlight that enters the building. By adjusting the density, orientation, and configuration of the lattice, architects can precisely control solar exposure, ensuring that areas which need shade receive it, while areas which can benefit from natural light are exposed to it. The dynamic interplay between light and shadow reduces the need for artificial cooling systems by mitigating solar heat gain while also improving the esthetics of the building. Moreover, the lattice structure can be integrated with photovoltaic cells in order to capture solar energy, transforming what is typically a challenge for architects into a potential source of renewable energy. The lattice truss building skin, as an embodiment of parametric design, provides a multifaceted approach to addressing the challenge of solar energy while also being environmentally friendly.

Author Contributions: Conceptualization, M.A., N.A., L.A., T.A. and L.B.; methodology, M.A., N.A., L.A., T.A. and L.B.; validation, M.A., N.A., L.A. and T.A.; formal analysis, M.A., N.A., L.A., T.A., L.B., A.A., O.A.K., F.F. and Y.K.K.; investigation, M.A., N.A., L.A. and T.A.; resources, L.B.; data curation, L.B., A.A., O.A.K., F.F. and Y.K.K.; writing—original draft preparation, M.A., N.A., L.A. and T.A.; writing—review and editing, L.B., A.A., O.A.K., F.F. and Y.K.K.; visualization, M.A., N.A., L.A. and T.A.; supervision, L.B., O.A.K., F.F. and Y.K.K.; project administration, L.B.; funding acquisition, L.B. All authors have read and agreed to the published version of the manuscript.

Funding: This research was funded by United Arab University with Grant Code UPAR 12N116.

Institutional Review Board Statement: Not applicable.

Informed Consent Statement: Not applicable.

Data Availability Statement: The original contributions presented in the study are included in the article, further inquiries can be directed to the corresponding author.

Acknowledgments: The authors thank the United Arab Emirates university for the funds provide.

Conflicts of Interest: The authors declare no conflicts of interest.

References

1. Mahmoud, I.; Saravanan, C. Prevalence of Mental Disorders and the Use of Mental Health Services among the Adult Population in United Arab Emirates. *Asian J. Epidemiol.* **2019**, *13*, 12–19. [CrossRef]
2. 2021 UAE Wellbeing Findings | Cigna ME. Available online: https://www.cigna-me.com/en/blog/2021-uae-wellbeing-findings (accessed on 9 February 2024).
3. Business Benefits of Boosting Employee Health and Well-Being | World Economic Forum. Available online: https://www.weforum.org/agenda/2023/07/business-benefits-of-boosting-employee-health-and-well-being/ (accessed on 9 February 2024).
4. Nassar, A.K.; Alan Blackburn, G.; Duncan Whyatt, J. Developing the desert: The pace and process of urban growth in Dubai. *Comput. Environ. Urban Syst.* **2014**, *45*, 50–62. [CrossRef]
5. Dubai Land Department—DLD Launches Well-Being Certificate to Enhance Investor Trust and Support Building Sustainability. Available online: https://dubailand.gov.ae/en/news-media/dld-launches-well-being-certificate-to-enhance-investor-trust-and-support-building-sustainability/#/ (accessed on 9 February 2024).
6. Department of Urban Planning and Municipalities—Estidama Services. Available online: https://pages.dmt.gov.ae/en/Urban-Planning/Estidama-Program (accessed on 9 February 2024).
7. Alobaidi, K.A.; Rahim, A.B.A.; Mohammed, A.; Baqutayan, S. Sustainability achievement and estidama green building regulations in Abu Dhabi vision 2030. *Mediterr. J. Soc. Sci.* **2015**, *6*, 509–518. [CrossRef]
8. Taleb, H.M. Using passive cooling strategies to improve thermal performance and reduce energy consumption of residential buildings in U.A.E. buildings. *Front. Archit. Res.* **2014**, *3*, 154–165. [CrossRef]
9. Brebbia, C.A.; Beriatos, Ē. *Sustainable Development and Planning V*; WIT Press: Southampton, UK, 2011.
10. Ahmad, A.; Bande, L. AI Application in Architecture in UAE: Optimization of Parametric Structure as a Retrofit Strategy of a Mid-rise Residential Building in Downtown Abu Dhabi. In *Proceedings of the HCI International 2022—Late Breaking Papers: HCI for Today's Community and Economy*; Rauterberg, M., Fui-Hoon Nah, F., Siau, K., Krömker, H., Wei, J., Salvendy, G., Eds.; Springer: Cham, Switzerland, 2022; pp. 301–313.
11. Energy and Water Statistics Content List. 2017. Available online: https://scad.gov.ae/documents/20122/2310471/Energy2520and2520Water252020172520EN.pdf/6b2afaff-c6bb-56b1-5a1d-9dc1924e1e62?t=1684163031388 (accessed on 9 February 2024).
12. Krarti, M.; Dubey, K. Review analysis of economic and environmental benefits of improving energy efficiency for UAE building stock. *Renew. Sustain. Energy Rev.* **2018**, *82*, 14–24. [CrossRef]

13. AlNaqbi, A.; AlAwadhi, W.; Manneh, A.; Kazim, A.; Abu-Hijleh, B. Survey of the Existing Residential Buildings Stock in the UAE. *Int. J. Environ. Sci. Dev.* **2012**, *3*, 491–496. [CrossRef]
14. Shanks, K.; Nezamifar, E. Impacts of climate change on building cooling demands in the UAE. In Proceedings of the SB13 Dubai: Advancing the Green Agenda Technology, Practices and Policies, Dubai, United Arab Emirates, 8–10 December 2013.
15. Wong, S.L.; Wan, K.K.W.; Li, D.H.W.; Lam, J.C. Impact of climate change on residential building envelope cooling loads in subtropical climates. *Energy Build.* **2010**, *42*, 2098–2103. [CrossRef]
16. Maksoud, A.; Mushtaha, E.; Al-Sadoon, Z.; Sahall, H.; Toutou, A. Design of Islamic Parametric Elevation for Interior, Enclosed Corridors to Optimize Daylighting and Solar Radiation Exposure in a Desert Climate: A Case Study of the University of Sharjah, UAE. *Buildings* **2022**, *12*, 161. [CrossRef]
17. Xuereb, K. The Al Bahr Towers—Enhancing Sustainability Through Innovation. In Proceedings of the Sustainable Energy 2014: The ISE Annual Conference, Qawra, Malta, 20 March 2014; pp. 16–27.
18. Lobaccaro, G.; Wiberg, A.H.; Ceci, G.; Manni, M.; Lolli, N.; Berardi, U. Parametric design to minimize the embodied GHG emissions in a ZEB. *Energy Build.* **2018**, *167*, 106–123. [CrossRef]
19. Hudson, R. Strategies for parametric design in architecture. *Civ. Eng.* **2010**, *274*, 77–81.
20. Milena, S.; Ognen, M. Parametric Modeling for Advanced Architecture. *Int. J. Appl. Math. Inform.* **2011**, *5*, 9–16.
21. Learn More about Dynamo for Revit: Features, Functions, and News | ArchDaily. Available online: https://www.archdaily.com/945566/learn-more-about-dynamo-for-revit-features-functions-and-news (accessed on 9 February 2024).
22. Ceylan, S. Artificial Intelligence in Architecture: An Educational Perspective. In Proceedings of the 13th International Conference on Computer Supported Education (CSEDU 2021), Virtual Event, 23–25 April 2021; Volume 1, pp. 100–107. [CrossRef]
23. Vectorworks Marionette—Python Scripting. Available online: https://forum.vectorworks.net/index.php?/topic/101776-marionette-to-python-script/ (accessed on 9 February 2024).
24. Wibranek, B.; Tessmann, O. Interfacing Architecture and Artificial Intelligence—Machine Learning for Architectural Design and Fabrication. In *The Routledge Companion to Artificial Intelligence in Architecture*; Routledge: London, UK, 2021; p. 14.
25. Taleb, H.; Musleh, M.A. Applying urban parametric design optimisation processes to a hot climate: Case study of the UAE. *Sustain. Cities Soc.* **2015**, *14*, 236–253. [CrossRef]
26. Myers, B.A. Taxonomies of Visual Programming P and Rogram Visualization g Taxonomies of Visual Programmin and n Program Visualizatio. 1989. Available online: https://www.cs.cmu.edu/~bam/papers/vltax2.pdf (accessed on 9 February 2024).
27. Heidegger, M. Building-Dwelling-Thinking. In *Poetry, Language, Thought*; Hofstadter, A., Translator; Harper & Row: New York, NY, USA, 1971.
28. Downey, E. Roofs and roofing. *Constr. Build. Mater.* **1988**, *2*, 176. [CrossRef]
29. Abass, F.; Ismail, L.H.; Wahab, I.A.; Elgadi, A.A. A Review of Green Roof: Definition, History, Evolution and Functions. *IOP Conf. Ser. Mater. Sci. Eng.* **2020**, *713*, 012048. [CrossRef]
30. Fátrai, G. Historic Roof Structures. pp. 1–13. Available online: https://www.scribd.com/document/659793154/Historic-roof-structures (accessed on 9 February 2024).
31. Schwinn, T.; La Magna, R.; Reichert, S.; Waimer, F.; Knippers, J.; Menges, A. Prototyping Biomimetic Structures for Architecture. In *Prototyping Architecture: The Conference Papers*; Building Centre Trust: London, UK, 2013.
32. Dzwierzynska, J.; Prokopska, A. Pre-rationalized parametric designing of roof shells formed by repetitive modules of Catalan surfaces. *Symmetry* **2018**, *10*, 105. [CrossRef]
33. Zanelli, A.; Monticelli, C.; Viscuso, S.; Mazzola, C. Parametric design and the manufacturing process of an ultra-lightweight roof system for humanitarian relief contexts. In Proceedings of the 1st Conference Transformables, Seville, Spain, 18–20 September 2013.
34. Imbert, F.; Frost, K.S.; Fisher, A.; Witt, A.; Tourre, V.; Koren, B. *Concurrent Geometric, Structural and Environmental Design: Louvre Abu Dhabi BT—Advances in Architectural Geometry 2012*; Hesselgren, L., Sharma, S., Wallner, J., Baldassini, N., Bompas, P., Raynaud, J., Eds.; Springer: Vienna, Austria, 2013; pp. 77–90.
35. Kure, J.; Manickam, T.; Usto, K.; Clausen, K.; Chen, D.; Pugnale, A. Parametric Design and Construction Optimization of a Freeform Roof Structure. In *Computational Design Modeling*; Springer: Berlin/Heidelberg, Germany, 2011; pp. 137–144. [CrossRef]
36. Turrin, M.; Van Den Ham, E.; Kilian, A.; Sariyildiz, S. Integrated design of a large span roof: A parametric investigation on structural morphology, thermal comfort and daylight. In Proceedings of the EG-ICE 2010—17th International Workshop on Intelligent Computing in Engineering, Leuven, Belgium, 30 June–3 July·2019.
37. Alvarado, R.G.; Gottlieb, A.L.; Cendoya, P.; Salcedo, P. Parametric Development of Variable Roof Structures with Central Supports (Tulips). *Nexus Netw. J.* **2013**, *15*, 257–269. [CrossRef]
38. Al Ain Climate: Average Temperature, Weather by Month, Al Ain Weather Averages—Climate-Data.org. Available online: https://en.climate-data.org/asia/united-arab-emirates/abu-dhabi/al-ain-1210/#temperature-graph (accessed on 30 June 2022).
39. Revit Software | Get Prices & Buy Revit 2022 Directly from Autodesk Middle East. Available online: https://www.autodesk.ae/products/revit/overview (accessed on 28 August 2021).
40. Grasshopper—Algorithmic Modeling for Rhino. Available online: https://www.grasshopper3d.com/ (accessed on 12 September 2021).

41. The Pearl Rating System for Estidama Community Rating System Design & Construction | Jose Niño Peralta—Academia.edu. Available online: https://www.academia.edu/41520072/The_Pearl_Rating_System_for_Estidama_Community_Rating_System_Design_and_Construction (accessed on 8 August 2022).
42. AHSRAE 90.1. Available online: https://www.ashrae.org/technical-resources/bookstore/standard-90-1 (accessed on 9 February 2024).
43. Utkin, S. Review and Analysis of Heat Loads Calculation in Autodesk Revit MEP. Bachelor's Thesis, Oulu University of Applied Sciences, Oulu, Finland, 2012.
44. Roudsari, M.S.; Pak, M. Ladybug: A parametric environmental plugin for grasshopper to help designers create an environmentally-conscious design. In Proceedings of the 13th Conference of International Building Performance Simulation Association, Chambéry, France, 26–28 August 2013.

buildings

Article

Analysis of Energy Reduction and Energy Self-Sufficiency Improvement Effects by Applying a Bidirectional Reflectance PV Array with Integrated External Shading at a School Building

Keon Ho Lee [1] and Young-Hak Song [2,*]

[1] Department of Architectural Engineering, Donga University, Pusan 49315, Republic of Korea; lee1ncdh@dau.ac.kr
[2] Department of Architectural Engineering, ERI, Gyeongsang National University, Jinju 52828, Republic of Korea
* Correspondence: songyh@gnu.ac.kr; Tel.: +82-55-772-1756

Abstract: In South Korea, the introduction of new and renewable energy in the building sector has been promoted through various policies since the early 2000s. As a result, solar photovoltaics (PV), which are mostly applied to the rooftops of buildings, and building-integrated photovoltaics (BIPV), which are installed on the elevated surfaces of buildings, have been applied to various sites with subsidies. Renewable energy will be mandatory for all buildings from 2025. In general, the power generation efficiency of PV panels varies depending on the installation angle. According to Korean standards, the power generation efficiency is 100% for an installation on a 30° slope, 90% for a horizontal installation, and 70% for a vertical installation. This study proposes a BIPV that improves the power generation efficiency using the unique reflectance of PV panel surfaces made of glass and a bidirectional reflectance PV array. This new type of BIPV structure improves the power generation efficiency and reduces the solar heat gain coefficient (SHGC) as it protrudes over the windows, providing external shading. It is defined as bidirectional reflectance PV (BRPV), and its performance is evaluated. The effects of applying BRPVs (48 KW by 160 PV Panels with 300 W) to a school building with a fixed slope PV on the rooftop were calculated based on annual measurement results, and it was found that the energy independence rate of the building increased from 34.1 to 65.8%.

Keywords: model-based simulation; annual measurement; energy independence rate; renewable energy; photovoltaics; bidirectional reflectance PV array; external shading

1. Introduction

1.1. Research Background

Korea's energy import dependence rate is 92.8%, and 18.5% of the GDP is spent on purchasing overseas energy (2020) [1]. In addition, the construction of new thermal power plants, such as coal-fired power plants, has been avoided to achieve the Nationally Determined Contribution (NDC) agreed upon in the Paris Agreement [2]. This situation has led to high-efficiency energy production and active development of energy-reducing technologies across all industries. As a result in the building sector, renewable energy is mandatory for all buildings from 2025 [3].

In Korea, subsidies and technical support have been provided to expand the dissemination of nine renewable energy sources, including solar power, solar heat, and wind power, and three new energy sources, including fuel cells, in addition to the government's support on energy production management (2022). Table 1 shows the energy produced through new and renewable energy sources in Korea over the last decade; solar thermal, ocean, and hydropower have undergone almost no changes [4]. In contrast, energy production by solar photovoltaics (PV), wind power, geothermal, and fuel cells have continuously increased, and the dissemination of hydrothermal and integrated gasification combined cycle (IGCC) have begun relatively recently. In particular, the proportion of the power generated by solar PVs increased from 13.2% in 2011 to 56.9% in 2020.

Table 1. New and renewable energy production in Korea (unit: thousand TOE (tons of oil equivalent)).

	2011	2012	2013	2014	2015	2016	2017	2018	2019	2020
Solar thermal	27.4	26.3	27.8	28.4	28.5	28.5	28.1	27.4	26.9	26.4
Solar PV	198.0	243.0	359.0	579.0	907.0	1183.0	1672.0	2194.0	3055.0	4156.0
Wind power	185.0	192.7	242.4	241.8	283.5	355.3	462.2	525.2	570.8	671.1
Hydropower	965.4	814.9	892.2	581.2	453.8	603.2	600.7	718.8	594.5	826.3
Ocean	11.2	98.3	102.1	103.8	104.7	104.6	104.3	103.4	101.0	97.4
Geothermal	47.8	65.3	87.0	108.5	135.0	162.0	183.9	205.5	224.7	241.0
Hydrothermal	0	0	0	0	4.8	6.0	7.9	14.7	21.2	21.2
Fuel cells	63.3	82.5	122.4	199.4	230.2	241.6	313.3	376.3	487.2	750.8
IGCC	0	0	0	0	1.3	76.1	273.9	362.5	219.7	506.4
Total	1498	1523	1833	1842	2149	2760	3646	4528	5301	7297

1.2. Zero Energy Building (ZEB) Certification

In Korea, a detailed roadmap for stepwise mandatory zero-energy construction was announced in June 2019, and the legal basis for mandatory zero-energy construction certification was developed by revising the Green Building Act. Since then, a program for mandatory ZEB certification has been implemented in stages for public buildings. ZEB is defined as a green building that minimizes the energy load required by the building and minimizes energy requirements through power generation using renewable energy. From 2020, new public buildings with a total floor area of more than 1000 m^2, public buildings with a total floor area of more than 500 m^2, general buildings with a total floor area of more than 1000 m^2, and apartment buildings with more than 30 households will be included in the program. From 2030, all buildings with a total floor area of more than 500 m^2 will be included. As the realization of 100% zero energy construction in high-density urban development leads to prohibitive costs, a stepwise certification from Class 1 (100% independence rate) to Class 5 (20% independence rate) has been performed according to the ratio of new and renewable energy production to energy consumption, i.e., the energy independence rate. This system is intended to consider the market situation and introduce the concept of ZEB, which has associated cost limitations compared to technology, into the social system. The following two conditions are required for certification in addition to the above energy independence rate. First, the building energy efficiency rating must be maintained at 1++ or higher, with a primary energy consumption of less than 90 kWh/m^2/year for residential buildings and less than 140 kWh/m^2/year for non-residential buildings (Table 2). Next, a building energy management system (BEMS) or electronic meters for remote readings must be applied at six or more locations. These criteria indicate that building energy consumption must be reduced and power generation must be simultaneously increased.

Table 2. Classification of the building energy efficiency standard in Korea (kWh/m^2 year).

Classification	Residential Building	Non-residential Building
1+++	~60	~80
1++	60~90	80~140
1+	90~120	140~200

Table 2. *Cont.*

Classification	Residential Building	Non-residential Building
1	120~150	200~260
2	150~190	260~320
3	190~230	320~380
4	230~270	380~450
5	270~320	450~520
6	320~370	520~610
7	370~420	610~700

1.3. Research Purpose

As can be seen from Table 1, solar PVs have been the most widely applied new and renewable energy source in Korea. As the floor area is smaller than the total facade area for high-rise buildings, applying PV panels only on rooftops imparts limitations in increasing power generation. As it is easy to acquire solar radiation for high-rise buildings due to the large facade area and fewer surrounding obstacles, installing building-integrated photovoltaics (BIPV) on building facades effectively improves the energy independence rate. The acquisition of high solar radiation through the facade windows of high-rise buildings, however, increases the cooling energy consumption of the buildings. Therefore, it is necessary to install efficient shading for transparent windows to control the inflow of high solar radiation. However, the g-value, the solar heat gain coefficient (SHGC = τ(Transmittance) + q_i(secondary heat transfer)), significantly varies depending on the position and type of shading installed. When the g-values are 0.12 and 0.6 under the same outdoor air temperature conditions, the indoor temperatures are 30 °C and 35 °C, respectively, indicating the significance of the g-value on the indoor temperature. As the value increases, cooling energy consumption increases [5]. Therefore, achieving both an increase in solar PV power generation and a reduction in the inflow of solar radiation can be viewed as essential in improving the energy independence rate of a building.

The power generation efficiency of PV panels varies greatly depending on the installation angle. This study proposes an L-shaped BIPV, which recycles the inherent reflectance of the PV panel surface to improve the power generation efficiency due to the installation angle. Solar PVs are generally installed on the front of a window's upper part and protrude as external sunshades [6]. On the other hand, the new concept, the L-shaped solar PV installed on the upper part of a window, is referred to as a bidirectional reflectance photovoltaic (BRPV). It serves two functions: it increases solar PV power generation and blocks direct solar radiation blocking by shading. The protruding structure of BRPVs can secure additional power generation areas beyond the rooftop as it is applied to the upper part of a window and is expected to contribute to a reduction in the g-value by generating shadows, thereby imparting external shading. This study aims to quantitatively analyze the effects of an increase in power generation by new and renewable energy and a reduction in cooling energy consumption on the energy independence rate of a school building.

A school was selected for the simulation to verify the technology's effectiveness because buildings of this type usually have very short cooling and heating periods due to school vacations during the summer and winter. Therefore, the building's annual energy consumption per unit area is low. A school was selected because power generation from renewable energy significantly impacts the energy independence rate in these types of buildings.

2. Research Method

This study sought to accurately reflect the special characteristics of school buildings. Simulation data, among the building's energy usage, cooling, heating, and ventilation, apply calculated values. The lighting and hot water consumption apply actual measured values, and the power generation by PVs and BRPVs apply measured values. Simulations using this approach provide more accurate information on changes in the energy independence rate. In general, the impact of fixed external awnings on building energy consumption can be evaluated through simulations as it is an evaluation of the shading due to the geometric structure of the awning. However, no current simulation model can evaluate the power generation of reflective solar power systems, such as BPRVs. Accordingly, it was necessary to present a new evaluation methodology for BRPVs.

In order to verify the solar-radiation-blocking effect of BRPVs on the external shading, the building's cooling/heating load and thermal performance were simulated using the DesignBuilder tool. Due to issues with the appropriateness of input values for lighting and hot water supply energy in the school building, actual measurements were applied. Electricity production by PVs and BRPVs used actual measurements of the power generation of the PV array at every minute for one year starting in May 2019.

Applying this approach, three stages of simulation were possible. Case 1 (Stage 1) is a simulation of a universal school building in Korea. The simulation result reflects performance in accordance with the latest Korean building energy conservation design standards for 2022, and renewable energy and external shading were not applied. Case 2 (Stage 2) applies Case 1 standards but reflects the result of solar PVs installed at a 30-degree slope on the rooftop. It is a simulation of the most common school building design in Korea. Lastly, Case 3 (Step 3) applied Case 2 standards and reflected the result of applying the BRPV proposed in this study. As a result, the building's solar power generation capacity increases, and at the same time, the building's energy use is reduced by reducing the cooling load due to the shading effect of BRPVs. The difference in the application range for each simulation stage can be seen in Figure 1.

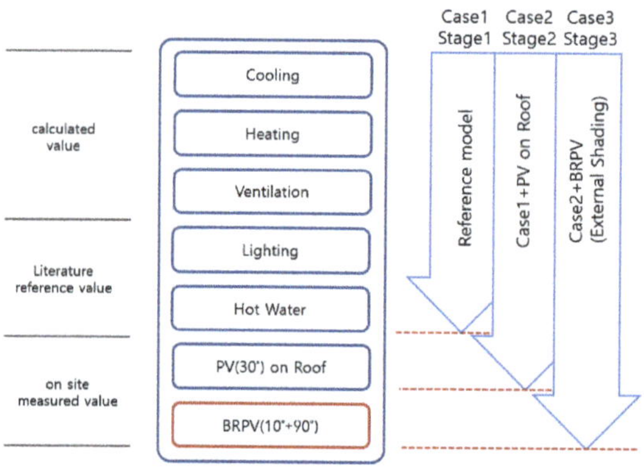

Figure 1. Simulation stage.

3. Previous Studies Related to the Improvement in Solar PV Power Generation and BRPV

3.1. Basic Knowledge Concerning Solar Incidence Angles and Power Generation

The most important parameters for the efficiency of PV modules are the solar altitude, which varies by season and time, and the PV module installation angle [7,8]. The power generation efficiency of the horizontally and vertically installed panels also differs depending on the latitude [8]. Seoul is geographically located at 37°33′ N, 126°59′ E. Thus, 100% maximum efficiency occurs at an installation angle of 30° to 35° toward the south. Based on this, 89% for horizontal installation and 63% for vertical installation efficiency is achieved. In a city where the latitude is approximately 4° lower with a similar longitude, the maximum power generation efficiency is 95% for horizontal installation and 54% for vertical installation with a 30° installation angle toward the south.

However, the underlying reason for the difference in the power generation efficiency between the two cities is the angle of incidence (qi). At the same longitude, the latitude difference between the two cities is 4°09′. Therefore, based on the optimal installation angle of 30°–35°, cities with lower latitudes have a higher solar incidence angle, which is why the efficiency of horizontal panels is higher than that of Seoul, and the efficiency of vertical panels is relatively higher in Seoul, where the solar incidence angle is lower.

Another reason is the reflectance of the panel glass surfaces. The reflectance of the glass surface increases as the qi increases. As the qi decreases, the reflectance of the glass surface decreases while the transmittance increases. In other words, the angle of incidence, qi, is highly related to the reflectance of the glass surface. Danks evaluated the reflectance of the most common 6 mm clear glass according to the angle of incidence and observed that the surface reflectance ranged from 8% to 10% for installation at 40°, over 15% at 60°, and over 50% at 80° [9]. This phenomenon also occurs on smooth PV surfaces. Therefore, as the solar incidence angle increases, the power generation efficiency of the installed PV panel decreases due to increased surface reflectance.

3.2. Effect of the Surface Reflectance According to the Angle of Incidence on Power Generation

The surface of every material has a reflectivity value for incoming light, and as the angle of incidence of light decreases, the reflectance increases. The difference in PV efficiency, depending on the installation angle, is attributed to the change in the absolute area of solar incidence according to the angle of incidence (qi) and the reflectance of the PV surface depending on the solar altitude [10]. In particular, the surface of the PV panel that absorbs 90% sunlight at 90° reflects up to 60% depending on the angle of incidence. The decrease occurs due to the decrease in the solar incidence area. This generally occurs during the sunrise and sunset hours. Therefore, the reflectance is less than 5%, and the solar incidence area reaches its maximum when the fixed panel directly faces the sun. However, this phenomenon lasts for only a few minutes. Therefore, the reflectance according to the solar incidence angle significantly affects the PV panels' efficiency.

Consequently, maximum power generation is possible when the angle of incidence (qi) is close to 0°. As the angle increases, the power generation of the PV panel rapidly decreases due to the increase in the reflectance and reduction in the incidence area. This depends on the time of day. The maximum power generation occurs at noon when the angle of incidence is close to 0°; however, the amount of power generated is small in the early morning and late afternoon hours when the angle of incidence is large. When PVs are installed on the rooftop of a building, the angle of incidence and surface azimuth can be optimized. However, in the case of BIPVs installed on facades, such as vertical building walls, it is difficult to optimize the installation angle and surface azimuth for power generation because of the geometry of the existing exterior walls.

3.3. Cases of Improving Power Generation Efficiency Using Concentrators

Glass and aluminum are commonly used for exterior curtain walls in high-rise buildings. Using the refractive index data (refractiveindex.info) that provides the reflectance characteristics of general materials, the reflectance characteristics of glass according to the angle of incidence can be identified. In the case of 5 mm single-pane glass, the reflectance is 6.3% at an incidence angle of 90° and 59.6% at an incidence of 10°, indicating a significant difference in the reflectance depending on the incidence angle. On the other hand, aluminum's reflectance is 91.2% at an incidence angle of 90° and 98.4% at an incidence of 10°, indicating very high reflectance, regardless of the angle of incidence. Absorption or reflection occurs on metallic surfaces, such as aluminum, as they are non-permeable materials, but the transmittance is relatively high for glass, as it is a permeable material.

The high reflectance of aluminum has been applied to improve the power generation efficiency of PV panels. Various reflective PV systems that maximize the power generation of PV panels have been developed by designing aluminum concentrators to be installed around the panels so sunlight can be concentrated on the PV panel. The installation of such aluminum concentrators increased the amount of power generated by approximately 20%. When a tracking function was applied together with concentrators, the power generation efficiency improved by more than 40%.

Examples of using a fixed concentrator include Jeong, who improved the output of a PV panel by approximately 20% by installing aluminum concentrators on both sides of the panel [11]. Palaskar increased the amount of power generated by more than 21% by installing aluminum concentrators on both sides of a PV panel [12], and Bahaidarah also proved that electrical power generation output could be improved by 22.8% by introducing a concentrating system similar to Palaskar [13]. Khan achieved an output improvement of 28% using aluminum concentrators on both sides of a PV panel [14,15]. When such fixed concentrators were installed, maximum output could be achieved during the daytime.

Improved results were derived when applying to track concentrators rather than with fixed concentrators. Kim introduced aluminum concentrators on both sides by integrating them with a tracking PV system that operated within 45° in the left and right directions at a slope of 15° and improved power generation efficiency by up to 42% by maintaining the concentration of light according to the position of the sun [16]. Using the concentrating PV tracking module (CPVT), Naseer improved the daily average power generation by up to 91% compared to PV installed at a slope of 30° by applying two concentrators and an x-y tracking system [17].

These new systems to increase the efficiency of PV panels using concentrators improved the actual power generation efficiency by 20% to 28% for the fixed type and by 42% to 91% for the tracking type compared to conventional PV systems; however, it is not easy to find application cases. The main causes are difficulty in standardization and a global rise in the price of raw materials. For new PV systems, it is necessary to use commercially available products, but most have unique structures or designs and thus have limitations in standardization.

Moreover, the price of PV panels reached grid parity in 19 countries in January 2014, and the price is continuously decreasing [18]. In contrast, the price of aluminum showed no significant change over the same period. In fact, aluminum prices have been continuously increased recently [19]. The additional coating on the aluminum surfaces to maintain high reflectance further increases the price of aluminum concentrators. In other words, utilizing relatively expensive aluminum concentrators to improve the efficiency of cheap PV panels is not economically feasible. Therefore, PV systems that used aluminum concentrators with high reflectance were unsuccessful in terms of commercialization.

3.4. *BRPV Overview and Characteristics*

The BRPV proposed in this study comprises two PV panels (top vertical and bottom horizontal panels), as shown in Figure 1. The angle between the two panels ranges from 90° to 100°, and the panels are installed on the spandrel in the upper part of a window on the building facade. Existing reflective PVs are comprised of a PV panel and an aluminum plate with high reflectance to improve efficiency through the concentration of the reflected light. For BRPVs with two PV panels, on the other hand, mutual reflection is generated according to the constantly changing incidence of sunlight, increasing power generation efficiency, and the horizontal panel provides external shading.

The BRPV system shows the highest power generation efficiency at a solar incidence angle of approximately 45°, the median value of the angle between two PV panels. Because the solar altitude is high in summer in Korea, located in the northern hemisphere's middle latitude, the power generation efficiency of the horizontal PV panel is higher than that of the vertical panel. In winter, the power generation efficiency of the vertical panel increases because the solar altitude decreases. Therefore, in the BRPV structure, the sunlight reflected from the horizontal panel improves the power generation efficiency of the vertical panel in summer, whereas the sunlight reflected from the vertical panel increases the power generation efficiency of the horizontal panel in winter. Additionally, the geometrical structure allows an increase in the number of PV panels. In particular, for high-rise buildings, the number of PV panels applied as BIPVs increases due to an increase in the facade area. Finally, when BRPVs are applied to the upper part of the outside of a window, the protruding horizontal panel provides external shade that reduces the cooling load caused by solar radiation in summer. Consequently, when applied to the upper part of a window facing south, as shown in Figure 2, the BRPV increases power generation by increasing the number of PV panels and decreases the cooling energy consumption by reducing the indoor cooling load.

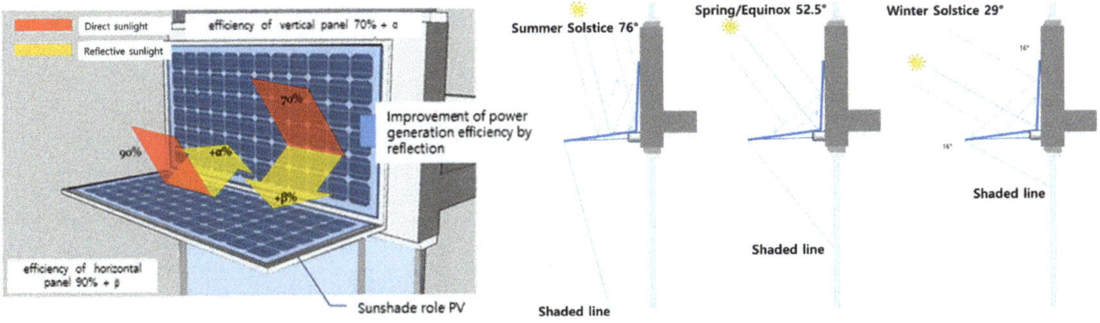

Figure 2. Conceptual diagram of BRPV power generation improvement and window shading effect.

4. Examination of Energy Consumption and Power Generation

4.1. Overview and Modeling of the Target Building

The simulation target is a middle school located in Seoul, Korea. This area corresponds to 4A in the ASHRAE climate zone classification. A 30-day summer vacation (July 27 to August 25), a 35-day winter vacation (January 2 to February 5), and a 10-day spring vacation (February 19 to 28) were reflected in the calculations. During the semester, the entire building, including the general classrooms, was measured from 09:00 to 16:00, and a load generation rate of 25% was considered from 16:00 to 18:00, considering the presence of the teaching staff. During vacation periods, it was assumed that cooling and heating were provided to approximately 20% of the building, including support facilities, such as administration offices and teachers' rooms. Figure 3 shows the school building's first floor and typical floor plans. They are large in the east–west aspect, and most classrooms are on the south side. Table 3 shows the overview of the building. The building has five above-ground floors and a total floor area of approximately 6219 m^2. The variable refrigerant flow (VRF) system was applied for cooling and heating.

Table 3. Overview of the building and HVAC.

Category			Content
Building outlines	Location		Seoul, South Korea
	Use		Middle school
	Total floor area		6219 m^2
	Building scale		Five floors above ground
	Window-to-wall ratio		30%
Skin Load	Heat transmittance (W/m^2 K)	Outer wall	0.24
		Roof	0.15
		Ground	0.29

Table 3. *Cont.*

Category		Content
Internal Heating Gain	Equipment power density	10 W/m^2
	Occupancy	0.3 person/m^2
	Light	500 lux
HVAC	VRF system Outdoor unit capacity	580 kW

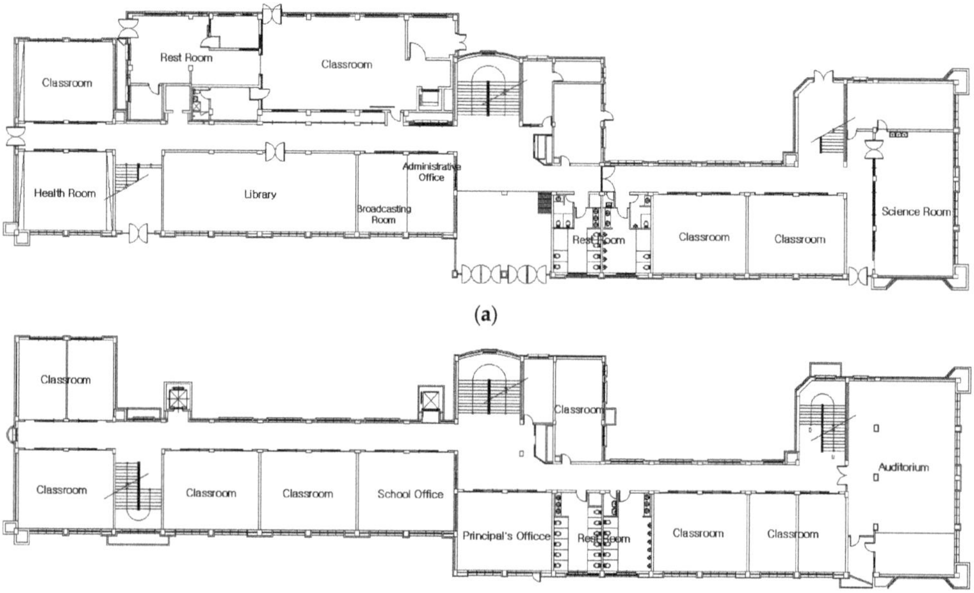

(a)

(b)

Figure 3. Floor plans of the target building. Floor plans of the first floor (**a**) and a typical floor (**b**).

Figure 4 shows the isometric plan of the target building. In the planning stage, external shading was not installed on the southern façade, as shown in Figure 4a (Cases 1 and 2). In this study, however, external shading that protruded by 1 m was added to the upper part of the windows to examine the effect of applying BRPVs (Case 3), as shown in Figure 4b.

4.2. Cooling/Heating Energy Consumption Calculation Results and Examination

Figure 5 shows the cooling and heating loads for representative weeks during the summer and winter periods, including the presence and absence of shading. The peak time cooling load without shading ranged from approximately 350 kWh to 400 kWh; however, shading reduced the load by approximately 5% to 8% by decreasing the SHGC. The heating load in the morning hours of the representative week ranged from 15 kWh to 328 kWh in the model without shading. However, shading increased the load by approximately 15% to 18% by decreasing the SHGC. Meanwhile, the cooling energy consumptions of the representative week were 3818 kWh and 3560 kWh, depending on the presence/absence of shading, indicating that shading saved 257.9 kWh (approximately 6.7%). Without shading, the heating energy consumption increased by 186.8 kWh (approximately 13.7%) from 1356 kWh to 1542 kWh.

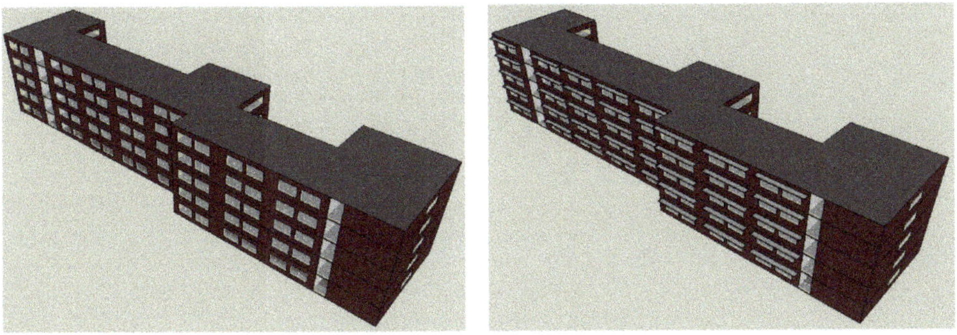

(a) (b)

Figure 4. Isometric plan of the target building. (**a**) No shading (Cases 1 and 2), (**b**) Window shading installed (Case 3).

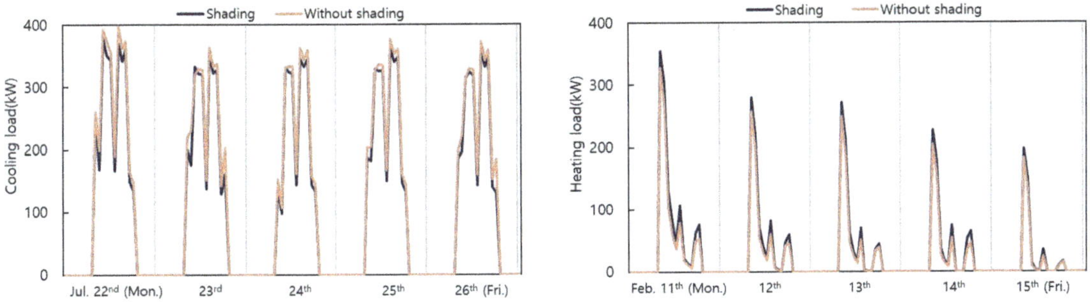

(a) (b)

Figure 5. Cooling and heating loads of the representative weeks in summer and winter. (**a**) Weekly cooling load, (**b**) Weekly heating load.

Table 4 shows the annual heating and cooling energy consumption by month, considering ventilation. The annual cooling and heating energy consumptions were found to be 65,177 kWh and 11,393 kWh without shading and 55,225 kWh and 12,762 kWh with shading due to the application of BRPV, respectively. Therefore, external shading decreased the cooling energy consumption by 15.3% but increased the heating energy by 12.0%. When the cooling, heating, and ventilation energy consumptions were combined annually, BRPVs were found to save approximately 11.2%, as the results were 76,060 kWh without shading and 67,987 kWh with shading.

Table 4. Comparison of annual cooling and heating energy consumption (kWh) depending on the presence/absence of shading.

	Cooling	Reduction	Heating	Reduction
Without shading	65,177	N/A	11,393	N/A
Shading	55,225	9952 (15.3%)	12,762	−1369 (−12.0%)

From the perspective of building energy consumption, internal shading is favorable in cold regions, and external shading is preferred in hot regions. However, in the case of warm climate zones, external shading must be applied in summer and internal shading in winter [20]. When PV panels integrated with fixed external shading are applied in Korea, which belongs to a warm climate zone but has a large amount of solar radiation, a reduction in the inflow of solar radiation leads to a decrease in cooling energy consumption but an increase in heating energy consumption. Because heating energy consumption has decreased over the last 40 years due to the reinforcement of insulation standards, an overall reduction in cooling energy consumption is assumed to be more effective.

4.3. Lighting and Hot Water Supply Energy Consumption Calculations

To examine the feasibility of a zero-energy school, this study referred to the lighting and hot water supply energy consumption of a previous study [21] that performed energy analysis modeling of a school. This study, which pursues a zero-energy school, can provide more realistic results; the building energy analysis was based on the actual energy consumption observations. The annual lighting energy consumption of the school was 5.1 kWh/m^2, and the hot water supply energy consumption was 7.7 kWh/m^2, excluding the cafeteria's hot water supply energy consumption, which has distinct characteristics as a separate facility. When these were reflected in the area of the target school building in this study, the annual lighting energy consumption was calculated to be 31,779 kWh/m^2, and the annual hot water supply energy consumption was 47,886 kWh/m^2.

4.4. Examination of Solar PV Power Generation

The installation location (37.67° N, 126.74° E) of the PVs for actual measurements was near Seoul, Korea, and the power generation was monitored after installing five PV modules on the building rooftop [22]. Each panel had a 300 W single crystal type with a power generation efficiency of 17.5% and an area of 1.65 m^2. It was confirmed in advance that the output difference of each panel was 1% or less. The power amount of the five PV modules was measured every minute by five Wattman's HPM-300A devices. Measurements were performed for one year starting in May 2019. The number of days with valid data was 234 due to the omission of measurements and equipment malfunctions, and the extraction interval of the raw data was 1 min. The five PV panels were installed facing south. The installation angle of each panel was set to 30° (PV 30), which is the most common installation angle in the mid-latitude regions, Horizontal 10° (PV 10), vertical 90° (PV 90), horizontal 10° (BRPV 10), and vertical 90° (BRPV 90) were applied to identify the change in power generation due to reflection. As shown in Figure 6, each panel was separated by more than 1 m to avoid interference. In addition, the surroundings were arranged so vegetation or other nearby obstacles could not affect power generation.

Figure 6. Actual measurement environment of solar PV.

Figure 7 shows the daily power generation in ascending order. The *y*-axis represents the sum of daily power generation, and its value is determined by the total solar radiation of the day. The daily power generation of the horizontal BRPV panel (BRPV 10) was larger than that of the conventional horizontal panel (PV 10) on most days, and it was approximately 8.6% larger on average. The same pattern was observed from the vertical BRPV panel (BRPV 90) and the conventional vertical panel (PV 90); the daily power generation of BRPV 90 was 32.8% on average. The amount of power generated by the horizontal and vertical panels constituting the BRPVs was larger because of the bidirectional reflection. This proves that the design intent was appropriate. Meanwhile, the conventional PV panel (PV 30), installed on the horizontal surface (rooftop) of a building at a fixed angle of 30° in mid-latitude regions, showed the largest power generation on cloud-free days.

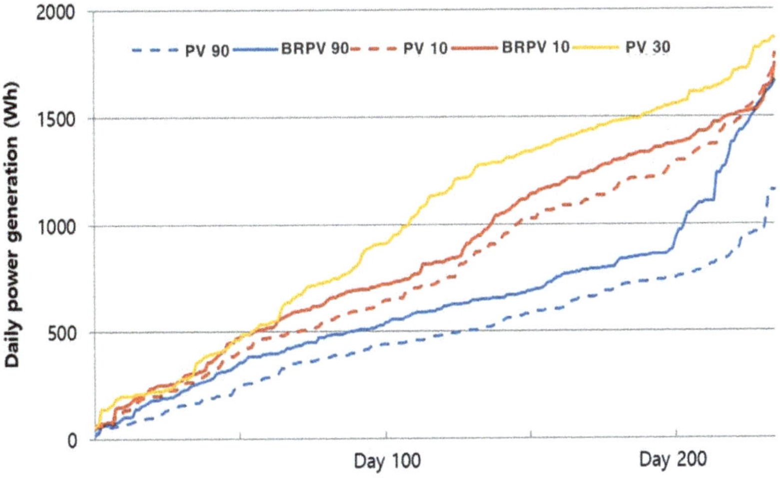

Figure 7. Daily power generation by PV array in ascending order.

Power generation during the valid measurement period of 234 days was found to be 353.6 kWh for the BRPV set (BRPV 10 + BRPV 90) and 247.8 kWh for one unit of PV 30. In addition, it is judged that correcting the power generated during the measurement period will not be significantly difficult as there is no distinction between the rainy and dry seasons in Korea. When the power generation is converted based on 365 days, the expected power generation is 549.2 kWh for one BRPV set and 384.9 kWh for one unit of PV 30.

5. Examination of the Energy Independence Rate of the Building

Based on the above building energy consumption simulation results, the effects of PV and BRPV applications on the energy consumption of the entire building were compared and analyzed, as shown in Table 5. Case 1 is the default value of the school building that reflected only the building envelope thermal performance of Korea in 2022. Case 2 is the result of installing 138 PV panels (41.4 KW) at a slope of 30° on the rooftop, while the building envelope performance of Case 1 was reflected. The number of PV panels corresponds to 20% of the rooftop area, which is reasonable considering the outdoor unit area of VRF and basic rooftop landscaping. In Case 3, 80 BRPV sets (160 panels, 48 KW) were installed on the outer wall above the windows on the south side of the building in addition to the conditions of Case 2.

First, the total building energy consumption was calculated to be 156,235 kWh for Cases 1 and 2. In Case 3, the total building energy consumption decreased by approximately 5.5% to 147,652 kWh with the reduction in cooling energy consumption and the increase in heating energy consumption due to the external shading effect of the BRPV application in Case 3.

In the simulation results, it was observed that the insulation performance of the building was similar to that of a passive house and that the cooling and heating energy consumptions were lower due to the usage characteristics of the school facility (vacations during the cooling and heating periods). Accordingly, the proportion of hot water supply and lighting energy consumption in the total energy consumption was relatively large, 51.0% in Cases 1 and 2 and 53.9% in Case 3.

Table 5. Energy consumption, power generation, and independence rate by case (kWh).

		Case 1	Case 2	Case 3	Remarks
a. Cooling		65,177	65,177	55,225	Including ventilation energy
b. Heating		11,393	11,393	12,762	
c. Hot water		47,886			
d. Lighting		31,779			
e. Total building energy consumption (a + b + c + d)		156,235	156,235	147,652	In Case 3, where the external shading effect was reflected, the total building energy consumption decreased by 5.5% compared to Cases 1 and 2.
f. Solar PV	f-1. PV 30	N/A	53,195	53,195	138 PV panels (41.4 KW) were installed on the rooftop.
	f-2. BRPV	N/A	N/A	43,933	80 BRPV sets (160 PV panels, 48.0 KW) were applied to the upper part of the windows on the south side.
g. Sum of Solar PV power generation (g − 1) + (g − 2)		N/A	53,195	97,128	
h. Energy independence rate ((g/e) × 100)		0%	34.1%	65.8%	

In addition, a significant difference was observed in the power generation improvement effect by the application of PV panels. In Case 2, the total power generation by the 138 PV panels installed at an angle of 30° on the rooftop was 53,195 kWh, and the energy independence rate (ratio of new and renewable energy production to building energy consumption) was 34.1%. In Case 3, a total power generation of 97,128 kWh was obtained as the sum of the power generated by the PV panels installed on the rooftop and 80 BRPV sets (annual power generation of 549.2 kWh per set), and the energy independence rate increased to 65.8%.

6. Conclusions

In this study, bidirectional reflectance photovoltaic (BRPV) panels with a shading effect were installed in a school building, and the impact of applying BRPVs on the annual energy consumption and power generation was examined. The main conclusions are as follows.

(1) The sum of cooling, heating, and ventilation energy consumptions was 76,060 kWh in Cases 1 and 2 with internal shading and 67,987 kWh in Case 3, indicating that the external shading of BRPVs reduced energy consumption by approximately 11.2%. The total energy consumption of Case 3 (147,652 kWh) is expected to be approximately 5.5% lower than Cases 1 and 2 (156,235 kWh). This appears to be because the proportion of hot water supply and lighting energy consumption is relatively large due to the reinforced insulation level of the envelope.

(2) The power generation of BRPVs (BRPV 10 + BRPV 90) was identified through on-site measurements. Compared with conventional horizontal and vertical panels (PV 10 + PV 90), the power generation of BRPVs was improved by 10.7% in winter, 13.4% in spring, and 10.2% in summer.

(3) For the target school building, the power generation area can be increased by approximately 75% through the installation of BRPVs on windows compared to the installation of conventional PV30 on the rooftop only. Based on this, the power generation of the BRPVs (549.2 kWh) is expected to be 143% higher than that of PV30 (384.9 kWh).

(4) When the new and renewable energy power generation rate was reflected based on the actual measurement results of solar PVs, the energy independence rate was found to be 34.1% in Case 2, where only rooftop PVs were applied, and 65.8% in Case 3 with rooftop PVs and BRPVs. This means the building's ZEB rating can be improved when BRPVs are additionally applied.

In future climate scenarios, increased electricity consumption in buildings is predicted. Therefore, increasing power generation through the application of new and renewable energy to future buildings will be key in realizing zero-energy buildings. In this study, the power generation performance of BRPV modules for maximizing new and renewable energy application areas in buildings was evaluated, and the effects on the energy consumption and energy independence rate of a school building were verified. The results are expected to be the basis for the future commercialization of BRPVs. However, in order to commercialize BRPVs, various design developments and the standardization of specifications must be secured. The PV modules used in this study were silicon-type single-crystal PV modules; however, additional experiments on the amorphous silicon type and the dye-sensitized type (Dye-TiO_2), which is mainly applied to building-integrated photovoltaics (BIPV), will be performed in the future at a newly constructed long-term monitoring experiment facility. Based on this, the BRPV structure that can be applied regardless of the PV module type will be analyzed in various forms.

Author Contributions: Conceptualization, K.H.L.; methodology, K.H.L.; experiment, K.H.L.; simulation, Y.-H.S.; validation, Y.-H.S.; formal analysis, K.H.L. and Y.-H.S.; investigation, K.H.L. and Y.-H.S.; data curation, K.H.L. and Y.-H.S.; writing—original draft preparation, K.H.L. and Y.-H.S.; writing—review and editing, K.H.L. and Y.-H.S.; visualization, K.H.L. and Y.-H.S.; supervision, K.H.L.; project administration, Y.-H.S.; funding acquisition, K.H.L. All authors have read and agreed to the published version of the manuscript.

Funding: This research was funded by the Dong-A University research fund.

Data Availability Statement: The original data are not publicly available because the experiment was funded by Korean state research fund.

Acknowledgments: The authors would like to thank Dong-A University for their financial support.

Conflicts of Interest: The authors declare no conflict of interest.

References

1. Yang, E.S. *Yearbook of Energy Statistics*; Korea Energy Economics Institute: Ulsan, Republic of Korea, December 2022; Available online: https://www.kesis.net/sub/sub_0003_eng.jsp (accessed on 18 October 2021).
2. 2030 National Greenhouse Gas Reduction Goal, Republic of Korea Government. 2021. Available online: https://www.ctis.re.kr/en/downloadBbsFile.do?atchmnflNo=7394 (accessed on 18 October 2021).
3. Carbon Neutrality Committee. 2050 Carbon Neutral Scenario, Republic of Korea. October 2021. Available online: http://webarchives.pa.go.kr/19th/www.2050cnc.go.kr/base/board/read?boardManagementNo=4&boardNo=101&searchCategory=&page=1&searchType=&searchWord=&menuLevel=2&menuNo=15 (accessed on 18 October 2021).
4. 2020 New Renewable Energy Supply Statistics, Korea Energy Agency, Center of Renewable Energy. 2021. Available online: https://www.knrec.or.kr/biz/pds/statistic/view.do?no=120 (accessed on 18 October 2021).
5. Daniels, K. *Technology in Ecological Building: The Fundamentals and Approaches, Examples and Ideas*; Birkhaeuser: Basel, Switzerland, 1997.
6. Appelbaum, J.; Aronescu, A.; Maor, T. Shading by Overhang PV collectors. *Appl. Sci.* **2019**, *9*, 4280. [CrossRef]
7. Reale, A.; Cinà, L.; Malatesta, A.; De Marco, R.; Brown, T.M.; Di Carlo, A. Estimation of Energy Production of Dye-Sensitized Solar Cell Modules for Building Integrated Photovoltaic Applications. *Energy Technol.* **2014**, *2*, 531–541. [CrossRef]
8. Kim, D.S.; Shin, U.C.; Yoon, J.H. Annual energy yield prediction of building added PV system depending on the installation angle and the location in Korea. *KIEAE J.* **2014**, *14*, 67–74. [CrossRef]
9. Danks, R.; Good, J.; Sinclair, R. Avoiding the Dreaded Death Ray: Controlling Facade Reflections Through Purposeful Design. In Proceedings of the 2016 Facade Tectonics Institute World Congress, Los Angeles, CA, USA, 10–11 October 2016.
10. Yellowhair, J.; Ho, C.K. Assessment of Photovoltaic Surface Texturing on Transmittance Effects and Glint/Glare Impacts. In Proceedings of the ASME 2015 9th International Conference on Energy Sustainability, San Diego, CA, USA, 28 June–2 July 2015.
11. Jeong, H.J.; Lee, Y.W.; Ju, S.M.; Lee, H.J.; Boo, S.J. Fabrication and Evaluation of Low Concentrator Photovoltaic Modules with Aluminium Reflectors. *Korean Soc. New Renew. Energy* **2008**, *4*, 17–22.

12. Palaskar, V.N.; Deshmukh, S.P.; Pandit, A.B. Deshmukh, Design and Performance Analysis of Reflectors Attached to Commercial PV Module. *Int. J. Renew. Energy Res.* **2014**, *4*, 240–245.
13. Bahaidarah, H.M.; Tanweer, B.; Gandhidasan, P.; Rehman, S. A Combined Optical, Thermal and Electrical Performance Study of a V-Trough PV System—Experimental and Analytical Investigations. *Energies* **2015**, *8*, 2803–2827. [CrossRef]
14. Khan, M.A.; Ko, B.; Alois Nyari, E.; Park, S.E.; Kim, H.J. Performance Evaluation of Photovoltaic Solar System with Different Cooling Methods and a Bi-Reflector PV System (BRPVS): An Experimental Study and Comparative Analysis. *Energies* **2017**, *10*, 826. [CrossRef]
15. Khan, M.A.; Zeb, K.; Uddin, W.; Sathishkumar, P.; Ali, M.U.; Hussain, S.; Ishfaq, M.; Himanshu; Subramanian, A.; Kim, H.-J. Design of a Building-Integrated Photovoltaic System with a Novel Bi-Reflector PV System (BRPVS) and Optimal Control Mechanism: An Experimental Study. *Electronics* **2018**, *7*, 119. [CrossRef]
16. Kim, Y.S.; Kang, G.H.; Sim, S.Y.; Lee, H.R.; Lee, J.S.; Hong, J.K. Demonstration Research of Photovoltaic System with Solar Reflectors. *J. Korean Sol. Energy Soc.* **2009**, *29*, 64–69.
17. Kasim, N.K.; Atwan, A.F.; Oleiwi, F.M.; Hameed, J.S. Study the Performance of V-through PV solar System with Two Axis Tracking. *AIP Conf. Proc.* **2019**, *2144*, 030008.
18. Nemet (2009); International Renewable Energy Agency (IRENA). Available online: https://ourworldindata.org/grapher/solar-pv-prices?time=earliest..latest (accessed on 10 January 2022).
19. Henry Sanderson, 9 August 2021. Available online: https://www.ft.com/content/a07c556d-2aa4-4b25-b00f-fdd42a2cf9b8 (accessed on 9 August 2021).
20. Lee, K.H. Investigation into the Possibility of Using Double Facades in Highrise Office Buildings with Glass Facades in Extreme Climate Areas. Doctoral Thesis, Technical University Berlin, Berlin, Germany, December 2002.
21. Yoon, J.H.; Shin, U.; Cho, J.I.; Park, J.W.; Kim, H.J.; Lee, C.S. Energy Evaluation of *A* Primary School Building for Zero Energy School, The Korean Solar Energy Society. In Proceedings of the KSES 2009 Spring Annual Conference, Daegu, Republic of Korea, 4 October 2009.
22. Lee, K.H.; Kim, S.J.; Jeong, J.W.; Song, Y.H. Measurement Analysis of Power Generation Change with Bidirectional Reflectance PV Array. *J. Korean Inst. Archit. Sustain. Environ. Build. Syst.* **2021**, *15*, 190–199.

buildings

MDPI

Article

Solar-Assisted Heat Pump with Electric and Thermal Storage: The Role of Appropriate Control Strategies for the Exploitation of the Solar Source

Stefania Perrella [1], Fabio Bisegna [2,*], Piero Bevilacqua [1], Daniela Cirone [1] and Roberto Bruno [1]

[1] Department of Mechanical, Energy and Management Engineering (DIMEG), University of Calabria, P. Bucci 46/C, 87036 Rende, Italy; stefania.perrella@unical.it (S.P.); piero.bevilacqua@unical.it (P.B.); daniela.cirone@unical.it (D.C.); roberto.bruno@unical.it (R.B.)

[2] Department of Astronautic, Electrical and Energy Engineering, Sapienza University, Via Delle Sette Sale 12/B, 00184 Roma, Italy

* Correspondence: fabio.bisegna@uniroma1.it

Abstract: In the EU, the building sector is responsible for 40% of the global energy consumption for final uses and 36% of the carbon dioxide (CO_2) emissions. Heat pumps allow for the replacement of conventional systems based on fossil fuels with the perspective of combining PV and solar thermal collectors. In order to rationalize the use of the solar source, this paper examined the self-consumption electricity share, the CO_2 equivalent emissions, and the domestic hot water demand covered by renewable sources which were determined in two opposite climatic conditions. These involved both electric and thermal storage systems and considered two different control strategies. The first is commonly used for the management of air-conditioning systems, the second was specifically conceived to maximize the exploitation of the solar source. Results showed that the latter significantly reduced grid dependence in both locations, determining the direct satisfaction of 76% of the thermal and electric loads through the PV self-consumption, determined by 18 kWp of installed PV and a battery capacity of 24 kWh. In terms of equivalent CO_2 emissions, when the two control strategies were compared, a remarkable reduction in emissions was registered for the latter, with percentages ranging between 8% and 36% as a function of PV surface and battery capacity. The analysis of domestic hot water supplies revealed disparities between the two localities: the colder first, relied more on heat pumps for water heating, while the warmer second, benefitted from the large availability of solar radiation.

Keywords: solar-assisted heat pump; active solar systems; thermal storage; thermal storage management

1. Introduction

The global environmental landscape has been undergoing profound changes in recent decades [1]. With the growing recognition of the adverse impacts of climate change, the need to transition towards sustainable and environmentally responsible energy sources has become increasingly urgent. As the world grapples with the consequences of greenhouse gas emissions, the pursuit of cleaner and more efficient energy systems has emerged as a central focus of research, policy, and innovation. At the heart of this transformation lies the imperative to harness renewable energy sources. Renewable energy, characterized by its sustainability and low environmental impact, offers a promising avenue for reducing carbon footprint and mitigating the effects of climate change. Solar, wind, hydro, and geothermal energy, among others, have all emerged as key players in the global effort to shift away from fossil fuels. As these sources become more accessible and cost-effective, they present unprecedented opportunities to transform the energy landscape and combat the environmental challenges.

Environmental sustainability has become a central focus, particularly in the building sector and the related energy consumption. This is evidenced by the European Union's recent "Fit for 55%" package, which set ambitious targets, including a 55% reduction in greenhouse gas emissions by 2030, compared to 1990 levels. This legislative commitment underscores the growing urgency to address climate-related challenges. Specifically, the Fit for 55% package underlines the pivotal role of HVAC systems in achieving these goals. The legislation encourages the optimization of existing HVAC systems to bolster energy efficiency [2].

Buildings represent one-third of the world's energy consumption and have the potential to significantly reduce carbon emissions [3,4]. In order to achieve this goal, "smart buildings" equipped with innovative air conditioning systems and renewable sources need to be implemented [5].

Heat pumps combined with solar energy (solar-assisted heat pumps) offer high sustainability. Recent research and field applications have made the use of heat pumps necessary as they are efficient, reliable, and environmentally friendly generators for annual air conditioning. Additionally, combining solar power with heat pumps offers a dual advantage. It maximizes the use of solar energy and increases the efficiency of the heat pump, thus optimizing and rationalizing the use of renewable sources [6,7]. The versatility in usage that allows for the provision of both heating and cooling as well as the production of domestic hot water with a single device makes heat pumps particularly suitable for the conditioning of civil buildings. The emission-mitigating potential of heat pumps is contingent upon various factors, including the specific technology employed, geographical location, and the composition of the electricity source [8,9].

In recent years, many researchers have focused on integrating heat pumps with solar collectors, photovoltaic panels, and hybrids for domestic heating. M. S. Bucker et al. and X. Wang conducted reviews on systems that combined heat pumps with solar collectors, photovoltaic panels, and hybrids. They classified these systems based on their components, the type of use, and the performance achieved in various case studies [10].

Dikici et al. conducted an experimental analysis to evaluate the performance of heat pumps at different sources for domestic heating. They subsequently analyzed the performance of multi-heat pump sources to provide a comparison between the various configurations [11]. In another study, F. Huide et al. [12] investigated the performance of an integrated system consisting of a helios-assisted heat pump with photovoltaic/thermal collectors made of heat pipes. The system was used to heat water directly or to act as a heat pump evaporator, depending on different weather conditions [12].

For solar-assisted heat pumps, thermal and electric energy storage systems are pivotal for enhancing self-consumption, narrowing the gap between energy demand peaks and troughs, and increasing the stability of the grid. Over recent years, the widespread adoption of thermal energy storage has been driven by its ability to address the dual requirements of electricity and space heating, while also ameliorating the irregularities in energy demand [13].

Storage systems contribute to enhancing the integration of renewable energy sources into the power grid while boosting system flexibility, all without compromising the security of the transmission and distribution network [14].

Aneli et al. conducted a study to assess how well an energy system, which included an electric heat pump powered by a photovoltaic generator, as well as thermal and electric energy storage, performed. The analysis results showed that systems with around 1000 L of thermal storage and 5.0 kWh of electrical storage can reach self-consumption and self-sufficiency rates of about 80%, which is three times greater than what an energy system lacking storage can achieve. Additionally, this system configuration substantially decreased grid power exchange [15].

Pinamonti et al. [16] conducted a study on a system that integrated a water-water heat pump with short- and long-term storage, using a dynamic simulation carried out with TRNSYS. The purpose of the long-term storage was to store the surplus solar energy collected by solar collectors during the summer season, which could then be utilized for space heating during the winter months. As a result, the utilization of this system increased the solar fraction by five percentage points [16].

Bellos et al., conducted an energy and cost comparison between a solar-assisted heat pump heating system utilizing flat plate collectors, an insulated tank, and a standard air source heat pump system across twenty European cities through the TRNSYS simulation tool for the entire winter season. The solar-assisted heat pump system demonstrated significant electricity savings, ranging from 30% to 40%, across various insulation scenarios. In these systems, the coefficient of performance (COP) was approximately 4, while for conventional air source systems, it was about 2.5. Notably, the solar-assisted heating system emerged as the financially optimal choice for most of the studied cities, particularly in areas characterized by low insulation levels in buildings [17].

Miglioli et al. [18] provided the state-of-the-art photovoltaic-thermal (PV/T) solar-assisted heat pumps designed for fulfilling thermal energy requirements in buildings. The primary emphasis was given to integration methodologies, potential system configurations, utilization of diverse energy sources, and the design of subsystem components. General guidelines suggested sizing the PV/T area to cover peak thermal demand during the sunniest month and the heat pump according to peak thermal load. Energy storage choices should align with project objectives, favoring water thermal storage (WTS) and advanced control logic for enhanced PV electricity self-consumption and reduced grid dependency. Integrating WTS into the solar circuit for direct user or heat pump requirements is advised for optimal solar energy utilization [18].

Furthermore, the integration of heat pumps with thermal storage systems coupled with phase change materials has also been extensively studied. Ning et al. analyzed the combination of phase change thermal storage technology with air-source heat pumps, improving the performance coefficient and stability of air-source heat pumps operating in a low-temperature environment. The findings indicated that incorporating phase change materials (PCMs) led to a substantial enhancement in heating time by up to 30.6% and an increase in COP of up to 33.9% at a temperature of 35 °C. However, at 55 °C, this inclusion may trigger an adverse effect [19]. Furthermore, the operation time of the compressor can be reduced by as much as 16.5% at 35 °C, but it increases by up to 15.7% at 55 °C. The study emphasized the importance of balancing sensible and latent heat storage. J. Yao et al. performed a numerical simulation to analyze the performance of a system that combined the use of a heat pump associated with solar/photovoltaic panels and thermal storage with PCM [20].

Instead, E. Georges et al. developed an optimization system that allowed load modulation of heat pumps installed in residential buildings to quantify their flexibility of use by adjusting consumption to electricity prices [21]. Finally, L. Kreuder et al. analyzed the combination of heat pump use with demand side management (DSM) strategies [22].

While previous research primarily focused on the effectiveness of the technologies employed, this work aimed to explore the simultaneous impact of combined electrical and thermal storage systems with a control logic aimed at maximizing the utilization of excess renewable energy on the overall performance of solar-assisted heat pump systems. In order to overcome this gap, this paper focused on the possibility of combining an air-to-water heat pump with other renewable sources to increase the use of renewable energy and self-consumption. The proposed system included an air-to-water heat pump, a field of photovoltaic panels with electrical storage, a thermal solar collector, and an insulated tank as thermal storage. The domestic hot water (DHW) is also provided by the tank supplied via thermal solar collectors working in forced convection. The paper's aim was the development of suitable control strategies targeted to optimize the employment of the free solar source. In particular, two different control logics were compared: in the first, the heat pump kept the tank at the set-point temperature, regardless of the presence of renewable energy, whereas the second allowed it to surpass the set point even when the demand for renewable electricity was exceeded in order to store thermal energy in tanks via an appropriate management of renewable surpluses. Simulations were conducted, in the TRNSYS environment for two distinct Italian cities, Milan and Messina, with opposite climatic conditions (heating dominates the first, cooling dominates the second). This analysis also aimed to underscore any potential drawbacks associated with the proposed solution. To quantify the performances of the proposed system, efficiency was chosen as a reference parameter and determined through the analysis of:

- Self-consumption and electricity supplied by the national grid.
- CO_2 equivalent emissions.
- Domestic hot water (DHW) supplied by solar collectors.

2. Methodology

In this section, the reference building-plant system and numerical setup are presented. The main characteristics of the air-to-water heat pump, PV generator, thermal solar collectors, and thermal and electrical storage are introduced. Successively, the implemented control strategies are described. Finally, the method for determining performance parameters is outlined.

2.1. Case Study Building

The case study building was a one-story residential structure with a total area of 200 m^2. It was comprised only of thermal zones and had an inter-floor height of 2.7 m (Figure 1). The building envelope was designed to meet national energy efficiency regulations for buildings currently mandatory in Italy for new constructions [23]. The U-value of the external walls was 0.34 W/m^{2}K, with the stratigraphy outlined in Table 1.

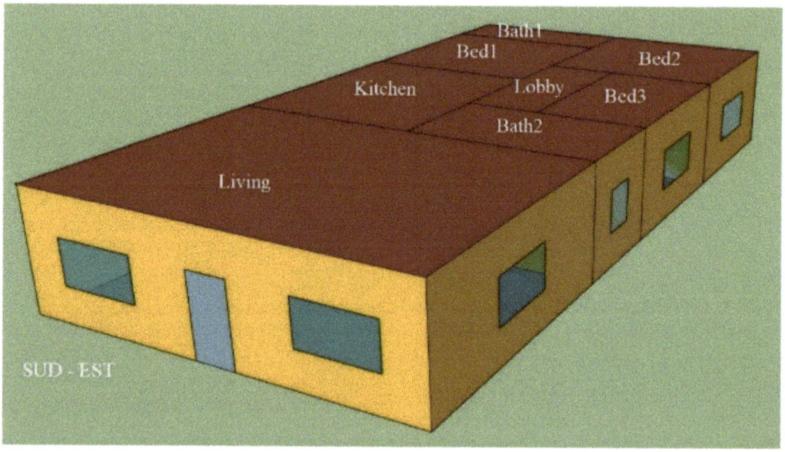

Figure 1. 3D view of the case study building.

Table 1. Thermal properties of the external wall layers.

		External Vertical Wall (U = 0.34 W/m^{2}K)		
Material	**Thickness [m]**	**Density [kg/m^3]**	**Specific Heat [kJ/kg K]**	**Thermal Conductivity [W/mK]**
Plaster	0.02	900	1.00	0.250
Bricks	0.33	1800	0.84	0.720
Expanded Polystyrene	0.09	20	1.20	0.041
Plaster	0.02	900	1.00	0.250

The ground floor and roof slabs had a thermal transmittance of 0.277 W/m^{2}K and 0.225 W/m^{2}K respectively. The layers are listed in Tables 2 and 3.

Table 2. Thermal properties of the ground floor slabs.

Ground Floor Slabs				
Material	**Thickness [m]**	**Density [kg/m³]**	**Specific Heat [kJ/kg K]**	**Thermal Conductivity [W/mK]**
Tiles	0.010	2.3×10^3	0.84	0.720
Adhesive	0.010	1.53×10^3	1.00	0.720
Sand	0.050	1.73×10^3	0.84	1.400
Expanded polystyrene	0.088	20	1.20	0.041
Bitumen	0.005	1.23×10^3	1.00	0.170
Concrete	0.084	1.93×10^3	0.84	1.060
Bricks	0.330	1.3×10^3	0.84	0.300
Plaster	0.010	2.33×10^3	0.84	0.720

Table 3. Thermal properties of the Roof slabs.

Roof Slabs				
Material	**Thickness [m]**	**Density [kg/m³]**	**Specific Heat [kJ/kg K]**	**Thermal Conductivity [W/mK]**
Plaster	0.015	1.83×10^3	1.00	0.700
Concrete and bricks	0.300	1.3×10^3	0.84	0.300
Concrete	0.084	1.93×10^3	0.84	1.060
Bitumen	0.010	1.23×10^3	1.00	0.170
Expanded polystyrene	0.080	203×10^3	1.20	0.041
Sand	0.050	1.73×10^3	0.84	1.400
Adhesive	0.330	13×10^3	0.84	0.300
Plaster	0.010	2.33×10^3	0.84	0.720

The windows consisted of a double-pane 4/15/4 configuration with an argon fill and a thermal transmittance of 2.2 W/m^2K. They had a frame-to-glazed surface ratio of 0.15. The building was intended for occupancy by 5 individuals, with a designated sensible heat rate of 75 W per person. An infiltration rate of 0.28 air changes per hour (h^{-1}) was specified. The winter set-point temperature was established at 21 °C, and in summer, it was set to 26 °C. The occupancy profile followed the schedule depicted in Figure 2.

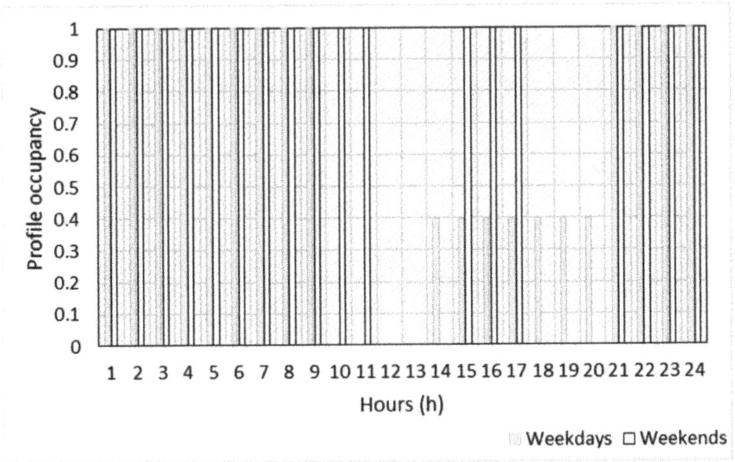

Figure 2. Occupancy profiles for weekdays (on the left) and weekends.

Initially, the building was assumed to be located in Milan, characterized by a continental climate classified as subtype 'CFb' (marine west coast climate) according to [19]. Successively, simulations were also conducted in Messina, which features a hot Mediterranean/dry-summer subtropical climate as per the same classification.

2.2. *Electrical and DHW Loads Profiles*

The DHW usage pattern was established based on data collected from residential buildings. Separate profiles were created for weekdays and weekends, and these profiles were used throughout the year. This choice was made as no significant variations were observed in the measured data. The DHW profile, represented in terms of the required thermal energy, is depicted in Figure 3.

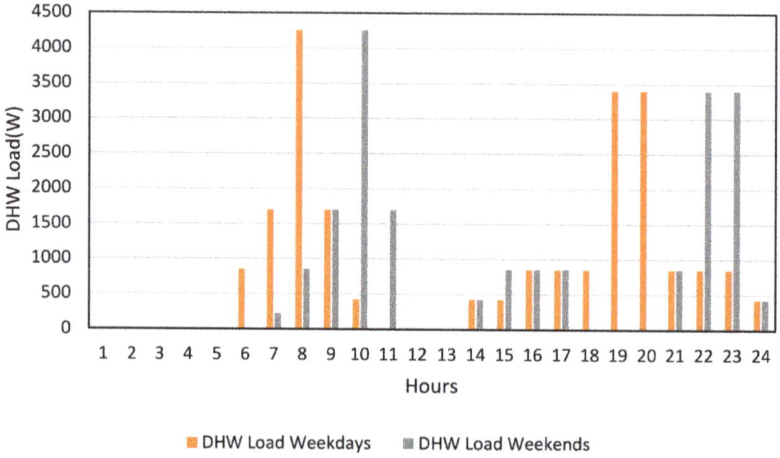

Figure 3. Profile of requested thermal power for DHW applications for weekday and weekends.

Similarly, measurements taken on the primary appliances in residential buildings were used to establish an appropriate schedule for the building's electric energy demand. Monitored appliances included the refrigerator, washing machine, dishwasher, phone, iron, personal computers, and television. Data analysis led to the definition of the profiles depicted in Figure 4, which were categorized for weekdays and weekends.

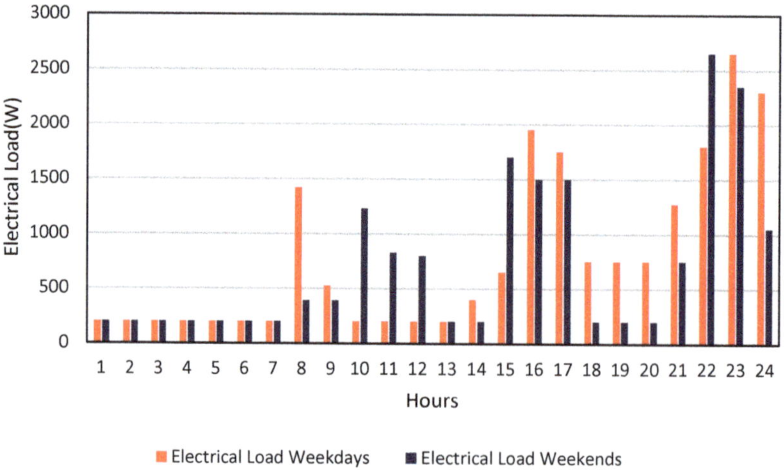

Figure 4. Profile of requested electric power on weekdays (left) and weekends (right).

2.3. Simulation Model and HVAC Plant

In accordance with recent European directives on energy efficiency, the HVAC system used in the case study relied on an air-to-water heat pump. The plant above was implemented in the TRNSYS 18 environment (Figure 5). TRNSYS 18 is widely recognized as a reliable and versatile tool for energetically simulating complex systems. Its capability to intricately model both the components of the system and their interactions has been crucial in achieving a precise evaluation of the dynamics of the studied system. Additionally, this program incorporates comprehensive libraries of system and technology components, encompassing various types of solar systems, buildings, and heating-cooling loads.

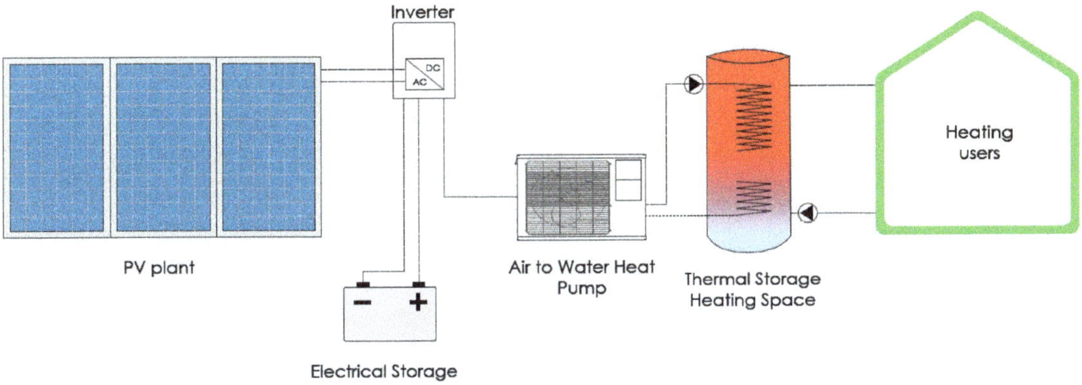

Figure 5. Scheme of the simulated plant in TRNSYS 18.

The construction of the model involved the following work-flow (Figure 6): construction of the 3D model, a thermophysical definition of the reference building and finally, the mathematical definition. The building's 3D model was created using Sketchup, incorporating the TRNSYS 3D plug-in to facilitate a comprehensive representation of its geometry. TRNBuild establishes the required characteristics of the building envelope and its regime, encompassing radiation mode, infiltration and ventilation, the number of zones/air nodes, as well as the thermal and moisture capacitance of the zone/air node. TRNBuild generates an information file that describes the outputs and inputs required by the Type 56 multizone building that in TRNSYS 18, is completely dedicated to modeling the behavior of even complex buildings. The plant, on the other hand, was primarily modeled using the following types: "type 941" for the air-to-water heat pump, "type 156" for the thermal storage tank, "type 987" to simulate the behavior of the emission system, "type 103b" for the photovoltaic system, "type 47a" for the electric storage battery, and "type 48b" employed to regulate energy flows.

A preliminary building simulation determined the necessary heating capacity of the heat pump, which was then set to 11 kW in set-point conditions for the colder locality.

The heat pump provided direct heating to the thermal storage tank. The key features of the thermal storage are detailed in Table 4. The tank was equipped with an immersed heat exchanger in the form of a coiled tube. The fluid in the storage tank interacted with the fluid in the heat exchanger (through heat transfer with the immersed heat exchanger), the environment (through thermal losses from the top, bottom, and edges), and up to two flow streams that passed into and out of the storage tank. The tank was divided into isothermal temperature nodes to model water temperature stratification. Each constant-volume node was assumed to be isothermal and interacted thermally with the nodes above and below through fluid conduction and movement.

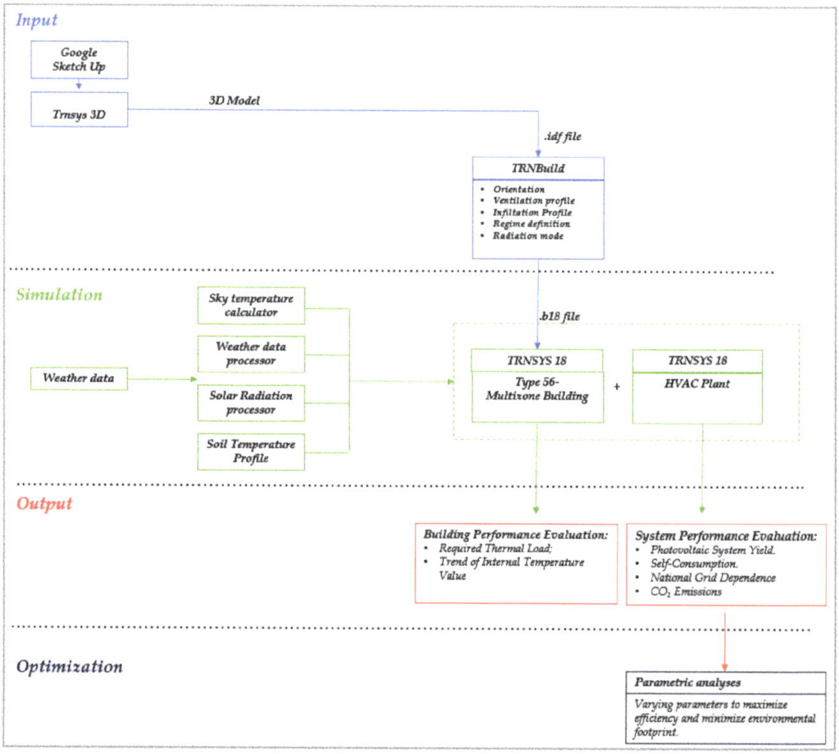

Figure 6. Structure of the simulation procedure.

Table 4. Main characteristics of thermal storage.

	Value
Tank height	1 m
Number of nodes	5
Thermal loss coefficient	0.924 W/m^{2}K
Height of inlet 1	0.8 m
Height of outlet 1	0.2 m
Height of inlet 2	0.2 m
Height of outlet 2	0.8 m
Height of HX inlet	0.5 m
Height of HX outlet	0.2 m

An on/off differential controller governed the operation of the heat pump based on the difference between two temperature thresholds (Th and Tl) relative to two hysteresis temperature differences (DTh and DTl). The controller incorporated hysteresis as its control function relied on the previous input function value. For winter operation, the controller permitted the heat pump operation between 50 °C and 55 °C at the outlet (first control logic). Additionally, an upper limit condition of 80 °C was in place to prevent the control function from activating if the maximum temperature was exceeded. The building's conditioning was performed by a fan coil heating or cooling the air as it passed over hot or cold liquid flow streams withdrawn from the thermal storage, allowing the outlet flowrate generated via the fan coil to be sent to the building for heating or cooling. Appropriate input variables, such as inlet temperature, humidity, and flowrate, were defined in the building type, where a maximum outlet temperature of 50 °C was allowed to account for real working conditions. It is achieved by mixing the outlet flowrate from the storage with a fraction of the fain coil returning water after passing through a tempering valve. In summer, the thermal storage volume stores chilled water for cooling services, so it cannot be used for DHW. A thermal storage of 300 L was used for DHW with solar collectors working in a forced convection. The solar collectors had an intercept efficiency of 0.8, a first-order loss coefficient of 3.61 W/m^2K, and a second-order loss coefficient of 0.139 W/m^2K^2. The circulation pump was activated when the temperature difference between the outlet of the solar collector and the storage was less than 20 °C and stopped when the difference reached 10 °C. The water withdrawn from the solar storage was then mixed with a colder flowrate from the supply network to achieve a set point temperature of 45 °C for domestic applications. The necessary area for the solar collectors was determined through a preliminary simulation to be 8 m^2 to fully meet the defined DHW requirements in Figure 2 for the summer period. In winter, when solar radiation was not sufficient to meet the DHW requirements, the additional energy needed was extracted from the thermal storage connected to the heat pump, giving priority to the solar collectors. To increase efficiency and the share of renewables, a grid-connected photovoltaic plant was included to provide part of the necessary electricity. The PV modules used were made of mono-crystalline silicon type and the main electric characteristics are reported in Table 5.

Table 5. Main characteristics of the PV modules.

	Value
Short-circuit current at ref. Isc	8.76 A
Open-circuit voltage at ref. Voc	37.9 V
Voltage at maximum power	31.0 V
Current at maximum power	8.07 A
Temperature coefficient of Isc	0.053 A/K
Temperature coefficient of Voc	−0.485 V/K
Number of cells in series	60
NOCT	47.5 °C
Area	1.64 m^2

The size of the PV plant was left as a parameter for the parametric analysis shown in the following section. Since there is not always synchronization between the required electric loads and the availability of solar radiation, the generator was fitted with an electric storage system—a traditional lead-acid storage battery—in order to better manage the surplus of PV power output. The simulated battery worked in conjunction with a solar cell array and power conditioning components, with the capacity of the battery being a variable in the parametric analysis discussed below. In DC source power systems—such as photovoltaic arrays—two power conditioning devices are necessary. The first is a regulator, which distributes DC power from the source to and from a battery (in energy storage systems) and the second component, an inverter. If the battery is fully charged or only requires a taper charge, excess power is either discharged or not collected by shutting off parts of the source. The inverter converts the DC power to AC, sending it to the load and/or sending it back to the utility. Type 48 simulates a power regulator and inverter which can be used with a PV array and a battery.

2.4. Management Logic

The paper investigated the influence of controllers on the rational use of renewable energy in air-conditioning systems with the aim of improving energy efficiency. Two main control management strategies were proposed. The first strategy activated the heat pump based on the temperature levels monitored on its outlet and the thermal storage, in order to maintain the tank at the desired set-point temperature, irrespective of the availability of renewable energy. This approach is currently the most commonly used logic in air-conditioning systems, although it gives little attention to the optimization of renewable sources. The second strategy, referred to as smart control, allowed for the tank temperature to exceed the set point, as the heat pump was activated even when the renewable electricity exceeded the demand. This meant that the tank could function as an electric storage unit for any PV surpluses in terms of hot/cold water produced through the heat pump. In other terms, PV surpluses were transformed into heating/cooling loads by the heat pump and stored in the tank for deferred uses. In this case, the combination of a heat pump and tank is considered an electrical storage system for managing excess photovoltaic electricity, therefore it can be exploited also when heating or cooling loads are not required. If the storage tank is saturated, thermal energy can be stored in the building structure by modifying the internal set-point values with strategies such as overheating in winter or undercooling in summer. To avoid eventual discomfort, indoor temperatures cannot exceed 22 °C in winter and 25 °C in summer. Thus, the implemented smart control strategy has the potential to optimize the use of renewable energy and improve energy efficiency in air-conditioning systems. In this paper, only the system's performance during the winter season was analyzed.

2.5. CO_2 Equivalent Emission

The equivalent emissions were calculated through the factors listed in Table 6 as a function of the energy carrier [24]. The calculation process involved multiplying the respective emission factor for each energy carrier by its corresponding energy consumption in kilowatt-hours.

Table 6. Emission factors for each energy carrier.

Energy Carrier	$kgCO_{2eq}/kWh$
Natural gas	0.21
Electricity energy from the grid	0.46
Electricity from PV	0.00

3. Results

A study was conducted to identify the optimal configuration for each management strategy using parametric analysis. Various parameters were tested, including thermal storage volume, number of photovoltaic panels, and battery capacity. Photovoltaic system simulations were conducted using 18, 36, and 72 panels for peak powers of 6, 12, and 24 kW, respectively. Electrical storage was tested with two different capacities, 1.2×10^4 Wh and 2.4×10^4 Wh. Tank volume was tested with 0.5 m^3 and 1 m^3 capacities. The results showed that energy efficiency was not significantly affected by tank volume variation and hence not reported in the text. Therefore, only the results of the other two analyzed parameters were reported and referred to a tank storage volume of 1 m^3. The efficiency of the system was measured by determining the self-consumption of photovoltaic electricity, the percentage of electricity supplied by the grid and CO_2 emissions. The study also examined the percentage of DHW supplied by solar collectors.

3.1. Calculation of the Heating Loads of the Reference Building in the Considered Climate

In Figure 7, the thermal loads for the building are illustrated, with Milan depicted in red and Messina in orange.

The monthly load data presented represents the dynamic simulation results obtained through TRNSYS for a building based on actual weather data provided by a typical meteorological year (TMY) that is a collation of selected weather data for a specific location, providing hourly records of solar radiation and meteorological elements over a one-year period. These values were derived from a database with a duration significantly longer than a year, typically spanning at least 12 years. The selection was specifically tailored to encompass the full range of weather phenomena for the given location, while still providing annual averages that aligned with the long-term averages for that particular geographic area.

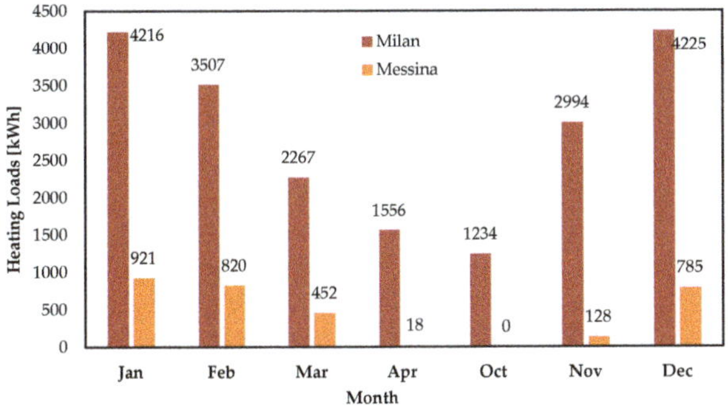

Figure 7. Heating Loads for Milan and Messina.

The load data for the building in Milan reveals significant variations throughout the year. Notably, the highest heating demands were observed during the winter months (January, February, and December), with substantial values reaching up to 920.70 kWh. These load patterns aligned with Milan's continental climate, characterized by cold winters and warm summers. In contrast, the load data for the building in Messina reflected a substantially different climatic profile. The heating demands in Messina were considerably lower than in Milan, with minimal winter loads that almost reached zero. This discrepancy was due to Messina's Mediterranean climate, featuring mild winters and hot summers.

The graphs in Figure 8 depict meteorological variations between Milan and Messina, presenting a comparison of air temperature (°C) and solar radiation on the horizontal plane (kJ/h·m^2). These data provide insights into the availability of solar energy in their respective regions.

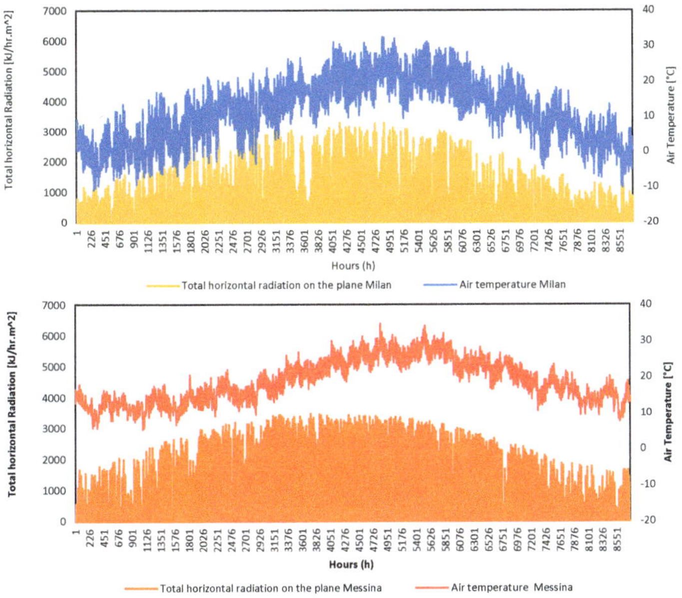

Figure 8. Weather conditions in selected localities: comparison of air temperature and total horizontal radiation on the plane in Milan and Messina.

This analysis underscores the importance of tailoring HVAC systems to the specific climatic conditions of each location. In Milan, the focus should be on effective heating solutions for winter, while in Messina, a strong emphasis on cooling systems is essential to address the summer heat. These insights were crucial for optimizing energy efficiency and ensuring occupant comfort in both regions.

3.2. *Self-Consumption and Electricity Supplied by the National Grid*

This analysis investigated the impact of varying peak installed power (PV panels) and battery capacity on the system's performance, with a focus on the heating season (from October to April). The key parameters evaluated include PV generation, electrical storage, electricity from the grid, PV self-consumption, and building power consumption. Six scenarios were considered: Scenario 1: 18 PV panels and 1.2×10^4 Wh battery capacity. Scenario 2: 36 PV panels and 1.2×10^4 Wh battery capacity. Scenario 3: 72 PV panels and 1.2×10^4 Wh battery capacity. Scenario 4: 18 PV panels and 2.4×10^4 Wh battery capacity. Scenario 5: 36 PV panels and 2.4×10^4 Wh battery capacity. Scenario 6: 72 PV panels and 2.4×10^4 Wh battery capacity.

Simulations were conducted for all the scenarios listed above, carefully examining both the application of the first and the second control logic. In these simulations, the behavior of the systems under different conditions was analyzed, considering both traditional management, based on the first logic, and the more advanced approach of the second control logic. This allowed us to evaluate how each system performed in terms of energy demand coverage, self-consumption, and the use of renewable sources in various scenarios and environmental conditions.

The following charts depict the percentage distribution of the electrical load provided by each component: the contribution from the direct photovoltaic system is shown in green, that of the electrical storage in light green, and finally, the national electrical grid in red.

3.2.1. Milan

From the graphs depicted in Figure 9 (from Scenario 1 to Scenario 3), it can be seen that as the peak photovoltaic power installed increased, the percentage of energy demand coverage directly from the photovoltaic system and battery increased, consequently the dependence on the national grid decreased. In particular, the maximum percentage of dependence on the grid, recorded in December with 18 panels, decreased to 76% in December with 72 panels installed. By doubling the battery storage capacity (from Scenario 4 to Scenario 6), the use of photovoltaics and batteries increased proportionally to the number of photovoltaic panels. It is particularly noticeable that the benefits were lower in colder months and higher in April, October, and November. In October, with 72 panels and a battery capacity of 2.4×10^4 Wh, almost complete energy self-sufficiency was achieved compared to the national grid.

The diagrams in Figure 10 (from Scenario 1 to Scenario 3) show the percentage of load managed by each component through the implementation of the second control logic, while the number of photovoltaic panels increased and a 1.2×10^4 Wh storage battery was used. Comparing it with the graph in Figure 9, it can be noticed that for 18 panels, the situation remained essentially unchanged. However, the situation was different for the configuration with 36 and 72 panels. In the latter case, it was evident that the storage of excess photovoltaic energy as thermal energy can halve the usage of the national grid, especially in March, April, and October.

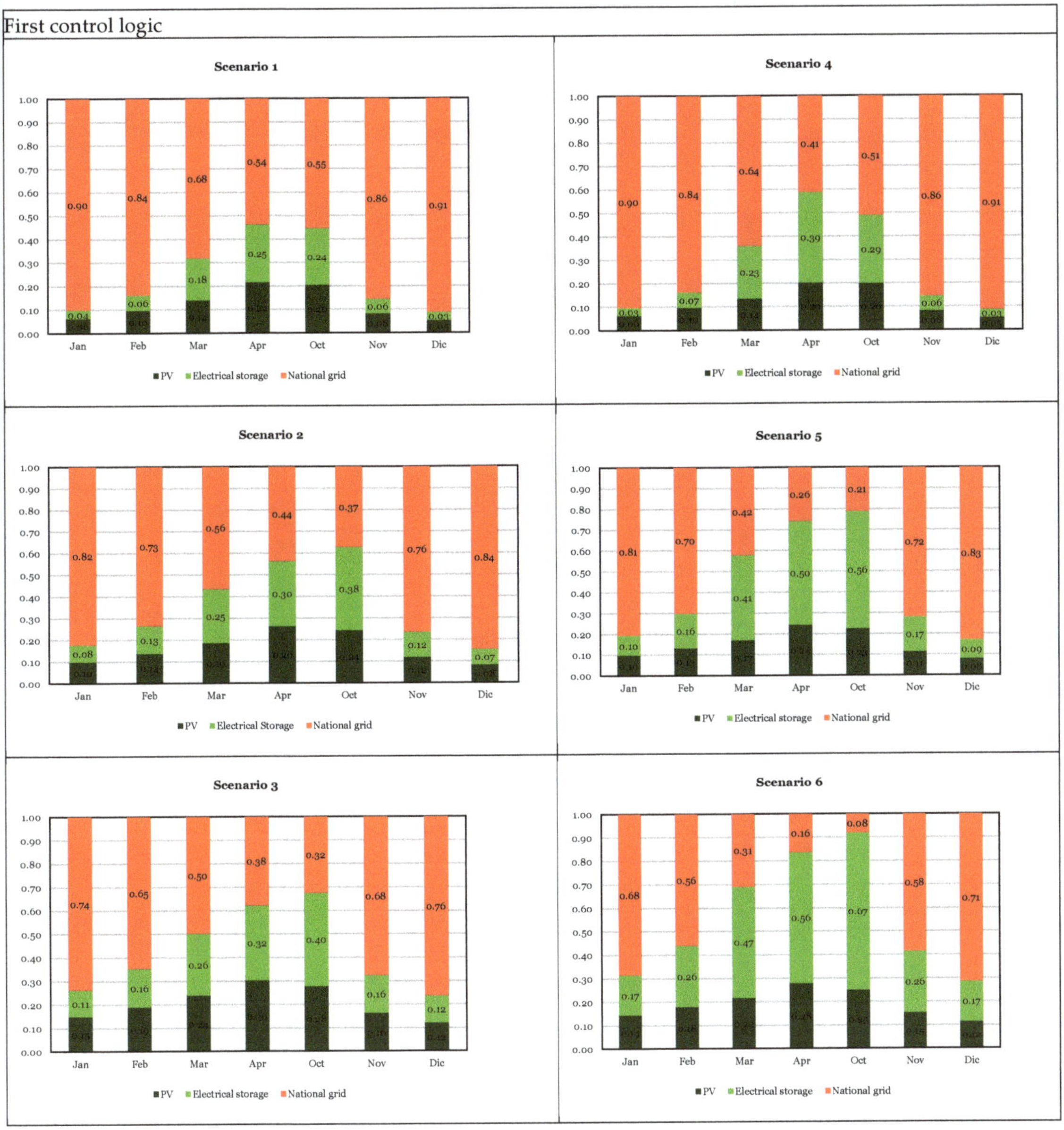

Figure 9. Percentage of load managed by each component varying the scenarios for the first control logic in Milan.

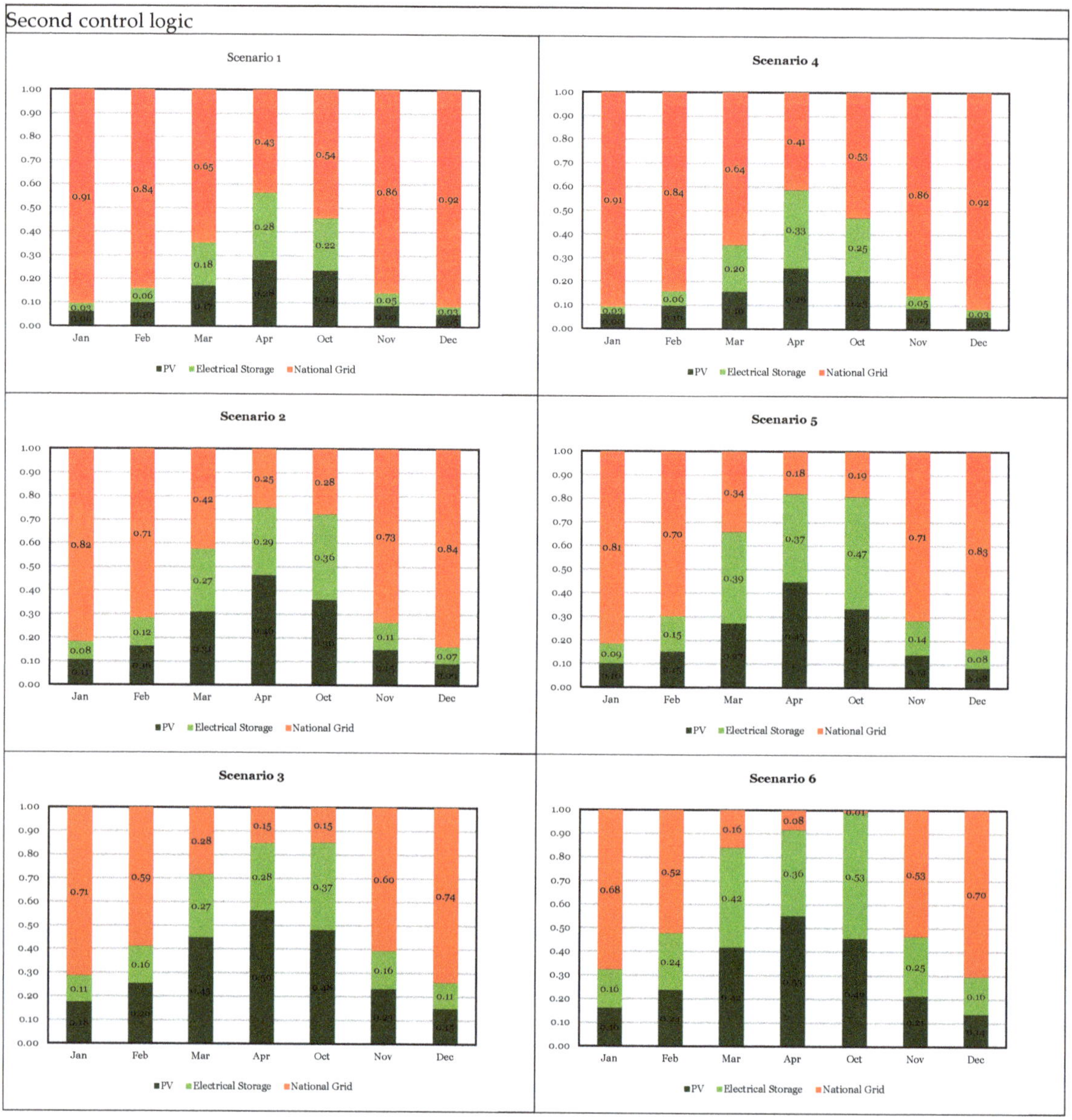

Figure 10. Percentage of load managed by each component varying the scenarios for the second control logic in Milan.

The charts in Figure 10 (from Scenario 4 to Scenario 6) depict the percentage of load regulated by each component via the second control logic, as the quantity of photovoltaic panels increased and a 2.4×10^4 Wh storage battery was employed. By comparing the percentages in the graph shown in Figure 10 with those in Figure 9 (the same combination but with a different control logic), it can be deduced that in this case as well, a control logic that was more focused on managing the surplus energy produced did not lead to improvements to the configuration with 18 panels. In fact, in this configuration, excess production compared to demand rarely occurred. On the other hand, the situation was different compared to the other two combinations. A smart control strategy allowed for almost always satisfying 50% of the thermal load using renewable energy sources.

In Figure 11, the annual percentages of self-consumed energy during the heating period are shown relative to the energy produced. It can be observed that this percentage decreased as the peak power installed increased. It was also evident that greater electrical storage capacity allowed for better utilization of solar radiation, resolving the problem of the mismatch between demand and solar radiation presence. Through the use of building envelope overheating strategies (second control logic), it was possible to increase the percentage of self-consumed energy by approximately 10%, maximizing the use of a clean energy source, except for a number of panels equal to 18.

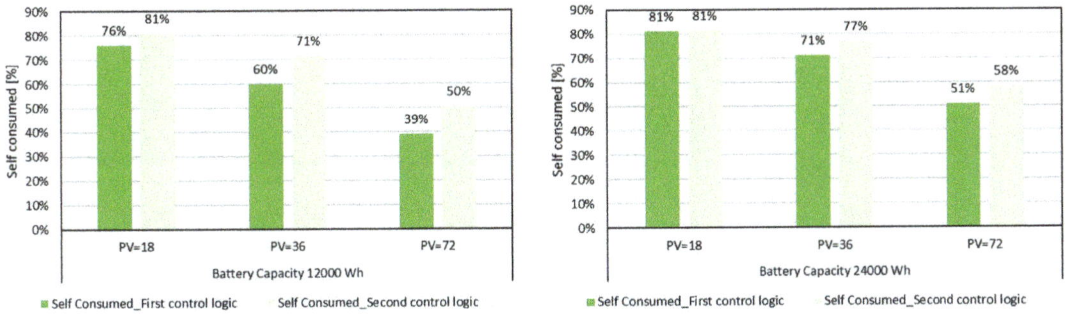

Figure 11. Annual percentages of total self-consumed electric energy in Milan.

Table 7 provides an annual breakdown of the percentage distribution of electrical load managed by various components within different scenarios. These scenarios were assessed under two distinct management logics: first control logic and second control logic. The results revealed notable variations in energy utilization patterns between the two logics. Under the first control logic, the systems exhibited a significant reliance on the national grid, ranging from 65% to 75% of energy sourced externally, highlighting substantial grid dependency. However, self-consumption rates were remarkable, surpassing 70% in certain scenarios, particularly during colder months. Conversely, the second control logic' showcased a substantial reduction in grid dependence across all scenarios. The percentage of grid-sourced energy significantly diminished, occasionally dropping as low as 38%, emphasizing the efficiency of this logic in optimizing local energy utilization. Moreover, self-consumption levels remained notably high, exceeding 70% in several scenarios. These findings underscored the reliability of the second control logic' in maximizing self-generated energy consumption while mitigating dependence on the national grid.

Table 7. Annual percentage distribution of electrical load managed by each component in different scenarios, comparing the two management logics in Milan.

Management Logic	Scenario	PV	Electrical Storage	National-Grid	Self-Consumed
First control logic	Scenario 1	12%	12%	75%	76%
	Scenario 2	16%	19%	65%	60%
	Scenario 3	21%	22%	57%	39%
	Scenario 4	12%	16%	72%	81%
	Scenario 5	15%	28%	56%	71%
	Scenario 6	12%	11%	76%	55%
Second control logic	Scenario 1	14%	12%	73%	81%
	Scenario 2	24%	19%	58%	71%
	Scenario 3	33%	21%	46%	50%
	Scenario 4	13%	14%	73%	81%
	Scenario 5	22%	24%	54%	77%
	Scenario 6	31%	30%	38%	58%

3.2.2. Messina

In the graphs of Figure 12, the percentages provided by each component of the first control logic are shown concerning the simulations conducted in the city of Messina.

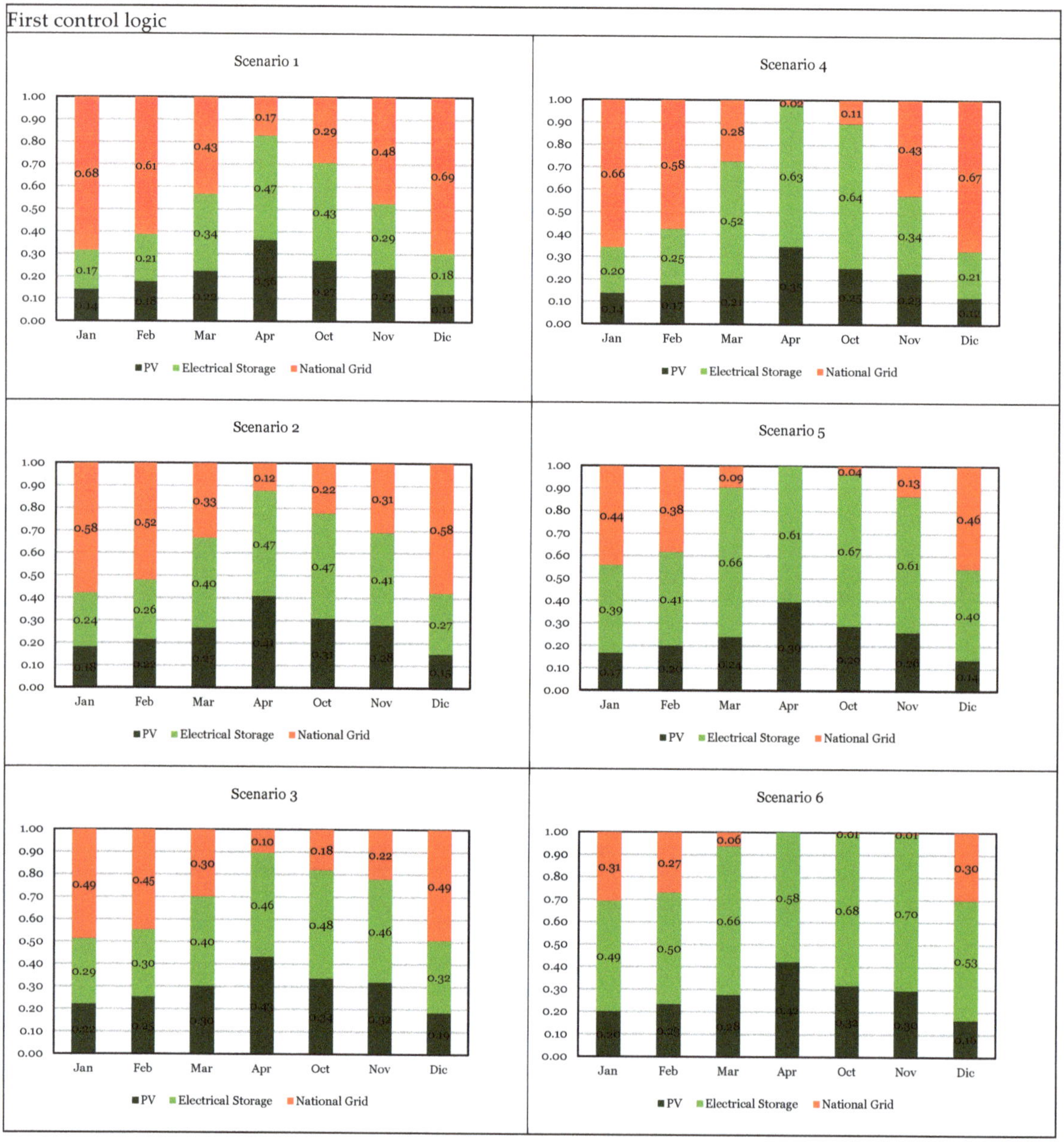

Figure 12. Percentage of load managed by each component varying the scenarios for the first control logic in Messina.

Similar to the simulations conducted in Milan, the data from the Messina simulations also revealed the impact of varying the number of photovoltaic panels while maintaining a constant 1.2×10^4 Wh battery capacity (Scenario 1 to 3) and the influence of doubling the battery capacity (Scenario 4 to 6) with an increasing number of panels.

In the Milan simulations, the results showed that as the peak photovoltaic power installed increased, the percentage of energy demand covered directly by the photovoltaic system and battery increased. Conversely, the reliance on the national grid decreased. For instance, the highest grid dependence in Milan was observed in December with 18 panels, covering 75% of the energy demand. However, with 72 panels installed, this dependence dropped to 49% in January, underscoring the potential for greater energy self-sufficiency as photovoltaic capacity increased.

Similarly to Milan, in the Messina simulations, the trend of increasing photovoltaic power leading to greater self-sufficiency was observed. The percentage of load directly supplied via the photovoltaic system (represented by the green bars) and the contribution of the electrical storage system (light green bars) both grew, while the percentage of grid-supplied energy (red bars) decreased. Notably, in Messina, with 72 panels and a 2.4×10^4 Wh storage battery, nearly complete energy self-sufficiency was achieved in October compared to the national grid, similar to the results obtained in Milan.

Regarding the simulations conducted in Messina during the heating period with the second control logic, the data (Figure 13) exhibited a similar trend to what was observed in the Milan simulations. Increasing the number of solar panels from Scenario 1 to Scenario 3 while maintaining a constant battery capacity of 1.2×10^4 Wh, the percentage of energy supplied directly via the solar panels rose from 16% to 22%, while the battery contribution varied from 16% to 28%. Concurrently, energy drawn from the national grid decreased from 68% to 49%. These results underscore the second control logic's capacity to promote greater grid independence, in parallel with the increase in solar panels.

Comparing these simulations with those of the first control logic in Messina revealed some significant differences. While the second control logic demonstrated reduced grid dependency, especially with a growing number of solar panels, the first logic maintained a higher percentage of energy supplied by the grid. Furthermore, the effect of doubling the number of panels and battery capacity, from Scenario 4 to Scenario 6, was less pronounced in the first logic, with a more modest variation in the energy drawn from the grid, ranging from 68% to 38%.

Table 8 provides a summarized presentation of the detailed outcomes derived from an analysis of the percentage distribution of electrical load across various scenarios, considering the implementation of two management logics. These data distinctly showcased the disparities between the two logics and their impact on the integration of renewable energy sources in the construction sector. Specifically, the second control logic emerged as the more efficient approach, demanding less energy from the national grid. For instance, in Scenario 1, it allowed for the direct satisfaction of 76% of the load through self-consumption, thereby reducing reliance on the national grid to 47%. This outcome holds profound significance in fostering enhanced integration of renewable energy sources within the building context, thereby aiding in diminishing the reliance on the national grid and optimizing the energy system's efficiency.

However, it is noteworthy that the results obtained in Messina with the second control logic differed from those in Milan during the heating period. Despite an identical number of solar panels, regional and climatic variables led to considerable variations in the outcomes. This variation underscored the importance of adapting energy management strategies to local specifics to maximize self-consumption and reduce grid dependence.

Within the context of the two control logics under examination, these climatic differences dictated how energy was generated, managed, and utilized within the system. The first control logic was more effective in the mild climate of Messina, where heating demands necessitated lower thermal capacity and thermal storage could be employed more efficiently. Conversely, the second control logic was advantageous in Milan, where lower temperatures required greater thermal capacity, and thermal storage became vital for heating.

These differences hold considerable significance in the design and optimization of the energy system, as they impact efficiency, self-consumption, and grid dependence in a climate-specific manner.

Figure 13. Percentage of load managed by each component varying the scenarios for the second control logic in Messina.

Table 8. Annual percentage distribution of electrical load managed by each component in different scenarios, comparing the two management logics in Messina.

Management Logic	Scenario	PV	Electrical Storage	National Grid	Self-Consumed
First control logic	Scenario 1	22%	30%	48%	68%
	Scenario 2	26%	36%	38%	42%
	Scenario 3	29%	39%	32%	24%
	Scenario 4	21%	40%	39%	78%
	Scenario 5	24%	54%	22%	54%
	Scenario 6	27%	59%	14%	31%
Second control logic	Scenario 1	24%	29%	47%	76%
	Scenario 2	35%	35%	30%	54%
	Scenario 3	44%	36%	20%	34%
	Scenario 4	23%	35%	41%	80%
	Scenario 5	33%	47%	20%	63%
	Scenario 6	42%	50%	9%	39%

3.3. Calculation of CO_2 Equivalent Emissions

In conclusion, the annual equivalent CO_2 emission percentages were calculated using the emission factors reported in Table 6 for all analyzed configurations. The results are presented in Figure 14 for Milan.

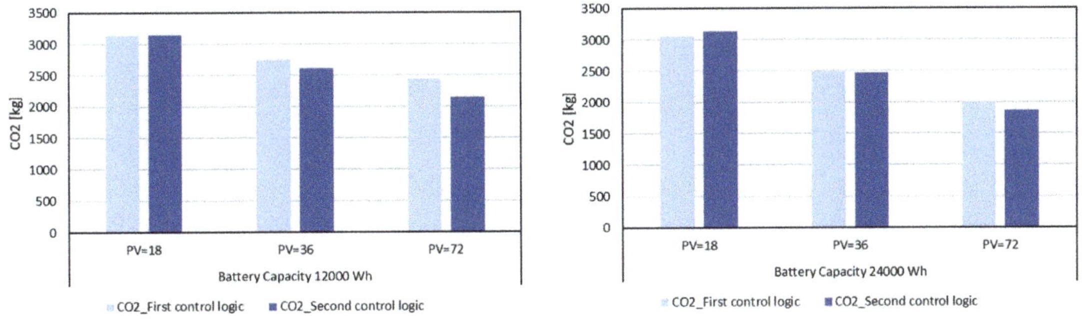

Figure 14. Annual amount of equivalent CO_2 emissions for Millan.

It was found that the highest CO_2 production was achieved with 18 photovoltaic panels, regardless of battery capacity and control logic, with an amount of approximately 3.0×10^3 kg of CO_2. However, increasing the battery capacity and peak power installed led to a reduction in this production. On the other hand, the lowest CO_2 production was observed with 72 photovoltaic panels and a 2.4×10^4 Wh battery, using the second control logic. In this case, the CO_2 production was around 1.86×10^3 kg. This represents a significant reduction compared to a traditional heating system with a gas boiler, which produces approximately 5.0×10^3 kg of CO_2.

The graphs in Figure 15 display CO_2 emissions observed during simulations conducted in Messina across various scenarios managed using two distinct control logics: the first traditional logic and the second "smart" logic. A comparative analysis of these two strategies revealed that the second "smart" logic led to a substantial reduction in CO_2 emissions compared to the traditional logic across all considered scenarios. In the initial three scenarios (Scenario 1, 2, and 3), the adoption of the second "smart" logic resulted in reductions of CO_2 emissions by 8%, 23%, and 36%, respectively, in comparison to the traditional logic. This trend continued in subsequent scenarios (Scenario 4, 5, and 6), with reductions in CO_2 emissions amounting to 0.1%, 13%, and 37%, respectively.

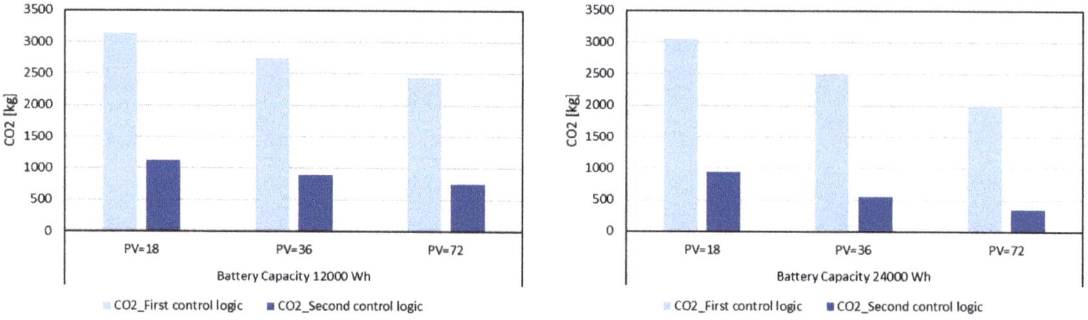

Figure 15. Annual amount of equivalent CO_2 emissions for Messina.

The data highlighted that the second "smart" logic had a significant impact on reducing CO_2 emissions, particularly in the first three scenarios. These findings underscored the importance of implementing advanced control strategies to enhance energy efficiency and mitigate environmental impact, especially in climatic contexts akin to Messina.

3.4. Domestic Hot Water Supplied by the Solar Collectors

During the summer season, the thermal storage system was used to store chilled water for cooling purposes and cannot be utilized for DHW. For this reason, a supplementary 300 L thermal storage unit was employed with solar collectors to provide DHW. The solar storage output was combined with a cooler flow rate from the supply network to attain a temperature of 45 °C for household usage. In the winter season, when solar radiation was insufficient to meet DHW demands, the heat pump took energy from the thermal storage system, with the solar collectors being given priority.

In Figure 16, the graph presents the coverage percentages for each component throughout every month of the year in the city of Milan. Furthermore, in Figure 17, an analogous graph is provided, which pertains specifically to the city of Messina.

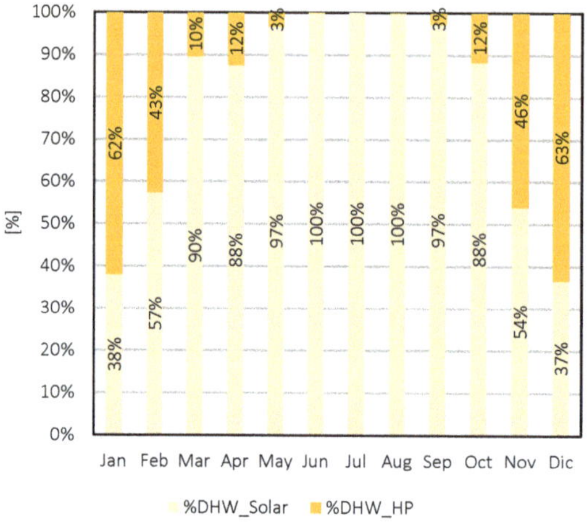

Figure 16. Coverage percentages for each component throughout every month of the year in the city of Milan.

In Milan, the data in the table showed that the use of solar collectors covered approximately 38% to 57% of the demand for DHW. However, the city relied more on heat pumps, which satisfied between 43% to 62% of the demand. This suggested that in Milan, the efficiency of solar collectors in providing hot water was relatively lower compared to

Messina. The colder climate in Milan led to a higher dependence on heat pumps to heat water, resulting in less reliance on solar collectors.

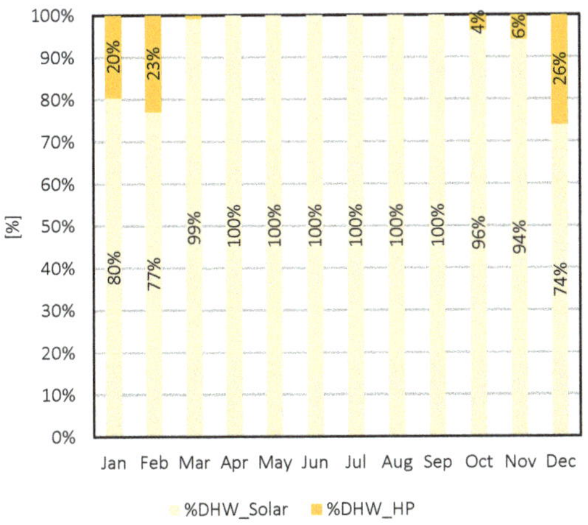

Figure 17. Coverage percentages for each component throughout every month of the year in the city of Messina.

Turning our attention to Messina, the data indicated that solar collectors were significantly more effective in covering the demand for hot water. In Messina, the use of solar collectors ranged from 74% to 100% in meeting the demand, highlighting the city's ability to harness solar energy efficiently. The use of heat pumps in Messina remained notably low, ranging from 0% to 26%. This outcome underscored the favorable climate conditions in Messina, which allowed for extensive utilization of solar energy and a reduced reliance on heat pumps for water heating.

4. Conclusions

In this study, the energy and environmental implications of a system that combines the use of an air-water heat pump with photovoltaic panels and electric and thermal storage systems were analyzed. Performances were compared between two different control strategies, the first commonly used for the management of air-conditioning systems, and the second one specifically conceived to maximize the employment of renewable solar sources.

A comprehensive parametric analysis was conducted to determine the optimal configuration for energy management strategies, considering various parameters such as thermal storage volume, the number of photovoltaic panels, and battery capacity. Increasing peak photovoltaic power enhanced direct coverage of energy demand by the photovoltaic system and battery, reducing reliance on the national grid. Comparative analysis of traditional and advanced "smart" control logics underscored the latter's efficiency in optimizing local energy utilization, significantly reducing grid dependence. The "smart" logic maximized self-generated energy consumption, mitigating national grid dependence. CO_2 equivalent emissions analysis revealed substantial reductions, especially in initial scenarios, emphasizing the advanced control's role in enhancing energy efficiency and environmental sustainability in climates akin to Messina.

Evaluation of domestic hot water supply highlighted differences between Milan and Messina. In Milan, solar collectors covered 38–57% of demand, while heat pumps satisfied 43–62%. Conversely, Messina exhibited higher solar efficiency, covering 74–100% of demand with less reliance on heat pumps. Results underscored local climatic influence on solar energy effectiveness for water heating.

In conclusion, the study emphasized tailoring energy management to regional climates. Advanced control logic benefits energy efficiency, reduces carbon emissions, and promotes sustainable energy use in Milan and Messina. These insights can guide optimal energy systems and enhance environmental sustainability in diverse climatic contexts.

5. Future Research Directions

Building on the findings of this study, several potential avenues for future research emerge:

Performance in Cooling Demand: investigating the system's performance concerning cooling demand is an intriguing area for future research. Given the substantial variations in heating and cooling requirements across different climates, understanding how these systems can efficiently provide cooling, especially in regions with high cooling demands, is essential.

Integration of Renewable Energy Sources: exploring the integration of various renewable energy sources, such as wind and geothermal energy, in addition to photovoltaic systems, can provide insights into further reducing the carbon footprint and enhancing energy efficiency. Assessing the interplay between these sources and their impact on system performance is a promising research direction.

Seasonal Energy Storage: the development of advanced energy storage solutions tailored to seasonal variations in energy supply and demand is critical. Future studies could investigate the feasibility and effectiveness of such long-term storage options, ensuring a reliable and continuous energy supply throughout the year.

User Behavior and Adoption: understanding user behavior and the factors influencing the adoption of renewable energy systems is essential for broader implementation. Future research could focus on user-centric studies to identify barriers, motivations, and strategies for increasing the adoption of sustainable energy technologies.

Smart Grid Integration: as smart grid technologies continue to evolve, assessing the impact of integrating residential renewable energy systems into these grids is crucial. Investigating the benefits, challenges, and optimal strategies for enhancing grid resilience and energy management is a promising area of study.

Economic and Policy Analysis: conducting economic analyses, including cost-benefit assessments and payback periods, can provide valuable insights into the financial feasibility of renewable energy systems. Additionally, examining policy incentives and regulations that promote the adoption of sustainable technologies is of great relevance.

Author Contributions: Conceptualization, S.P. and P.B.; methodology, S.P., P.B. and R.B.; software, S.P.; validation, S.P. and D.C.; formal analysis, S.P.; investigation, D.C.; resources, F.B.; data curation, D.C.; writing—original draft preparation, S.P.; writing—review and editing, R.B.; visualization, D.C.; supervision, F.B. and R.B. All authors have read and agreed to the published version of the manuscript.

Funding: This research received no external funding.

Data Availability Statement: The data presented in this study are available on request from the corresponding author. The data are not publicly available due to privacy.

Conflicts of Interest: The authors declare no conflict of interest.

References

1. IEA. Data. Statistics. Available online: https://www.iea.org/data-and-statistics (accessed on 1 January 2021).
2. Available online: https://www.europarl.europa.eu (accessed on 13 December 2023).
3. Korol, E.; Shushunova, N. Analysis and Valuation of the Energy-Efficient Residential Building with Innovative Modular Green Wall Systems. *Sustainability* **2022**, *14*, 6891. [CrossRef]
4. IEA. *Tracking Buildings 2020*; IEA: Paris, France, 2020.
5. Shushunova, N.; Korol, E.; Luzay, E.; Shafieva, D. Impact of the Innovative Green Wall Modular Systems on the Urban Air. *Sustainability* **2023**, *15*, 9732. [CrossRef]
6. Bee, E.; Prada, A.; Baggio, P.; Psimopoulos, E. Air-source heat pump and photovoltaic systems for residential heating and cooling: Potential of self-consumption in different European climates. *Build. Simul.* **2019**, *12*, 453–463. [CrossRef]

7. Baggio, P.; Bee, E.; Prada, A. Demand-side management of air-source heat pump and photovoltaic systems for heating applications in the italian context. *Environments* **2018**, *5*, 132. [CrossRef]
8. Popescu, L.L.; Popescu, R.; Catalina, T. Improving the energy efficiency of an existing building by dynamic numerical simulation. *Appl. Sci.* **2021**, *11*, 12150. [CrossRef]
9. Gaur, A.S.; Fitiwi, D.Z.; Curtis, J. Energy Research & Social Science Heat pumps and our low-carbon future: A comprehensive review. *Energy Res. Soc. Sci.* **2021**, *71*, 101764. [CrossRef]
10. Wang, Z.; Guo, P.; Zhang, H.; Yang, W.; Mei, S. Comprehensive review on the development of SAHP for domestic hot water. *Renew. Sustain. Energy Rev.* **2017**, *72*, 871–881. [CrossRef]
11. Dikici, A.; Akbulut, A. Exergetic performance evaluation of heat pump systems having various heat sources. *Int. J. Energy Res.* **2008**, *32*, 1279–1296. [CrossRef]
12. Huide, F.; Tao, Z. Performance Analysis of an Integrated Solar-assisted Heat Pump System with Heat Pipe PV/T Collectors Operating under Different Weather Conditions. *Energy Procedia* **2017**, *105*, 1143–1148. [CrossRef]
13. Zhao, H.; Wu, Q.; Hu, S.; Xu, H.; Nygaard, C. Review of energy storage system for wind power integration support. *Appl. Energy* **2015**, *137*, 545–553. [CrossRef]
14. Carvalho, A.T.G.M.L. Life Cycle Assessment of Stationary Storage Systems within the Italian Electric Network. *Energies* **2021**, *14*, 2047. [CrossRef]
15. Aneli, S.; Arena, R.; Tina, G.M.; Gagliano, A. Improvement of energy self-sufficiency in residential buildings by using solar-assisted heat pumps and thermal and electrical storage. *Sustain. Energy Technol. Assess.* **2023**, *60*, 103446. [CrossRef]
16. Pinamonti, M.; Beausoleil-morrison, I.; Prada, A.; Baggio, P. Water-to-water heat pump integration in a solar seasonal storage system for space heating and domestic hot water production of a single-family house in a cold climate. *Sol. Energy* **2021**, *213*, 300–311. [CrossRef]
17. Bellos, E.; Tzivanidis, C. Energetic and fi nancial sustainability of solar assisted heat pump heating systems in Europe. *Sustain. Cities Soc.* **2017**, *33*, 70–84. [CrossRef]
18. Miglioli, A.; Aste, N.; Del Pero, C.; Leonforte, F. Photovoltaic-thermal solar-assisted heat pump systems for building applications: Integration and design methods. *Energy Built Environ.* **2023**, *4*, 39–56. [CrossRef]
19. Ning, Z.; Zhang, X.; Ji, J.; Shi, Y.; Du, F. Research progress of phase change thermal storage technology in air-source heat pump. *J. Energy Storage* **2023**, *64*, 107114. [CrossRef]
20. Long, J.Y.; Zhu, D.S. Numerical and experimental study on heat pump water heater with PCM for thermal storage. *Energy Build.* **2008**, *40*, 666–672. [CrossRef]
21. Georges, E.; Cornélusse, B.; Ernst, D.; Lemort, V.; Mathieu, S. Residential heat pump as flexible load for direct control service with parametrized duration and rebound effect. *Appl. Energy* **2017**, *187*, 140–153. [CrossRef]
22. Kreuder, L.; Spataru, C. Assessing demand response with heat pumps for efficient grid operation in smart grids. *Sustain. Cities Soc.* **2015**, *19*, 136–143. [CrossRef]
23. UNI/TS 11300-1; Building Energy Performance—Part 1: Evaluation of the Energy Need for Space Heating and Cooling. 2014. Available online: https://www.uni.com (accessed on 1 October 2014). (In Italian).
24. Italian Higher Instutute for Environmental Protection and Research (ISPRA). Available online: https://www.isprambiente.gov.it/en/istitute (accessed on 1 January 2022).

MDPI AG
Grosspeteranlage 5
4052 Basel
Switzerland
Tel.: +41 61 683 77 34

Buildings Editorial Office
E-mail: buildings@mdpi.com
www.mdpi.com/journal/buildings

www.ingramcontent.com/pod-product-compliance
Lightning Source LLC
LaVergne TN
LVHW070457061126
851858LV00011B/252